Maladapting Minds
Philosophy, psychiatry,
and evolutionary theory

International Perspectives in Philosophy and Psychiatry

Series editors

Bill (K.W.M.) Fulford
Katherine Morris
John Z Sadler
Giovanni Stanghellini

Volumes in the series:

Mind, Meaning, and Mental Disorder 2e
Bolton and Hill

What is Mental Disorder?
Bolton

Delusions and other irrational beliefs
Bortolotti

Postpsychiatry
Bracken and Thomas

Unconscious knowing and other essays in psycho-philosophical analysis
Brakel

Psychiatry as Cognitive Neuroscience
Broome and Bortolotti (eds.)

Free will and responsibility: A guide for practitioners
Callender

Reconceiving Schizophrenia
Chung, Fulford, and Graham (eds.)

Nature and Narrative: An Introduction to the New Philosophy of Psychiatry
Fulford, Morris, Sadler, and Stanghellini (eds.)

Oxford Textbook of Philosophy and Psychiatry
Fulford, Thornton, and Graham

The Mind and its Discontents
Gillett

Dementia: Mind, Meaning, and the Person
Hughes, Louw, and Sabat (eds.)

Talking Cures and Placebo Effects
Jopling

Schizophrenia and the Fate of the Self
Lysaker and Lysaker

Responsibility and Psychotherapy
Malatesti and McMilan

Body-Subjects and Disordered Minds
Matthews

Rationality and Compulsion. Applying action theory to psychiatry
Nordenfelt

Philosophical Perspectives on Psychiatry and Technology
Phillips (ed.)

The Metaphor of Mental Illness
Pickering

Mapping the Edges and the In-between
Potter

Trauma, Truth, and Reconciliation: Healing Damaged Relationships
Potter (ed.)

The Philosophy of Psychiatry: A Companion
Radden

The Virtuous Psychiatrist
Radden and Sadler

Feelings of Being
Ratcliffe

Values and Psychiatric Diagnosis
Sadler

Disembodied Spirits and Deanimated Bodies: The Psychopathology of Common Sense
Stanghellini

Essential Philosophy of Psychiatry
Thornton

Empirical Ethics in Psychiatry
Widdershoven, McMillan, Hope and Van der Scheer (eds.)

Maladapting Minds
Philosophy, psychiatry, and evolutionary theory

Edited by
Pieter R. Adriaens and Andreas De Block

OXFORD
UNIVERSITY PRESS

Great Clarendon Street, Oxford OX2 6DP

Oxford University Press is a department of the University of Oxford.
It furthers the University's objective of excellence in research, scholarship,
and education by publishing worldwide in

Oxford New York

Auckland Cape Town Dar es Salaam Hong Kong Karachi
Kuala Lumpur Madrid Melbourne Mexico City Nairobi
New Delhi Shanghai Taipei Toronto

With offices in

Argentina Austria Brazil Chile Czech Republic France Greece
Guatemala Hungary Italy Japan Poland Portugal Singapore
South Korea Switzerland Thailand Turkey Ukraine Vietnam

Oxford is a registered trade mark of Oxford University Press
in the UK and in certain other countries

Published in the United States
by Oxford University Press Inc., New York

© Oxford University Press, 2011

The moral rights of the author have been asserted

Database right Oxford University Press (maker)

First published 2011

All rights reserved. No part of this publication may be reproduced,
stored in a retrieval system, or transmitted, in any form or by any means,
without the prior permission in writing of Oxford University Press,
or as expressly permitted by law, or under terms agreed with the appropriate
reprographics rights organization. Enquiries concerning reproduction
outside the scope of the above should be sent to the Rights Department,
Oxford University Press, at the address above

You must not circulate this book in any other binding or cover
and you must impose the same condition on any acquirer

British Library Cataloging in Publication Data
Data available

Library of Congress Cataloging in Publication Data
Data available

Typeset in Minion by Glyph International, Bangalore, India
Printed in Great Britain
on acid-free paper by
the CPI Antony Rowe, Chippenham, Wiltshire

ISBN 978–0–19–955866–7

10 9 8 7 6 5 4 3 2 1

To my son, Amos; my father, Luk; and my grandfather, Louis
- men of books

—PRA

Foreword

Psychiatry is a mess. Nobody seems to know how to distinguish normal behavior from mental disorders, or how to treat most mental disorders. Nobody seems to know how to integrate cutting-edge science (experimental psychopathology, community comorbidity studies, brain imaging, genome-wide association studies, multivariate behavior genetics) into a profession whose basic terms, concepts, empirical standards, professional institutions, funding sources, and intervention methods are decades old. There are strong, angry, and unresolved debates over how to revise the 5th edition of psychiatry's core reference work, the *Diagnostic and Statistical Manual of Mental Disorders* (DSM-V), to be published in 2013. There are continual tensions between research psychiatrists and clinical psychiatrists, between psychiatrists and clinical psychologists, and between mental health professionals and health insurers.

Evolutionary psychiatry promised to bring order to this chaos. In its two-decade history, it has made good progress in clarifying some terms, such as "disorder," "normal," "defence," and "emotion." It has yielded new insights into some mental disorders, notably depression, autism, phobias, anorexia, and psychopathy. It has promoted a bit more cross-fertilization among psychiatry, evolutionary psychology, behavior genetics, and biological anthropology. Yet it has left the bulk of psychiatry untouched.

Why has evolutionary psychiatry's impact been so limited, despite the impeccable Darwinian logic of basing the study of mental disorders on the study of evolved mental adaptations? There are the usual suspects—individual and institutional conservatism, the pre-Darwinian state of most medical school curricula, the vested interests of Big Pharma, the peculiarities of the American health insurance system, the vicious cycle between DSM categorizations and funding categories for research grants, the disappointments of psychiatric genetics, and the mindless but well-funded stampede towards neuroscience. No doubt these account for much of evolutionary psychiatry's limited impact.

Yet there may be deeper conceptual problems in evolutionary psychiatry. This is where philosophy might promote evolutionary psychiatry's progress and impact as a science.

Modern philosophy, I think, aims to analyze and clarify the terms, concepts, findings, and standards of evidence relevant to some domain of human discourse.

Psychiatry is one domain of human discourse with especially high stakes, such as trying to prevent suicide, rape, murder, despair, psychosis, and other forms of avoidable suffering. Following Nietzsche's demolition of grand philosophical systems—whether metaphysical, moral, or epistemological—much philosophy has become applied philosophy of some sort—philosophy of subject X, rather than Philosophy as an autonomous subject.

So, at its best, philosophy of science is pretty humble in its aspirations and methods. It largely means reviewing and critiquing scientific literatures with an eye towards unexamined assumptions, unclear concepts, slippery terms, internal contradictions, cultural prejudices, and historical amnesia. All of these problems are endemic to mainstream psychiatry, and remain fairly common in evolutionary psychiatry. Philosophy strives to do this concept-clarification work in a mindful, ruminative, deliberate, historically informed way, in contrast to the slapdash theorizing of many working scientists, who see literature reviews as onerous burdens to be finished quickly before the fun work of reporting methods and results in scientific papers. Insofar as philosophers gain specialist training in thinking clearly, debating sharply, knowing history, identifying counter-examples, and chasing implications imaginatively, they bring something useful and distinctive to science.

This book is a good example. Although half of the chapters are written by non-philosophers, most authors adopt the philosophical stance in relation to their particular issues. Of the 11 chapters after the introduction, some address general issues such as the nature of harmful dysfunctions, mechanistic versus evolutionary views of dysfunction, defenses versus disorders, generality versus modularity, and the role of human ethology in psychiatry. Other chapters focus on particular disorders: four on depression, and one each on phobias, sexual fetishes, autism, and schizophrenia. Yet even these disorder-specific chapters strive to gain insights that can be applied to other disorders, such as the difference between normally and abnormally regulated emotions, the interplay of evolved preferences and individual experience, and the differences between ancestral and modern environments. Philosophers often look for such cross-domain generalizations where concept-clarifications in one domain can be exported to other domains, whereas working scientists are usually more concerned with getting the theory right in just one domain.

Also, throughout all chapters, it is clear that the philosophy of evolutionary psychiatry has one huge advantage over the philosophy of nonevolutionary psychiatry: it can draw upon all the hard work that philosophers of biology have done since the 1970s to clarify concepts such as gene, trait, species, function, adaptation, selection, fitness, instinct, learning, and ancestral environment. It can also draw upon the hard work that evolutionary psychologists have done since around 1990 in refining such evolutionary concepts as they apply to human behavior, with all its distinctive complexity, flexibility, emotionality, sociability, culture-dependence, and consciousness.

So far, so good—good authors, good insights, good book. Yet future progress is uncertain. In the rest of this short foreword, I want to highlight a few ideas that I think might strengthen the next generation of evolutionary psychiatry theories.

The first point concerns science versus intuition. Philosophy can go wrong when it tries to reconcile our human intuitions about some domain with the technical concepts and findings in that domain. This often proves impossible. Philosophers of physics have learned that our intuitive physics—lay concepts of time, space, gravity, and impetus—are impossible to reconcile with core ideas in relativity, quantum mechanics, and cosmology. Philosophers of biology have learned that our intuitive biology—lay concepts of species essences and teleological functions—are impossible to reconcile with evolutionary population genetics and adaptationist teleonomy. Even moral philosophers such as Peter Singer are doubting whether our intuitive morality—self-deceptive, nepotistic, clannish, anthropocentric, and punitive—can be reconciled with any consistent set of moral values, whether deontological, consequentialist, or virtue-ethical. Yet philosophers of psychiatry still often make the argument that if some principled new way to think about mental illness has implications that seem intuitively unacceptable, the new idea must be rejected as absurd.

For example, my view of mental disorders as typically arising from mutation load—a view that I have defended elsewhere—suggests that there is no principled distinction between maladaptive disorder and "normal variation", because most "normal variation" reflects maladaptive deviations from optimal, species-typical design. The concept of "normal variation" in a mental trait makes evolutionary-genetic sense only if one assumes that the trait has been selectively neutral or under balancing (e.g., frequency-dependent) selection.

The implication is that almost all living humans have many mental disorders, mostly minor but some major, and these include not just DSM disorders like depression and schizophrenia, but diverse forms of stupidity, irrationality, religiosity, vices, and personality quirks. As the new positive psychology acknowledges, we are all very far from optimal mental health, and we are all more or less insane in multiple ways. Yet traditional psychiatry, like human intuition, resists calling anything a disorder if its prevalence is higher than about 10%.

My point here is not that the mutation load view is necessarily right, but that a mature philosophy of psychiatry may lead to insights so contrary to common sense that they compel us to rethink how psychiatry is taught, practiced, and researched. In other words, a twenty-second century psychiatry may fit no more comfortably with our evolved and acculturated intuitions than does twenty-first century M-theory in physics, with its 11 dimensions, P-branes, and supergravity. Indeed, we might hope that psychiatric theory eventually becomes so sophisticated, quantitative, and technical that it is no more comprehensible to working psychiatrists than M-theory is to engineers.

My second point concerns the mismatch of research topics between evolutionary psychiatry and evolutionary psychology. Most evolutionary psychology so far has focused on mate choice, sexual strategies, person perception, family conflict, reciprocity, aggression, decision heuristics, status, and emotions. Excepting the last two topics, very little of this work has informed evolutionary psychiatry. For example, there is almost no research connecting mate choice research to sexual dysfunctions such as dyspareunia, anorgasmia, vaginismus, or premature ejaculation, or to disorders that might promote short-term mating success, such as bipolar disorder and psychopathy. There is a gap between evolutionary personality psychology and the study of antisocial, borderline, narcissistic, or schizotypal personality disorders. There is a gap between the evolutionary psychology of aggression, warfare, rape, and conflict, and the study of post traumatic stress disorder. There is almost no evolutionary psychiatry work on any of the dissociative, impulse-control, somatoform, sleep, substance use, learning, neurological, or culture-specific disorders, not to mention other Axis III and IV issues. Also, evolutionary psychology identifies many mental adaptations that may have characteristic break down patterns and failure modes that constitute harmful dysfunctions, but that have been neglected by psychiatrists, notably the adaptations concerning food selection, habitat choice, mate choice, mate retention, sexual rivalry, ovulation, parental care, kinship, reciprocity, friendship, status-seeking, risk-taking, and decision-making. Evolutionary psychologists and evolutionary psychiatrists are learning important things from each other, but there's not the expected overlap between research on the normal and research on the abnormal in most domains of life functioning.

More generally, the evolutionary focus on differential reproductive success leads directly to an emphasis on conflicts of interest between genes, individuals, and groups. Yet evolutionary psychiatry has neglected many key conflicts of interest that may impose a heavy burden of suffering on people. These include evolutionary conflicts between pathogens and hosts, predators and prey, nuclear and mitochondrial genes, somatic and germ-line cells, parents and offspring, males and females, young and old, and rival groups, populations, and species. For example, few evolutionary psychiatrists yet take seriously Paul Ewald's suggestion that some mental disorders might reflect viruses or other pathogens influencing human behavior in their own interests. An evolutionary perspective could also highlight institutional conflicts of interest between psychiatrists and clients, psychiatrists and health insurers, psychiatrists and pharmaceutical companies, psychiatrists and clinical psychologists, and the mental health versus criminal-justice systems. These conflicts shape many debates in psychiatric research, practice, and policy, yet are rarely acknowledged.

These concerns about science versus intuition, and gaps between evolutionary psychiatry and evolutionary psychology, can be viewed positively. They identify low-hanging fruit—places where evolutionary psychiatry, and the philosophy thereof, can make substantial progress quickly and easily. This book reflects the state of the art now, but each chapter is also pregnant with potential to guide future research. We can look forward in coming years to evolutionary psychiatry growing more philosophically astute, psychologically informed, evolutionarily sophisticated, empirically fruitful, and clinically applicable in promoting mental health.

Geoffrey Miller

Contents

Contributors *xvii*

Introduction Why philosophers of psychiatry should care about evolutionary theory *1*
Andreas De Block and Pieter R. Adriaens

    I.1 Psychiatric Darwinism versus Darwinian psychiatry *3*

    I.2 Explaining the evolution of mental disorders *6*

    I.3 Sociobiology, evolutionary psychology, evolutionary psychiatry: philosophical criticism *10*

    I.4 Evolution, dysfunction, and mental disorders *19*

    I.5 *Das kranke Tier*: evolution, psychopathology, and human nature *23*

    I.6 Conclusion *27*

    References *27*

Part 1: Evolutionary psychiatry and its critics *33*

    1 Fearing new dangers: phobias and the cognitive complexity of human emotions *35*
    Luc Faucher and Isabelle Blanchette

        1.1 The case of phobia *36*

        1.2 A module for fear *42*

        1.3 Snakes and spiders vs syringes and guns *46*

        1.4 Further problems with the evolutionist's explanation of phobias *52*

        1.5 An alternative conception of emotion *55*

        1.6 Conclusion *59*

        Acknowledgments *60*

        References *60*

    2 Sexual imprinting and fetishism: an evolutionary hypothesis *65*
    Hanna Aronsson

        2.1 The science of fetishism: a history *66*

- 2.2 The evolution of human sexual preferences *68*
- 2.3 Sexual imprinting in animals *71*
- 2.4 Sexual imprinting in humans *78*
- 2.5 Adaptationism and sexual imprinting *82*
- 2.6 Conclusion *84*
  - Acknowledgments *86*
  - References *86*

3 Developmental disorders and cognitive architecture *91*
Edouard Machery
- 3.1 Psychopathologies and cognitive architecture *93*
- 3.2 Why developmental psychopathologies provide no evidence for modularity *98*
- 3.3 The epistemology of developmental dissociations *105*
- 3.4 Evaluation of the strong reading of Premise 2 *109*
- 3.5 Conclusion *112*
  - References *113*

4 On the role of ethology in clinical psychiatry: what do ontogenetic and causal factors tell us about ultimate explanations of depression? *117*
Erwin Geerts[†] and Martin Brüne
- 4.1 Introduction *117*
- 4.2 Ontogenetic processes: early attachment relationships, parental rearing styles, and their relationship with depression *121*
- 4.3 The ethological analysis of deviant behavior *122*
- 4.4 Association between observable behavior and depression *124*
- 4.5 Disturbed interpersonal behavior as a possible causal factor in depression *128*
- 4.6 Are causal factors of depression linked to adverse early experiences? *131*
- 4.7 Possible evolutionary explanations of depression *132*
- 4.8 Discussion: why psychiatry needs ethology *133*
  - References *136*

Part 2: Evolutionary theory and the concept of mental disorder *141*

    5 Darwin, functional explanation, and the philosophy of psychiatry *143*
    Jerome C. Wakefield

        5.1 Functional explanation: Aristotle, Lucretius, Darwin *144*

        5.2 Culver and Gert on distinct sustaining causes *152*

        5.3 The designed-defense objection *161*

        5.4 Nordenfelt's critique of evolutionary approaches to disorder *165*

        References *171*

    6 Evolutionary foundations for psychiatric diagnosis: making DSM-V valid *173*
    Randolph M. Nesse and Eric D. Jackson

        6.1 Diagnosis and its discontents *174*

        6.2 From clinical diagnosis to the DSM *176*

        6.3 The price of progress *179*

        6.4 The basic fault *181*

        6.5 Evolution and emotions *183*

        6.6 Emotional disorders *186*

        6.7 The importance of analyzing motivational structure *190*

        6.8 Towards an evolutionary foundation for psychiatric nosology *191*

        References *194*

    7 Normality, disorder, and evolved function: the case of depression *198*
    Daniel Nettle

        7.1 Introduction *199*

        7.2 Inductive evidence for a categorical depression/normality distinction *200*

        7.3 Evolved functions, dysfunctions, and depression *203*

        7.4 The challenge of individual variation *208*

        7.5 Disorder versus complaint as the basis for identifying depression *210*

        7.6 Conclusion *213*

        References *213*

8 Function, dysfunction, and adaptation? *216*
Kelly Roe and Dominic Murphy

    8.1 Introduction *217*

    8.2 The two-stage view *217*

    8.3 Theories of function *219*

    8.4 Dysfunction *228*

    8.5 Dysfunction and the role of science *231*

    8.6 Conclusion *235*

    Acknowledgments *236*

    References *236*

Part 3: Psychopathology, evolution, and human nature *239*

9 Mirroring the mind: on empathy and autism *241*
Farah Focquaert and Johan Braeckman

    9.1 Introduction *241*

    9.2 Autism spectrum conditions: a lack of "mirroring" and empathy *245*

    9.3 The genetics of autism spectrum conditions *248*

    9.4 Evolution of autistic traits: low empathy *252*

    9.5 Conclusion *258*

    Acknowledgments *259*

    References *259*

10 The role of mood change in defining relationships: a tribute to Gregory Bateson (1904–1980) *264*
John Price

    10.1 Introduction *264*

    10.2 The overthrown tyrant: a clinical case illustration *266*

    10.3 Darwin, Huxley, and sexual selection *268*

    10.4 Ritual agonistic behavior and ritual losing *270*

    10.5 A triune mind in a triune brain *272*

    10.6 Gregory Bateson: defining the relationship *276*

    10.7 Conclusion *281*

    Appendix 10.1 *283*

    References *286*

11 From "evolved interpersonal relatedness" to "costly social alienation:" an evolutionary neurophilosophy of schizophrenia *289*
Jonathan Burns

- 11.1 Introduction *289*
- 11.2 A philosophy of embodiment *292*
- 11.3 The evolution and development of the social brain *294*
- 11.4 Schizophrenia and the evolutionary paradigm *298*
- 11.5 "Interpersonal alienation" from the social world *299*
- 11.6 Schizophrenia as a social brain disorder *301*
- 11.7 Resolving the "schizophrenia problem" in evolutionary terms *302*

    References *304*

Index *309*

Contributors

Pieter R. Adriaens
Institute of Philosophy
University of Leuven
Belgium

Hanna Aronsson
Department of Zoology
Stockholm University
Sweden
and
Centre for the Study of
Cultural Evolution
Stockholm University
Sweden

Isabelle Blanchette
Département de
Psychologie
Université du Québec à
Trois Rivières
Canada

Johan Braeckman
Department of Philosophy and
Moral Sciences
Ghent University
Belgium

Martin Brüne
Research Department of
Cognitive Neuropsychiatry
and Psychiatric Preventive
Medicine,
LWL University Hospital
Ruhr-University Bochum
Germany

Jonathan Burns
Department of Psychiatry,
Nelson Mandela School
of Medicine
University of KwaZulu Natal
South Africa

Andreas De Block
Institute of Philosophy
University of Leuven
Belgium

Luc Faucher
Département de Philosophie
Université du Québec à Montréal
Canada

Farah Focquaert
Department of Philosophy and
Moral Sciences
Ghent University
Belgium

Erwin Geerts[†]
Department of Scientific Research
and Education
Mental Health Care Friesland
The Netherlands
and
Wenckebach Institute
Department of Communication,
Leadership, Assessment, and
Cooperation
University Medical Center
Groningen
The Netherlands

Eric D. Jackson
Vanderbilt University Law School
USA

Edouard Machery
Department of History and Philosophy of Science
University of Pittsburgh
USA

Geoffrey F. Miller
Department of Psychology
University of New Mexico
USA

Dominic Murphy
Unit for History and Philosophy of Science, Faculty of Science
University of Sydney
Australia

Randolph M. Nesse
Department of Psychiatry,
Department of Psychology,
Institute for Social Research,
Evolution and Human Adaptation Program
The University of Michigan
USA

Daniel Nettle
Centre for Behaviour and Evolution
Newcastle University
UK

John S. Price
Hackmans
Plumpton Lane
Plumpton
UK

Kelly Roe
Philosophy Program
Research School of the Social Sciences
Australian National University
Australia

Jerome C. Wakefield
Silver School of Social Work and Department of Psychiatry
New York University
USA

Introduction

Why philosophers of psychiatry should care about evolutionary theory

Andreas De Block and Pieter R. Adriaens

One of the most striking features of organs and organisms is their adaptive complexity. The traits of living organisms seem to be designed to solve problems that are posed by the environment. From Darwin, we know that adaptations only give the appearance of design. In fact, they are the result of a causal process dubbed natural selection—a process that makes some heritable traits more common in a population because these traits give their bearers a reproductive advantage over those who don't have them. Because of the importance of the Darwinian revolution in biology, contemporary philosophy of biology largely coincides with the philosophy of evolutionary theory. The pivotal concepts of evolutionary theory, such as "fitness" and "design", attract much more philosophical attention than the concepts of, say, cell biology or any other biological subdiscipline.

Yet evolutionary theory is also present in other areas of philosophy. Many philosophers have argued that Darwinian hypotheses and concepts have important implications for various issues in philosophy. In their view, Darwin's theory of evolution explains how science works (Hull, Popper) and how knowledge is acquired (Campbell, Bradie). Some philosophers even believe that evolutionary theory provides the basis for morality (Ruse) and for a genuine naturalization of philosophy of mind (Papineau, Carruthers). Yet the philosophical community has its skeptics too. Thomas Nagel, for instance, rejects evolutionary epistemology as a "reductionist dogma" (Nagel 1986, p. 81), thereby echoing Wittgenstein's famous claim that "Darwin's theory has no more to do with philosophy than any other hypothesis in natural science" (Wittgenstein 1922, 4.1122).

But how about philosophy of psychiatry? Why would Darwinian theory matter specifically to philosophy of psychiatry? We believe that there are three reasons why philosophers of psychiatry have taken an interest in evolutionary theory.

First of all, there is the nascent field of evolutionary psychiatry. "Evolutionary psychiatry" and "Darwinian psychiatry" are umbrella terms used to refer to various attempts to make sense of mental disorders within the general framework of evolutionary theory. While biological psychiatrists have always been interested in the causation of dysfunctional behavior, and while psychoanalytic psychiatrists have taken a distinctly developmental perspective, evolutionary psychiatrists engage with ultimate, rather than proximate, questions about mental illnesses. Being a young and youthful new discipline, evolutionary psychiatry allows for a nice case study in the philosophy of science. Thus, philosophers have asked questions about the scientific status of evolutionary explanations of mental disorders, as well as about the conceptual and empirical assumptions underpinning these explanations. Many hypotheses in evolutionary psychiatry are indebted to human sociobiology and evolutionary psychology, and therefore confront us again with the plethora of philosophical criticisms that have been leveled against these controversial disciplines.

Secondly, philosophers of psychiatry have engaged with evolutionary theory because evolutionary considerations are often said to play a role in defining the concept of mental disorder. The basic question here is: Can the concept of mental disorder be given an objective definition, or is it rather a normative concept? The most influential "objectivist" proposals rely heavily on evolutionary theory. Wakefield, for instance, argues that mental disorders are disorders because people suffering from them fail to meet a natural norm that is brought about by natural selection. Other "objectivists" do not necessarily agree with Wakefield's "selectionist" approach of biological function, but some of them do maintain that other key concepts of evolutionary theory, such as adaptation and fitness, are necessary to understand what mental disorders are.

Thirdly and finally, evolutionary thinking in psychiatry has often been a source of inspiration for a philosophical analysis of human nature. Many philosophers have claimed that psychopathology can give us a unique perspective on different aspects of human nature. In their view, mental disorders would (partially) reveal what it is like to be a human being. Evolutionary psychiatrists have taken up this line of thought in suggesting, for example, that man's vulnerability to mental disorders may well be one of the defining features of our species.

These three reasons for philosophers of psychiatry to engage with evolutionary theory provide the backbone of the themes and chapters of the present volume. In the last three sections of this introductory chapter we will briefly elaborate on these themes: the many philosophical critiques aimed at evolutionary explanations of mental disorders, the importance of evolutionary theory in analysing the concept of mental disorder, and the relevance of evolutionary psychiatry for various issues in contemporary philosophical anthropology. The first two sections of the present introduction are devoted to an overview of the main hypotheses of contemporary evolutionary psychiatry and a very brief history of evolutionary thinking in psychiatry. We consider this historical section to be necessary because we strongly believe that the best philosophy of psychiatry is always informed by the history of psychiatry.

I.1 Psychiatric Darwinism versus Darwinian psychiatry[1]

"Light will be thrown on the origin of man and his history" (Darwin 1859, p. 488). This seemingly innocent suggestion in the final chapter of *The Origin of Species* gave the green light to the diffusion of Darwin's thinking in the human sciences, including psychology and behavioral biology. It was only a matter of time before psychiatrists also yielded to this temptation. In fact, psychiatrists have been asking evolutionary questions about mental disorders since evolutionary theories became known. It is important, however, to distinguish between nineteenth-century and early twentieth-century evolutionary thinking about mental disorders, which we have labeled "psychiatric Darwinism" (Adriaens and De Block 2010, p. 135), and contemporary evolutionary psychiatry. While the latter goes back to a number of pivotal theoretical developments in evolutionary theory, such as ethology and the "modern synthesis", psychiatric Darwinism was imbued by non-Darwinian "evolutionisms", including social Darwinism, Lamarckism, and all sorts of degeneration theories.

I.1.1 Psychiatric Darwinism

The courtship between evolutionary theory and psychiatry goes back to the close collaboration between Darwin and the later doyen of British psychiatry, James Crichton Browne, on the occasion of Darwin's work for *The Expression of the Emotions in Man and Animals* (1872). Continuing one of the main

---

[1] In this introduction we take the term "Darwinian psychiatry" to be synonymous with "evolutionary psychiatry". A comprehensive overview of the many trends and schools in the history of evolutionary psychiatry can be found in an introductory paper in a recent special issue of *History of Psychiatry* (Adriaens and De Block 2010).

themes of *The Descent of Man* (Darwin 1871), the chief purpose of *Expression* was to further narrow the supposed gap between humans and other animals by showing that they share many of the skeletal, muscular, and behavioral elements involved in the expression of emotions. In Darwin's view, psychiatric patients closely resemble children (and "savages" too) in being able to experience the purity of emotions. A study of the emotional life of such patients was therefore simply essential in preparing his monograph. Darwin wrote:

> [I]t occurred to me that the insane ought to be studied, as they are liable to the strongest passions, and give uncontrolled vent to them. I had, myself, no opportunity of doing this, so I applied to Dr Maudsley and received from him an introduction to Dr J. Crichton Browne, who has charge of an immense asylum near Wakefield, and who, as I found, had already attended to the subject. This excellent observer has with unwearied kindness sent me copious notes and descriptions, with valuable suggestions on many points; and I can hardly over-estimate the value of his assistance.
>
> (Darwin 1872, p. 13)

Much of the detail on which Darwin based his arguments in *Expression* was indeed provided by James Crichton Browne (Pearn 2010). Beyond this, Darwin was not particularly interested in the etiology and classification of mental disorders, but at one point he mentions one of Maudsley's discussions of the causes of insanity:

> Dr Maudsley, after detailing various strange, animal-like traits in idiots, asks whether these are not due to the reappearance of primitive instincts – 'a faint echo from a far-distant past, testifying to a kinship which man has almost outgrown'. He adds, that as every human brain passes, in the course of its development, through the same stages as those occurring in the lower vertebrate animals, and as the brain of an idiot is in arrested condition, we may presume that it 'will manifest its most primitive functions, and no higher functions'.
>
> (Darwin 1872, p. 245)

Similarly, Maudsley (1916, p. 267) once wrote that man "is living his forefathers essentially over again". Darwin probably included this hypothesis in *Expression* because he thought that it echoed one of his own interests. Already in *The Variation of Animals and Plants Under Domestication* (Darwin 1868), he had devoted a complete chapter to regressions and atavisms, that is, the reappearance of ancestral traits in the individual. Madness, so he seems to suggest, may well be another example of such regression. The theme of regression and atavism was only one of the many melodies in a song that is often referred to as "degeneration theory", but it was certainly a popular one. It need not surprise us, then, that it has influenced many early attempts to understand and explain mental illnesses within an evolutionist framework.

Nineteenth-century degenerationist views were often imbued with another popular biological theory of the day: recapitulationism, or Haeckel's biogenetic law. In his *Natürliche Schöpfungsgeschichte* (1868), the most successful work of popular science in the nineteenth century, Haeckel explored the "laws of evolution" that had led to our species. Some of these laws were Darwinian in nature, others were of Lamarckian descent, but the biogenetic law was clearly Haeckel's own "discovery". The biogenetic law states that the individual's development somehow summarizes the evolution of our species ("ontogeny recapitulates phylogeny"), implying that every member of *Homo sapiens* goes through unicellar, multicellar, invertebrate, amphibian, reptilian, mammalian, and primate phases before the child becomes truly human. Moreover, Haeckel explicitly argued that his law held for both human physiology and psychology. Part of the biogenetic law's popularity was due to its apparent implications for psychiatry and criminology, as it was thought to show that criminals and psychiatric patients were either fixations or regressions to earlier evolutionary stages of development. Comparing these unfortunate individuals to "lower" animals, the coupling of degeneration theory and Haeckel's speculations proved to be a very explosive cocktail. Some have argued that it provided a breeding ground for eugenic practices in psychiatry (Kevles 1985, pp. 90–1), such as the large-scale sterilization of psychiatric patients in early twentieth-century America, and the genocide of more than 100 000 psychiatric patients in Nazi Germany (e.g., see Torrey and Yolken 2010)—a most unsavory chapter in the history of (evolutionary) psychiatry.

Freud's evolutionist speculations on mental illnesses are remarkably similar in content to those of degeneration theorists, including Maudsley. Moreover, Sulloway (1979) extensively documented that Freud was heavily indebted to both recapitulationist and Lamarckist thinking. In fact, Freud hoped to lend his own theory some credibility by grounding it in such evolutionist theories. This is nowhere more obvious than in a posthumously published paper, "A Phylogenetic Fantasy: Overview of the Transference Neuroses" (Freud 1987 [1915]), where Freud attempts to link up our ancestor's vicissitudes during the last Ice Age with man's present-day vulnerability to a series of mental illnesses. "Neurosis," he concludes, "must bring back the primeval picture (Freud 1987 [1915], p. 13). Unfortunately for Freud, his whole theory was built on faulty phylogenetic suppositions, thus providing a major motivation for biologists and other scientists to dispute the overall scientific status of Freudian psychoanalysis. To others, however, Freud's plan to examine "how much the phylogenetic disposition can contribute to the understanding of the neuroses" (Freud 1987 [1915], p. 13–14) should be cherished and continued. Freud may have got the details wrong, they argue, but surely his basic ideas are worth pursuing. Many contemporary evolutionary psychiatrists even credit Freud as being the "founding father" of evolutionary psychiatry (McGuire and Troisi 1998; Stevens and Price 2000).

I.1.2 Evolutionary psychiatry

Increasingly throughout the first half of the twentieth century a number of key developments in evolutionary theory separate "orthodox" Darwinian thinking from competing evolutionist theories, such as degeneration theory and recapitulationism. The rise of ethology, and the fusion of Darwin's ideas with Mendelian genetics (the so-called "modern synthesis"), revolutionized evolutionary thinking to the point where it became a solid science of evolution. Inevitably, these developments also affected evolutionary thinking in psychiatry, and perhaps it should not surprise us that the very first attempt to make sense of mental disorders within this modern evolutionary framework comes from two of the leading architects of the modern synthesis, Huxley and Mayr. Jointly with two psychiatrists, Abram Hoffer and Humphrey Osmond, they co-authored a *Nature* paper in 1964, in which they argue that schizophrenia is an evolutionary puzzle, at least to the extent that it is a heritable disorder with a fairly high prevalence. As such, it should have been weeded out by natural selection a long time ago—unless, of course, it confers important fitness advantages to schizophrenic patients and/or their relatives, such as enhanced resistance to contagious diseases, or perhaps enhanced fertility of female patients (Huxley *et al.* 1964).

Apart from explaining the evolutionary persistence of schizophrenia, and thereby "resolving" a Darwinian paradox, Huxley and Mayr also aimed at breaking the dominance of then fashionable environmentalist (and particularly psychoanalytic) conceptions of schizophrenia (De Bont 2010). Other evolutionary biologists were less critical of psychoanalytic thinking. Niko Tinbergen, for example, believed that his Nobel-prize winning work in ethology underscored the value of a psychodynamic approach to autism (Tinbergen and Tinbergen 1972; Vicedo 2010). As it happens, neither Huxley and Mayr's, nor Tinbergen's theories have stood the test of time. Yet it can easily be argued that, for various reasons, they are to be considered as the first "modern" evolutionary psychiatrists (rather than psychiatric Darwinists), whose work was firmly grounded in a truly modern, theory-based biological science. In the following section we explain what contemporary evolutionary psychiatry is by providing an overview of its many conceptualizations and explanations of mental illnesses.

I.2 Explaining the evolution of mental disorders

Most psychiatric Darwinists, including Darwin himself, were not very interested in detailed functional accounts of the emotions underlying mental disorders.

In *Expression*, for example, Darwin says very little about the function of the emotions, and much more about their phylogenetic origins and development. The adaptationist logic in understanding and explaining the adaptive value of mental disorders is indeed largely a new feature of contemporary evolutionary psychiatry.

Randolph Nesse, for instance, has argued that negative emotions should be seen as reactions to (real or imagined) situations with a negative cost-benefit outcome. Generally, negative emotions motivate the individual to do something about the current situation or to avoid similar situations in the future (Nesse 1990). For example, those ancestors who did not care about their sexual partner's infidelity had fewer offspring than those who reacted with jealousy. Jealousy might have helped our ancestors to cope with a series of reproductive threats because it elicits mate guarding and mate retention behaviors (Buss 2000). Of course there is an important difference between negative emotions and mental disorders, and no evolutionary psychiatrist would argue that all negative or aversive emotions are symptoms of a disordered mind. Someone who is angry about being fired is not considered mentally ill, and crying at one's best friend's funeral does not qualify as a symptom of depression. The reason why many evolutionary psychiatrists take an interest in aversive emotions is because such explanations teach us that natural selection did not design our minds for happiness or social harmony, but only for survival and reproduction. In short, feeling bad can be good for your evolutionary fitness.

But how about being depressed? Assuming that low mood is adaptive, is depression an adaptation, too? The idea that some mental disorders have some kind of functional significance is completely counterintuitive, but such claims have in fact been defended, notably for depressive disorders (e.g., see Hagen 1999). Less controversially, many evolutionary psychiatrists believe that mental disorders are enlargements of adaptive traits—"adaptive behavior gone wild", as Stevens and Price (2000, p. 94) have put it. Patients with paranoid personality disorder, for example, may be suffering from an overdose of suspicion which, in itself, is a very useful trait. Analogically, while the body's ability to produce fever is often said to be adaptive, for example in fighting bacterial infections, treating fever with Advil generally doesn't harm patients, which suggests that not all bouts of fever are useful. Interestingly, such "oversensitivities" or "excesses" need not be dysfunctional. In this context, Nesse (2001) has compared aversive emotions (and fever) with smoke detectors—their reliability requires them to express a number of false alarms, especially when the costs of such false alarms are low compared to the potential harm they protect against.

This set of evolutionary explanations of mental disorders, in which (core parts of) mental disorders are conceptualized as adaptations, rather than diseases, is only one of several nonexclusive explanatory models used in evolutionary psychiatry to account for man's vulnerability to disorder. We will call this set of explanations the *adaptationist model* and distinguish it, in the remainder of this section, from five other models in evolutionary psychiatry: the mismatch model, the trade-off (or balancing selection) model, the breakdown model, the displacement model, and the senescence model.

The *mismatch model* is still very close to the adaptationist model, as it builds on one of the central ideas in evolutionary psychology. Evolutionary psychologists claim that our ancestral environment, that is, the environment in which most of the evolution of our species took place, differs substantially from the modern cultural environment. Or, in the words of Tooby and Cosmides: "our modern skulls house a stone-age mind" (Tooby and Cosmides 1997). As a result, we are much better at solving the problems faced by our hunter-gatherer ancestors than the problems we encounter in modern cities. Many evolutionary psychiatrists consider this mismatch to be the hotbed of many of today's mental disorders. Although many seem to think of this model as quite homogeneous, it actually consists of three submodels. First of all, the environment we now live in probably frustrates many of our deep-seated evolved needs. As Stevens and Price (2000, p. 35) note: "If we are to understand the psychiatric disorders from which our contemporaries suffer, then we have to take into account the ways in which Western society frustrates the needs of paleolithic men or women still persisting as living potential within us in our present environmental circumstances." Secondly, many of the adaptive problems that we face now were not adaptive problems on the savannahs in Pleistocene East Africa. We do not fear guns the way we fear snakes, even though guns pose a much greater threat to our fitness today than snakes do (Öhman and Mineka 2001; but see Chapter 1). Thirdly, some objects or events in our contemporary environment resemble objects or events that were part of our ancestral environment. These modern objects or events trigger the very same reactions that evolved as adaptive responses to the Pleistocene objects and events, but the differences between these environments make the once adaptive reactions maladaptive:

> We enjoy strawberry cheesecake, but not because we evolved a taste for it. We evolved circuits that gave us trickles of enjoyment from the sweet taste of ripe fruit, the creamy mouthfeel of fats and oils from nuts and meat, and the coolness of fresh water. Cheesecake packs a sensual wallop unlike anything in the natural world because it is a brew of megadoses of agreeable stimuli which we concocted for the express purpose of pressing our pleasure buttons.
>
> (Pinker 1997, p. 524)

For some mental disorders, the *medical* or *breakdown model* seems to provide the most appropriate explanation. As Murphy notes, both mismatch and adaptationist models seem to assume that "none of our psychopathology involves something going wrong with our minds", while "nobody should deny that our evolved nature suffers from a variety of malfunctions and other pathologies" (Murphy 2005, p. 746). It may well be, for example, that low mood has some functional significance, but any organ can *dys*function, resulting, in the case of low mood, in "malignant" sadness or depression. Evolutionary psychiatrists using this model mention mostly proximate causes as the factors responsible for the *dys*function. Infections, lesions, and mutations are sometimes at the heart of psychiatric etiology.

The *trade-off model* (or balancing selection model) is one of the most popular models in contemporary evolutionary psychiatry. The classic example of a trade-off in evolutionary medicine is sickle-cell anemia, a disease commonly observed in people from African and Mediterranean descent. Geneticists have shown that heterozygote carriers of the sickle-cell gene do not suffer from anemia and, more importantly, they are much more resistant to malaria (Allison 1954). Malaria, of course, is very common in some parts of Africa and the Mediterranean, so one could say that sickle-cell anemia is the price some individuals pay for an unusual advantage in their relatives, that is, being resistant to a rampant and life-threatening infectious disease. Some evolutionary psychiatrists think that such a scenario may also apply to the evolution of schizophrenia, and perhaps other psychotic disorders. As Stevens and Price (2000, p. 146) note: "In a sense, schizotypic genes are like the genes responsible for sickle-cell anemia, which enhance the well-being of carriers by protecting them from malaria while impairing those with greater genetic loading by afflicting them with anemia." But what possible advantages could schizophrenia be associated with? While some researchers have focused on physiological, and particularly immunological, advantages (e.g., see Huxley *et al.* 1964; Erlenmeyer-Kimling 1968), most of them believe that schizophrenia (or bipolar disorder) owes its evolutionary persistence to a genetic association with a highly valuable and *typically human trait*, such as sociality (Burns 2007), language (Crow 2000), or creativity (Jamison 1993; Horrobin 2001; Nettle 2001). Horrobin (2001), for example, has suggested that minor mutations in the genetic code of the fat metabolism of our early ancestors, and the associated exponential increase of their cerebral capacities, have heralded the beginning of an amazingly creative new species: us. Yet the very same mutations also made us vulnerable for schizophrenia, therefore one could say that schizophrenia is the price we pay for our very humanity.

The *senescence model* is probably best used to explain certain neuropsychiatric diseases, such as Alzheimer dementia and Huntington's chorea. It may be true that the genes that code for these diseases have a pathological effect in the latter part of a lifespan, but have no negative effects before or during the reproductive period. In short, these disorders appear too late in our individual histories to come under negative selection pressures.

The *psychodynamic* or *displacement model* is rather underexposed in current evolutionary psychiatry. It states that some mental disorders can be defined by referring to the fixated and overactive use of defenses, cognitive mechanisms, and the like, in a context where no such defenses or mechanisms are needed, or where other defenses or mechanisms would be more appropriate (Tinbergen 1940; Demaret 1979). While similar to the breakdown model, the displacement model puts greater emphasis on the legitimate nature of these "out-of-context" reactions. For example, an individual (a patient with schizophrenia) uses a defense (suspiciousness) when confronted with problems (the hearing of voices) resembling the adaptive problems (gossip) that can be solved by using these defenses. Usually, such out-of-context defenses only aggravate the problem, thus creating a vicious looping effect.

As these models make abundantly clear, most evolutionary psychiatrists ask ultimate questions of the functional kind. Rather than reconstructing the phylogenetic vicissitudes of our mental and behavioral repertoire, the current trend in evolutionary psychiatry is to speculate about the past and current forces that undermine the adaptive value of our cognitive mechanisms and emotions. Ideally, such an evolutionary explanation should start from an accurate description of the condition and from a plausible proximate account of the condition. Getting the proximate mechanisms right often helps to find a correct answer to the ultimate questions. Conversely, evolutionary psychiatrists contend that good evolutionary explanations give us the heuristic means to refine proximate explanations. Other researchers are far more skeptical about the theoretical potential of evolutionary psychiatry. The following section lists and discusses some of the critiques that have been leveled against this new discipline.

1.3 Sociobiology, evolutionary psychology, evolutionary psychiatry: philosophical criticism

Philosophers of science sometimes try to assess whether or not a particular discipline as a whole meets the criteria of good science. There are good reasons to think that such an assessment is not valuable (nor feasible) in the case of evolutionary psychiatry. First of all, it will be clear by now that evolutionary psychiatry is not a homogenized field. The historical sketch above, for example, has revealed that different researchers (e.g., Tinbergen versus Huxley) work in somewhat different traditions (e.g., comparative ethology versus population genetics), even though they can be gathered under the umbrella of evolutionary psychiatry. These traditions are not necessarily guilty of the same sins. Secondly, it may be true that some evolutionary psychiatrists proceed in ways that are not exactly epistemologically disciplined (e.g., see Adriaens 2007), but it would be unfair to employ their way of working to dismiss every attempt to get an evolutionary grip on mental disorders. As one evolutionary psychiatrist once said, evolutionary psychiatrists are not all in one category, and they certainly do not all think the same things.

With this caveat in mind, we will devote this section to the most common philosophical critiques on evolutionary psychiatry. Firstly, we will try to evaluate to what extent the accusations against human sociobiology also have a bearing on contemporary evolutionary psychiatry. Then we will discuss whether the oft-heard criticisms of evolutionary psychology, one of the heirs of human sociobiology, can also be rightfully leveled against evolutionary psychiatry.

I.3.1 **Mere story telling?**

The upheaval around Wilson's *Sociobiology: The New Synthesis* (1975) marked the beginning of a deep and unresolved controversy that haunts evolutionary biology and the evolutionary social sciences to this day. Immediately after its publication, Stephen J. Gould, Richard Lewontin, and other prominent biologists took issue with Wilson's exploration of the biological basis of social behavior. So "the storm over sociobiology" (Segerstrale 2000) started in biological circles. Shortly thereafter, however, philosophers like Philip Kitcher and Michael Ruse joined the discussion because the debate also touched on the philosophical assumptions and moral implications of sociobiology as a science. While Lewontin, Rose, and the psychologist Leon Kamin rejected *Sociobiology* as a whole (Lewontin, Rose, and Kamin 1984), most other critics highlighted the problems only in the first and last chapter of the book. In these chapters, Wilson discussed the evolutionary underpinnings of human social behavior. The main accusations were that (human) sociobiology was the prototype of biological determinism, reductionism, adaptationism, and bad science. For the purposes of this volume, the latter two criticisms—adaptationism and bad science—are the most interesting, not least because they are still leveled against evolutionary psychiatry or closely related fields of research.

The attack on adaptationism was led by Lewontin and Gould. In their view, sociobiologists overemphasized the importance of natural selection and adaptation in the evolutionary process. Chance, drift, and history, to name just a few alternatives to selection, are simply ignored in sociobiological explanations, while Gould and Lewontin believed that these factors play a very important role in the evolutionary process. Assuming all aspects of human (social) behavior to be adaptations, sociobiological hypotheses are imbued with optimality assumptions, resulting in an overly optimistic ("panglossian") view of life: "if an organism has a trait, then it must be, in evolutionary terms, the best trait the organism is capable of having" (Sterelny and Griffiths 1999, 224). Another adaptationist sin, still according to Gould and Lewontin, is the construction of so-called just-so stories. Sociobiologists not only claim that all human traits are adaptations, they also claim to know what these traits have been selected for. Of course, critics of sociobiology wouldn't deny that eyes are for seeing, but they insist that most adaptations are far less obvious in this respect. The more creative individuals among us can easily come up with stories about the function of male baldness, being homesick, or athletic skills, but there is an important difference between storytelling and real science.

When Gould and Lewontin accused Wilson of adaptationism, they implied that sociobiology was bad science, for example because human sociobiologists took their speculations for reality. In sociobiological studies of human behavior, verbal arguments and rhetoric tricks replaced the gathering of detailed information in the field and the scrupulous testing of hypotheses. Similarly, the philosopher Philip Kitcher claimed that the conclusions of human sociobiology were not supported by the evidence and often started from unjustified assumptions. Kitcher draws a sharp line between this kind of pop sociobiology and the sociobiology of non-human animals, where the conclusions tend to be made with more caution and on the basis of vigorous testing. In his view, human sociobiology may have a future, but only if it learns from the faults Wilson and others have made. And these faults are many. Kitcher gives the example of Wilson's treatment of human (male) homosexuality:

> Wilson's argument for thinking that homosexuals might constitute a vertebrate 'caste' starts with an unscrutinized assumption to the effect that there is a single category of homosexual behavior, and that instances of this can be found in a wide variety of human and non-human contexts. He continues by inflating the credentials of studies in behavior genetics and offering some speculations about ways in which a propensity to have some homosexual offspring might boost one's inclusive fitness.
>
> (Kitcher 1987, p. 66)

Given that some "schools" in contemporary evolutionary psychiatry are indebted to human sociobiology, and given that at least some part of the above criticism of *Sociobiology* was justified, is evolutionary psychiatry guilty of the same sins that pop sociobiology committed in the 1970s? Or do we now have hard evidence to support evolutionary explanations of human behavior, including disordered behavior and mental illnesses?

As to the charge of adaptationism, evolutionary psychiatry need not plead guilty. As noted earlier, few evolutionary psychiatrists have argued that mental disorders are adaptations (and adaptationist accounts have in fact often been criticized from within the field; e.g., see Nettle 2004). However, nearly all evolutionary accounts of mental disorders are adaptationist in the sense that they do not discuss the evolutionary history of disorders or consider the role of drift in producing evolutionary change. To a certain extent, this is not too much of a problem because it is quite plausible that focusing on good design (or its absence) is generally thought to be a good way to study biological systems (Sterelny and Griffiths 1999). Much more problematic is that many evolutionary psychiatrists are not overly enthusiastic in testing the predictions that follow from their hypotheses—if they bother to formulate such predictions at all (but see Nesse 1999). Despite the philosophical storm over sociobiology, and despite the many critical voices repeating Kitcher's warnings,[2] little seems to have changed, and many of the hypotheses brought forward by evolutionary psychiatrists are not sufficiently supported by evidence.

That is not to say that evolutionary explanations of mental disorders are worthless and should be abandoned altogether. In psychiatry, many if not most etiologic explanations are speculative, so it would be a little unfair to reject evolutionary approaches because of their being speculative. Furthermore, evolutionary psychiatry is still in an embryonic state. Although the number of evolutionary psychiatrists and the number of papers on evolutionary psychiatry have increased over the last couple of years, evolutionary psychiatry still does not have its own journal, for example. Furthermore, there is change going on. Firstly, an interesting illustration of this change is given by Hanna Aronsson's chapter in this volume (Chapter 2), in which she argues that both normal and abnormal sexual preferences are better accounted for as the products of sexual imprinting, rather than as genetically determined adaptations. If correct, this hypothesis shows how evolutionary psychiatry can tackle adaptationism in other evolutionary social sciences. Secondly, some evolutionary psychiatrists have become more sophisticated in formulating and testing

---

[2] Daniel Dennett, for example, who is generally not very favorable to Gould's criticism of sociobiology (and evolutionary psychology), applauds Gould and Kitcher in *Darwin's Dangerous Idea*: "To the extent that adaptationists have been less than energetic in seeking further confirmation (or dreaded disconfirmation) of their stories, this is certainly an excess that deserves criticism." (Dennett 1995, p. 245)

hypotheses, and more cautious in drawing conclusions. In their attempts to demonstrate that low mood can be adaptive, Matthew Keller and Randolph Nesse predicted that different adverse life events would correlate with different patterns of low mood "symptoms" (Keller and Nesse 2005, 2006). With the help of questionnaires and a sophisticated experiment, they were able to confirm their predictions, even though it was not shown that different "symptom" patterns increase fitness in different adverse situations. As the authors note themselves, uncovering such a link is very difficult, for "fitness in modern environments, replete with birth control, medication, and other evolutionary novelties, may correlate poorly with ancestral fitness, which is the relevant criterion." (Keller and Nesse 2006, 328)

When Keller and Nesse refer to the difference between fitness in modern environments and fitness in the ancestral environment, they repeat one of the pivotal claims of evolutionary psychology—a claim which distinguished evolutionary psychology from human sociobiology and other, more recent, disciplines in the evolutionary social sciences. In the next subsection, we give a tentative assessment of the theoretical alliance between evolutionary psychology and evolutionary psychiatry.

I.3.2 Which brains in which environments?

The storm over sociobiology resulted in a fragmentation of the field of human evolutionary studies. To some, however, the fragmentation is only superficial:

> The lynching [of sociobiology] failed and the discipline still thrives, though many sociobiologists have been forced underground, traveling under disciplinary pseudonyms.
>
> (Queller 1995, p. 486)

These "disciplinary pseudonyms" include evolutionary psychology, human behavioral ecology and gene culture co-evolution (or dual inheritance theory) (for a comprehensive overview of the differences between these three programs, see Smith 2000, and below).

Unlike human sociobiology and human behavioral ecology, evolutionary psychology focuses on the evolved functions of the psychological mechanisms underpinning human behavior, rather than on human behavior as such or on behavioral strategies. Evolutionary psychologists argue that these psychological dispositions provide the evolutionarily relevant link between genes and behavior; they were naturally or sexually selected because they helped to solve the adaptive problems of our Pleistocene ancestors. Such a natural history of human psychology has two important implications. First of all, evolutionary psychologists insist that our minds were designed by the ancestral environment of our species, which is often referred to as the environment of evolutionary adaptedness (or EEA), and they were designed specifically to manage reproductively successful behavior in that particular environment. Secondly, evolutionary psychologists claim that the architecture of our minds is massively modular, meaning that they consist of a large number of innate and functionally specialized components. The many modules of our mind apply different processing mechanisms to different domains of information. As many evolutionary explanations of mental illnesses build on this so-called "massive modularity hypothesis", as well as on the supposed importance of the EEA, we need to scrutinize both assumptions and the philosophical critiques they attracted.

Much of the empirical support for the massive modularity hypothesis comes from neuropsychiatry (see also Chapter 3). Proponents of the massive modularity hypothesis argue that cases of selective cognitive impairment show that the breakdown of one component does not necessarily lead to the breakdown of other components, let alone to the breakdown of the whole system. For instance, the existence of patients with prosopagnosia (the inability to recognize faces) strongly suggests that humans have a separate module for face recognition (Boyer and Barrett 2005). Likewise, autistic people often have a deficit in mind-reading, while retaining normal linguistic abilities (Carruthers 2006). In the wake of these findings, a number of evolutionary psychologists have developed a special interest in evolutionary explanations of selective cognitive or emotional impairments. These accounts have now become part of evolutionary psychiatry, and it may come as no surprise to learn that an evolutionary psychologist-turned-evolutionary psychiatrist still adheres to many of the central ideas of the original discipline.

In recent years, evolutionary psychology's massive modularity hypothesis has been a major issue in numerous philosophical debates (e.g., see Carruthers 2006; Machery 2008). Some skeptics have argued that the brain is probably much more of a blank slate and much less specialized than many evolutionary psychologists and psychiatrists believe. On our view, this is largely an empirical issue, and it should be decided by empirical evidence. Yet it seems that the debates are also fed by conceptual confusion. Critics of the massive modularity hypothesis often build on a strong interpretation of modularity, while many evolutionary psychologists only subscribe to a much weaker interpretation. For instance, when critics argue that most cognitive subsystems draw on information that is held outside that subsystem (Currie and Sterelny 2000), they may have disproved a version of the massive modularity hypothesis, but not necessarily the version that is central to evolutionary psychology (Carruthers 2006, but see Sperber 2005). So despite the ongoing character of the discussion, we think it safe to say that the mind consists of a number of modules that vary in domain specificity, encapsulation, and isolation. The main issue seems to be of how many modular components the mind actually consists. So at this point, one may conclude that the alliance between evolutionary psychology and evolutionary psychiatry is rather unproblematic. But what about their joint interest in our ancestral environment?

The notion of the EEA is of psychiatric descent; it was first coined by the psychiatrist and attachment theorist John Bowlby. Contemporary evolutionary psychologists define the EEA as "a statistical composite of the adaptation-relevant properties of the ancestral environments encountered by members of ancestral populations" (Tooby and Cosmides 1990, 386). As noted earlier, there are important similarities and differences between the EEA and our contemporary environment. However, since, still according to evolutionary psychologists, the differences outweigh the similarities, there is a fundamental mismatch between these environments, resulting in all sorts of medical and mental illnesses. The basic assumption here is that our bodies and brains are still adapted to a savanna-like environment (because that is what our ancestral environment supposedly looked like during the bulk of our evolutionary history), as they could not keep up with the dazzling pace of cultural evolution.

However, there are good reasons to challenge this view. Gene-culture co-evolutionary theory, for instance, argues that we are a cultural species and that our brains have evolved to adapt to ever-changing cultural environments. Such adaptability explains how we were able to colonize the continents (except one) and to adapt ourselves to dramatically different climatologic, social, and ecological circumstances than those encountered in the African savanna (Boyd and Richerson 1985, 2005). Moreover, it is mistaken to think that, during the Pleistocene, all humans lived on the savanna or in savanna-like environments. Although much is still unknown about the EEA, research indicates that, during the evolution of early hominins there was quite some variation in the environments which they inhabited, ranging from forests to savannas to open-canopy woodlands (Potts 1998; Boaz and Almquist 2002). But even if all hominins lived on the African savanna during the Pleistocene, it is highly unlikely that humans are not adapted to any of the environments that have occurred since the end of the Pleistocene (Irons 1998). As human behavioral ecologists have pointed out repeatedly, 10 000 years (i.e. 300 to 400 human generations) is sufficient time for evolutionary change to occur, therefore we have to assume that more recent environments, shaped by agriculture, architecture, and other technologies, have had at least some impact on our genetic dispositions (Lumsden and Wilson 1981; Richerson and Boyd 2005). In short, the EEA is perhaps not as important in understanding human nature as evolutionary psychologists tend to believe.

How does this objection reflect on the credibility of evolutionary psychiatry? Firstly, it should be noted that not all evolutionary psychiatrists believe in a dramatic mismatch between the EEA and our contemporary environment. McGuire and Troisi (1998), for instance, seriously doubt the explanatory power of mismatch hypotheses. Secondly, there are some rather uncontroversial (but rough) explanations that make use of the mismatch model. Randolph Nesse's claim that "the rapid spread of alcohol-making technologies changed our world in ways our species has not yet adapted to" (Nesse 2005, p. 905) is a case in point. However, once hypotheses and predictions become more specific, evolutionary psychiatry pays a price for overrating the explanatory power of the EEA. Mismatch explanations of phobias are a nice illustration of this problem. One of the arguments supporting the claim that phobic fears were once adaptive, while they are a real nuisance today, is that we tend to be picky about the objects of our anxiety. Building on Seligman's famous concept of prepared learning, researchers have claimed that it is much easier to scare children with snakes and spiders, for example, than with guns and lorries, even though the latter are much more dangerous in our contemporary environment (Marks and Nesse 1994; Ohman and Mineka 2001). However, as Luc Faucher and Isabelle Blanchette argue in this volume, the effects used to point to the existence of a prepared mechanism of fear are also shown by "evolutionarily novel" objects, such as guns and syringes. After all, the EEA does not seem so important in understanding the nature of our fears.

Moreover, the prepared learning literature on fear and phobias also points at another problem with mismatch hypotheses. Most mismatch hypotheses assume that there is a point-by-point match between ancestral adaptive problems and our current psychological mechanisms, and, perhaps most importantly, that evolutionary theory is a powerful source of precise predictions about human nature. However, both assumptions are unsubstantiated. Evolutionary theory does not generate very specific predictions, partly because chance plays a decisive role in evolution. Eric Alden Smith and colleagues have aptly summarized this critique as follows:

> Because motor vehicles were not part of the EEA, does this mean that the evasive actions of wary pedestrians fall outside the purview of adaptive analysis? Or did the past existence of falling boulders and charging rhinos select for a cognitive module general enough to minimize the chances of collision with large moving objects, whether they be rhinos or Range Rovers? Or are the relevant evolved mechanisms even more general, having to do with motion detection, aversion to personal injury, and imitative learning? The verbal arguments of EEA proponents that seemed so incisive and revolutionary are compatible with any or all of the above, and hence rather lacking in predictive content.

(Smith *et al.* 2001, pp. 131–2)

1.3.3 Evolutionary psychiatry without evolutionary psychology?

The three "children" of sociobiology (evolutionary psychology, human behavioral ecology, and dual inheritance theory) differ in many ways. Hence the three approaches have sometimes been presented as complementary alternatives. In philosophy, Sterelny (2003) and Buller (2007) have argued against such an ecumenical position. In their view, evolutionary psychology is substantially flawed, while human behavioral ecology and dual inheritance theory are valuable research programs. Rejecting the ecumenical position, however, depends on very strong versions of both the massive modularity hypothesis and the concept (and importance) of the EEA. For instance, Buller interprets the massive modularity hypothesis as follows: "According to this hypothesis, *each* adaptive problem our lineage faced in its Pleistocene past was solved by a dedicated module." (Buller 2007, p. 271; our emphasis) But because the majority of evolutionary psychologists are committed to much weaker (but still informative) versions of both hypotheses, evolutionary psychology, human behavioral ecology, and dual inheritance theory may well be complementary. Obviously, such complementarity need not entail that all approaches are equally valuable at all times. What it does entail is that all evolutionary studies of human behavior should be informed by all three evolutionary approaches.

Evolutionary explanations of mental illnesses can build on some of the key ideas of evolutionary psychology (e.g., massive modularity), as long as these ideas are evaluated in the light of evidence brought forward by the other approaches. As such, we do not believe that evolutionary psychiatry would do much better without evolutionary psychology, as some have claimed (Panksepp and Panksepp 2000; Gerrans 2007). At the same time, we contend that evolutionary psychiatry can only mature if the concerns, methods, and basic findings of the other evolutionary styles are taken into account. To illustrate the importance of such pluralism, consider the "culture-boundedness" of at least some mental disorders (Hacking 1998) and the cultural relativity of ideas about mental health and mental disease (Lloyd 2007). In order to make evolutionary sense of these cultural variations, a good understanding of the evolutionary roots of social learning is indispensable. But since evolutionary psychology downplays the importance of social learning as a source of preferences, beliefs, and behaviors, it lacks a good evolutionary account of social learning. What is more, because it lacks such theory, evolutionary psychiatrists who theorize along the lines of evolutionary psychology often preclude the possibility that there are cultural variations in the domain of mental illness. Here dual inheritance theories, with their typical interest in social learning, could be of great value to evolutionary psychiatry.

I.4 Evolution, dysfunction, and mental disorders

Biological determinism is an important issue in the philosophical controversy surrounding human sociobiology. According to Lewontin and other left-wing critics, human sociobiology tries to legitimize the social and political status quo by explaining social inequalities as natural characteristics of our society. In *Not in our Genes*, Lewontin, Kamin, and Rose explored a great many aspects and implications of biological determinism, including its pervasive influence on psychiatry. In their view, Western psychiatrists were "willing or merely compliant agents of political oppression" (Lewontin *et al.* 1984, p. 167) because they biologize and medicalize all behaviors that pose a threat to the existing social order. Thus Western psychiatry would want us to believe, for example, that hyperactive children are diseased, rather than rattled and messed up by their parents, the school system, or the society at large.

Even though this and similar criticism, which was often highlighted by the antipsychiatry movement, has not gained widespread credence among psychiatrists, it has certainly generated many interesting questions in the philosophy of psychiatry. What, if anything, do we mean when we claim some or other condition to be a mental disorder? Are mental disorders biologically real? What about psychiatric classification? Should it be based on scientific facts, or is it always, and inevitably, a normative enterprise? The second part of the present volume shows that, in answering such questions, philosophers of psychiatry often appeal to evolutionary theory and to the function debate in the contemporary philosophy of biology. In this section, we will provide a brief introduction to this important theme in the philosophy of psychiatry.

The philosophical debate about the concept of mental disorder is mainly a discussion between normativists and naturalists. Naturalists hold that (mental) health is a natural concept, while normativists argue that it is a normative one. A third strand defends a hybrid concept, claiming that the concept of mental disorder involves a conjunction of facts and values. Evolutionary theory plays a prominent role in naturalist and hybrid approaches. The naturalist Christopher Boorse initiated the debate by arguing for a survival and reproduction account of the concept of disorder. Criticizing Boorse's account, Jerome Wakefield developed a hybrid approach, claiming that mental disorders are harmful failures of naturally selected functions.

It should be noted here that, by claiming the concept of mental disorder to be a nonnormative concept, naturalists are not claiming that there are no norms involved in determining whether a condition is a disorder. In fact, all defenders of naturalist approaches are convinced that a mental disorder is a deviation from a norm, but unlike normativists they argue that the relevant norm is a *biological* norm. The paradigmatic cases of mental disorders, such as schizophrenia and major depressive disorder, are understood as dysfunctions, and a dysfunctional psychological mechanism is a psychological mechanism that fails to conform to a natural norm. The main debate within the naturalist school is on the nature of this norm; while they all agree that the relevant norm can be scientifically discovered, they disagree on which science can do the job.

Boorse's bio-statistical account of the concept of mental disorder implies that mental disorders are failures of normal species functions. The normal species function of a mental mechanism or process is the statistically typical contribution by that mechanism or process to the individual's survival and reproduction. Because Boorse's analysis of function hinges on species-typical design and on survival and reproduction, its connection to evolutionary theory is obvious. In fact, Boorse himself has emphasized that his notion of species-typical design is completely consistent with contemporary evolutionary theory:

> The typical result of evolution is precisely a trait's becoming established in a species, only rarely showing major variations under individual inheritance and environment. On all but evolutionary time scales, biological designs have a massive constancy vigorously maintained by normalizing selection.
>
> (Boorse 1977, p. 557)

Since the function of a mechanism is its causal contribution to the biological goals of survival and reproduction, it follows that the criterion of lowered survival or lower reproductive fitness is proposed as the purely scientific means to identify disorder.

Even though Boorse's account is clearly influenced by evolutionary theory, he distinguishes his bio-statistical account from so-called "adaptation accounts" or "fitness accounts" of function, such as Wakefield's. Boorse argues that health is not identical to adaptation, and, hence, that disease is not the absence of such adaptations. The most important reason for Boorse to reject adaptation or fitness approaches is that such approaches would entail the environmental relativity of mental disorders. In his view, for example, conditions like myopia do not cease to be diseases in special environments where they would be advantageous. It is here that the views of Boorse and Wakefield diverge. Contrary to Boorse, Wakefield would claim that myopia would actually cease to be a disease in an environment where it would turn out to be adaptive. According to Wakefield, a person is only suffering from a disorder if, and only if, the dysfunction causes the person some kind of harm. For this reason, Wakefield adopts a hybrid view of the concept of mental disorder, rather than a purely naturalistic view. He argues that his harmful dysfunction account preserves the important insights of both naturalistic and normative accounts—the normativists' insight being that social values are always involved in identifying mental disorders "because disorders are negative conditions that justify social concern" (Wakefield 1992, p. 376).

The value component in Wakefield's harmful dysfunction analysis has been underexposed in the literature about the concept of mental disorder (but see De Block 2008). In our view, this component can be partially incorporated in Boorse's conceptual framework. Boorse calls an individual *ill* when the individual has a *disease* that is serious enough to be incapacitating or harmful (Boorse 1975), so he would say that an illness is a harmful dysfunction, whereas Wakefield would claim that a disorder is a harmful dysfunction. However, there are more important differences between their accounts. For one thing, even though Boorse and Wakefield agree that the concept of disorder/disease involves dysfunction, they clearly disagree on what the appropriate account of function is. In Boorse's account, the relevant effects of a mental organ that constitute the organ's natural function are those that contribute to the organism's survival and reproduction. Wakefield argues that the relevant effects are the effects that explain the organ's presence. Wakefield's so-called etiological theory of biological functions explains the function of an organ for which it is an adaptation (or for which it is selected). This means that, in Wakefield's view, quite a few mental disorders do not have an effect on mortality or fertility. However, these conditions are mental disorders because they fail to produce the effects that led to their selection.

In the present volume there are four chapters dealing with this issue, so we will not take it further here. Rather, in the remainder of this section we will to discuss the criteria to assess these naturalist or semi-naturalist accounts of mental disorder, and to indicate why the choice between them matters for psychiatric classification.

The first and most obvious criterion is whether the proposed approach can be applied to conditions that are uncontroversially considered to be mental disorders. For example, since most people agree that schizophrenia and major depressive disorder are paradigmatic mental disorders, any successful analysis of the concept of mental disorder must be able to account for these disorders.

A second and related criterion used to assess the value of an account is whether it is possible to come up with counterexamples and, perhaps more importantly, how many counterexamples there are. Proponents of the bio-statistical view have come up with some counterexamples that are claimed to disprove the harmful dysfunction approach, while Wakefield has produced examples of mental disorders that are not abnormal in a bio-statistical sense. However, the presence of counterexamples need not always be fatal for a theory or a definition. Ideally, a successful approach should be immune to them, but as long as the list of such counterexamples does not include paradigmatic mental disorders, opponents can duck the problem by denying that the counterexample in question is generally considered to be a disorder. Since there are many debates over the psychiatric nature of a condition, this strategy is often successful. Of course, another equally successful strategy is to argue that the proposed counterexample of a mental disorder is actually dysfunctional in the relevant sense, but that this is overlooked by the critic.

Thirdly, it is important to assess how the concepts of function and dysfunction are understood by psychiatrists and other mental health practitioners. For instance, some have argued that function talk in biology differs from function talk in the medical sciences. In this view, neuroscientists and psychiatrists would be interested not in the evolutionary history of a function, but rather in the activities an organ or an organism can perform (Boorse 2002). Philosophers like Godfrey-Smith believe that both functions are real and that both views can make the distinction between function and dysfunction. According to Godfrey-Smith, only one concept of function works well for behavioral ecology and evolutionary biology, while another concept of function works best for biochemistry, developmental biology, and the medical sciences (Godfrey-Smith 1993). A similar theoretical argument is developed by Roe and Murphy in their contribution to this volume (Chapter 8). They attempt to show that their systemic capacity view of function provides a better account of how functions are understood in medical practice in general and psychiatric practice in particular.

The debate about the concept of mental disorder is quintessentially philosophical, but at the same time it seems to be one of the few philosophical debates that actually matter to other academic disciplines too. A successful analysis of the concept of mental disorder leads us to a better understanding of the nature of mental disorders, and hence to a solid and reliable criterion to distinguish sound and unsound ascriptions of mental disorder. Together with the sociologist Allan Horwitz, Wakefield has appealed to his harmful dysfunction analysis to argue that many conditions are now diagnosed as disordered, while they are in fact completely normal. In *The Loss of Sadness* (Horwitz and Wakefield 2007), they claim that major depressive disorder is really a mental disorder because it is a harmful dysfunction of our evolved loss response mechanisms. However, because of the symptom-based approach of the *Diagnostic and Statistical Manual of Mental Disorders*, intense normal sadness following an adverse life event is nowadays confused with depressive disorder. In other words, if the *Diagnostic and Statistical Manual* would replace its symptom-based approach with a harmful dysfunction approach, psychiatrists would realize that low mood following an important loss is not abnormal: it is exactly what the loss response mechanism is designed (naturally) to produce in that context. In their chapter for this volume, Randolph Nesse and Eric Jackson argue along similar lines, claiming that evolutionary theory is the missing biological foundation for diagnosing and classifying mental disorders (Chapter 6).

It should be mentioned, however, that not everyone shares this optimism. Is it really true that a good naturalist account of the concept of mental disorder provides the necessary tools to revolutionize psychiatric classification? Skeptics admit that such an account would surely provide a rough kind of demarcation point between disorder and normality, but they argue that it would not enable us to close the case on controversial conditions, and, in their view, there is not much hope that this will ever be possible. Suppose that mental disorders are indeed harmful failures of naturally selected functions, as Wakefield would have it. In that case, a condition is a disorder if it can be shown that it lowers evolutionary fitness and has a selective history. However, providing the evidence for these assumptions poses enormous practical and scientific difficulties, as Nettle argues in his chapter for this volume (see also Murphy 2006). To some extent, Wakefield himself admits that we should not expect too much of this philosophical enterprise:

> Because of the complexity of the inferences involved in judgments of dysfunction and our relative ignorance about the evolution of mental functioning, it is easy to arrive at differing judgments about mental dysfunction even on the basis of the same data.
>
> (Wakefield 1992, p. 386)

1.5 *Das kranke Tier*: evolution, psychopathology, and human nature

In this final section we will examine the implications of evolutionary psychiatry for philosophical thinking about human nature. As the three chapters of the third part of this volume illustrate, evolutionary explanations of mental disorders have often been a source of inspiration for philosophers, especially in their reflections on the human predicament. Here we consider a number of traditions in philosophical anthropology that explicitly refer to mental disorders, and we attempt to uncover how these traditions have been informed by evolutionary theory.

On numerous occasions the famous philosopher Friedrich Nietzsche expressed his contempt for modern man by comparing the latter to an ailing animal ("das kranke Tier") and even to a mad animal ("das wahnsinnige Tier"). As he explained himself: "For man is more ill, uncertain, changeable and unstable than any other animal, without a doubt,—he is the sick animal"

(Nietzsche 1994 [1887], 88). Western civilization has seriously jeopardized the human animal's mental health. The only cure for this sickness of humanity is to let go everything that is all too human, and to become an animal again (Ham 1997). In Nietzsche's view, ignoring and denying our instincts has made us weak, sickly, and even mad. Of course, the idea that the domestication of our instincts has made us vulnerable in many ways is much older than Nietzsche's philosophy. Yet Nietzsche was certainly the first to link up this claim with a Darwinian and naturalist view of life (Richardson 2004).[3]

A highly similar analysis of the human condition is one of the key elements in George Estabrooks's Man: *The Mechanical Misfit* (Estabrooks 1941). Here the author sings of the perfect mind and body of the Cro-Magnon, the oldest known modern humans in Europe. The life of Cro-Magnon may not have been as bucolic as Rousseau would have wanted us to believe, but at least they lived their lives in accordance with their instincts. The rise of civilization soon brought an end to this natural state, and burdened Cro-Magnon's successor with a catalogue of complaints. The rising population density initiated an explosion of infections, such as syphilis, while urbanization led to various stress diseases, including stomach ulcers. In short, compared to the Cro-Magnon, modern man is a "degenerate mammal", a "mechanical misfit in modern civilization" (Estabrooks 1941, p. 105).

Following Nietzsche, it became common practice in philosophy to understand the vulnerabilities of human nature with the help of evolutionary principles and processes. Most often, philosophers have claimed that it is modern society that ruined us and that only a return to a more "natural" way of life could put an end to our chronic misery. To some extent, the work of Sigmund Freud is part of this tradition that ends with today's mismatch explanations in evolutionary psychology and evolutionary psychiatry. Freud agreed with Nietzsche in considering our vulnerability to be a defining characteristic of

---

[3] Conversely, Nietzsche also criticized Darwin in suggesting that he should give a central role to what he calls "ennoblement through degeneration". In more modern terms, Nietzsche thought that the evolvability of our species crucially depends on the presence of the weak: "Wherever progress is to ensue, deviating natures are of greatest importance. Every progress of the whole must be preceded by a partial weakening. The strongest natures retain the type, the weaker ones help to advance it. Something similar also happens in the individual. There is rarely a degeneration, a truncation, or even a vice or any physical or moral loss without an advantage somewhere else. In a warlike and restless clan, for example, the sicklier man may have occasion to be alone, and may therefore become quieter and wiser; the one-eyed man will have one eye the stronger; the blind man will see deeper inwardly, and certainly hear better. To this extent, the famous theory of the survival of the fittest does not seem to me to be the only viewpoint from which to explain the progress of strengthening of a man or of a race." (Nietzsche 1984 [1878], p. 224)

human nature, and in explaining this vulnerability by referring to the discrepancy between ancestral and contemporary environments. Yet Freud was more pessimistic than Nietzsche. Man's misery, he says, is ultimately inevitable because all major mental disorders, including the psychoses, are side-effects of highly useful human traits, such as puberty and high levels of parental care. As long as the advantages of these traits are bigger than the disadvantages of these disorders, humans will remain ailing animals (De Block 2005). Jonathan Burns's contribution to the present book fits in nicely with this tradition in philosophical anthropology, as he argues that schizophrenia—"this most human of maladies"—may represent a costly downside to the emergence of embodied social consciousness. In a highly similar vein, the famous German philosophical anthropologist Helmuth Plessner offered an "evolutionary" explanation that accounted both for our vulnerability to mental disorders and for other uniquely human features, such as the "eccentric" intentionality of human consciousness (Moss 2005). According to Plessner, the origins for these features have to be sought in the unfinished and helpless state in which the child is born.

In short, some of the greatest thinkers in philosophy have argued that our vulnerability to mental disorders distinguishes humans from other animals, and, to account for this vulnerability, many of them have embedded their claims in an explicitly Darwinian framework. The question is, however, whether or not these evolutionary explanations generate genuinely philosophical insights. Let us suppose, for instance, that Plessner's explanation is theoretically and empirically warranted. Then how could his evolutionary account steer the philosophical exploration of human intentionality? Does the origin of human intentionality matter because it tells us something about the structure of that intentionality? More generally, evolutionary biology seems to give us rather straightforward answers to the questions "What are we?" and "How did we come to be?", but it is still unclear if these scientific answers have the philosophical depth they are often thought to have. Many philosophical issues can be, and have been, illuminated by scientific findings, and most philosophers believe that the biological sciences should constrain philosophical theories about human nature. But not everyone agrees that there are controversies in philosophical anthropology that can be decided on the basis of evolutionary theory.

Psychiatric phenomena have also played an important role in another tradition in philosophical anthropology, particularly as it is laid down in the works of Karl Jaspers, Martin Heidegger, Maurice Merleau-Ponty, and many others. For these philosophers, mental disorders matter, not so much because they highlight the human–animal divide, but rather because they help us to elucidate certain aspects of normal human behavior, cognition, and experience. Part of this potential is due to the fact that some crucial human capacities go unnoticed until they happen to malfunction, but it is equally possible that mental illnesses show in an exaggerated and magnified way tendencies and mechanisms that constitute our humanity. Much in line with this tradition, Freud also once compared psychopathology to a broken crystal:

> We are familiar with the notion that pathology, by making things larger and coarser, can draw our attention to normal conditions which would otherwise have escaped us. Where it points to a breach or a rent, there may normally be an articulation present. If we throw a crystal to the floor, it breaks; but not into haphazard pieces. It comes apart along its lines of cleavage into fragments whose boundaries, though they were invisible, were predetermined by the crystal's structure. Mental patients are split and broken structures of the same kind.
>
> (Freud 1933, p. 58)

Interestingly, while this principle of understanding human nature by looking at its disordered versions is very characteristic of twentieth-century so-called "continental" philosophy, it is also central in much of contemporary cognitive science and evolutionary psychology. As Machery explains in Chapter 3, evolutionary psychologists often refer to cognitive impairments in patients suffering from developmental disorders to support their hypotheses about the architecture of normal human cognition. Similarly, John Price's chapter in this volume (Chapter 10) documents how an evolutionarily informed understanding of mood changes and mood disorders can shed a new light on the structure of human relationships, and, conversely, how Gregory Bateson's ideas about social dynamics have been a source of inspiration for a number of evolutionary psychiatrists. Apart from this remarkable methodological similarity, it is not immediately obvious how this tradition in philosophical anthropology can be informed by evolutionary thinking. After all, its primary focus is on a refined phenomenology of normal and abnormal cognition and experience. Evolutionary or other scientific accounts of those cognitions and experiences seem to add very little to such phenomenological work. For instance, if everything is indifferent to patients with depression, this emptiness might tell us something about how human beings experience time, and about how this experience differs from the temporality experienced by non-human animals. But it is difficult to see how an evolutionary account of depression could complement or complete the description of depressive emptiness (or of our species-typical experience of time).

Still, we think that there are a number of reasons why a Darwinian approach can be useful to understand specific key issues in philosophical anthropology and philosophical psychology.

Firstly, evolutionary theory has been of critical importance in the philosophical debate on essentialism. Insofar as philosophical anthropology is engaged in a search for the essential characteristics of our species, it should take note of evolutionary accounts of essentialism. According to David Hull, for example, Darwinian theory shows that our species lacks anything that might be termed an "essence" (Hull 1998). At the same time, evolutionary research into how people from all over the world conceptualize the living world has resulted in interesting hypotheses about why we often think along essentialist lines, for example about species and even about mental disorders (Atran 1999, Keller and Miller 2006; Adriaens and De Block, in press).

Secondly, nineteenth- and twentieth-century philosophical anthropology has consistently criticized the definition of man as a rational animal. Part of this criticism consisted of showing that human rationality wasn't as autonomous as Descartes and other modern philosophers had claimed. Marx, Nietzsche, and Freud all argued that human rationality depends on more basic drives, instincts, and urges, which they thought to be revealed by all sorts of "symptoms" that Freud described as a "psychopathology of everyday life". Today, evolutionary "takes" on mental disorders can be used for similar purposes. In principle, they could inform us about the natural origins and limits of human rationality. For example, as Focquaert and Braeckman argue in their contribution to this book, a natural history of autism might further our understanding of social intelligence and empathy, if only by showing its evolutionary underpinnings. Moreover, they contend that Simon Baron-Cohen's evolutionary explanation of autism also informs us to what extent some of the sex differences are innate.

1.6 Conclusion

The philosophy of psychiatry is one of the most exciting areas in philosophy. One of the great things about it is that it involves a high degree of cross-talk between philosophers and psychiatrists. We believe that it is of utmost importance that philosophical approaches to evolutionary trends in psychiatry adopt a similar interdisciplinary awareness, that is, awareness of the work done in evolutionary biology, philosophy of biology, philosophy of psychiatry, and evolutionarily inspired psychiatry. We hope that this volume and its contributors, who come from a variety of disciplines, give a sense of why interdisciplinary dialogue is both necessary and exciting.

References

Adriaens, P. (2007) Evolutionary psychiatry and the schizophrenia paradox: a critique. *Biology & Philosophy*, 22 (4), 513–28.

Adriaens, P. and De Block, A. (2010) The evolutionary turn in psychiatry: a history. *History of Psychiatry*, **21** (2), 131–43.

Adriaens, P. and De Block, A. (in press) Why we essentialize mental disorders. *Journal of Medicine & Philosophy*.

Allison, A. (1954) Protection afforded by sickle cell trait against subtertian malarian infection. *British Medical Journal*, **1**, 290–4.

Atran, S. (1999) The universal primacy of generic species in folk biological taxonomy: Implications for human biological, cultural, and scientific evolution. In R. Wilson (ed.), *Species: New Interdisciplinary Essays*. MIT Press, Cambridge, MA, pp. 229–61.

Boaz, N. and Almquist, A. (2002) *Biological Anthropology: A Synthetic Approach to Human Evolution*. Prentice Hall, New Jersey.

Boorse, C. (1975) On the distinction between disease and illness. *Philosophy and Public Affairs*, **5**, 49–68.

Boorse, C. (1977) Health as a theoretical concept. *Philosophy of Science*, **44**, 542–73.

Boorse, C. (2002) A rebuttal on functions. In A. Ariew, R. Cummins, and M. Perlman (eds), *Functions: New Readings in the Philosophy of Psychology and Biology*. Oxford University Press, Oxford, pp. 63–112.

Boyd, R. and Richerson, P. (1985) *Culture and the Evolutionary Process*. University of Chicago Press, Chicago.

Boyer, P. and Barrett, H. (2005) Evolved intuitive ontology: integrating neural, behavioral and developmental aspects of domain-specificity. In D. Buss (ed.), *Handbook of Evolutionary Psychology*. Wiley, New York, pp. 96–118.

Buller, D. (2007) Varieties of evolutionary psychology. In M. Ruse and D. Hull (eds), *Cambridge Companion to the Philosophy of Biology*. Cambridge University Press, New York, pp. 255–74.

Burns J. (2007) *The Descent of Madness: Evolutionary Origins of Psychosis and the Social Brain*. Routledge, London.

Buss, D. (2000) *The Dangerous Passion: Why Jealousy Is as Necessary as Love and Sex*. Free Press, New York.

Carruthers, P. (2006) *The Architecture of the Mind:* Massive Modularity and the Flexibility of Thought. Oxford University Press, Oxford.

Crow, T. (2000) Schizophrenia as the price that Homo sapiens pay for language: a resolution of the central paradox in the origin of the species. *Brain Research Reviews*, **31**, 118–29.

Currie, G. and Sterelny, K. (2000) How to think about the modularity of mindreading. *Philosophical Quarterly*, **50**, 145–60.

Darwin, C. (1859) *On the Origin of Species by Means of Natural Selection*. John Murray, London.

Darwin, C. (1868). *The Variation of Animals and Plants under Domestication*. John Murray, London.

Darwin, C. (1871) *The Descent of Man and Selection in Relation to Sex*. John Murray, London.

Darwin, C. (1872) *The Expression of the Emotions in Man and Animals*. John Murray, London.

De Block, A. (2005) Freud as an "evolutionary psychiatrist" and the foundations of a Freudian philosophy. *Philosophy, Psychiatry & Psychology*, **12** (4), 315–24.

De Block, A. (2008) Why mental disorders are just mental dysfunctions (and nothing more): some Darwinian arguments. *Studies in the History and Philosophy of the Biological and Biomedical Sciences*, **39** (3), 338–46.

De Bont, R. (2010) Schizophrenia, evolution and the borders of biology: on Huxley *et al.*'s 1964 paper in *Nature*. *History of Psychiatry*, **21** (2), 144–59.

Demaret, A. (1979) *Ethologie et Psychiatrie. Valeur de Survie et Phylogenèse des Maladies Mentales*. Mardaga, Bruxelles.

Dennett, D. (1995). *Darwin's Dangerous Idea. Evolution and the Meanings of Life*. Simon & Schuster, New York.

Erlenmeyer-Kimling, L. (1968) Mortality rates in the offspring of schizophrenic parents and a physiological advantage hypothesis. *Nature*, **220**, 798–800.

Estabrooks, G. (1941) *Man. The Mechanical Misfit*. The Macmillan Company, New York.

Freud, S. (1933) New Introductory Lectures on Psycho-analysis. In *New Introductory Lectures on Psycho-analaysis and Other Works. The Standard Edition of the Complete Psychological Works of Sigmund Freud*, vol. 22. Hogarth Press, London, pp. 3–182.

Freud, S. (1987 [1915]) *A Phylogenetic Fantasy. Overview of the Transference Neuroses*. Cambridge University Press, Cambridge.

Gerrans, P. (2007) Mechanisms of madness. Evolutionary psychiatry without evolutionary psychology. *Biology and Philosophy*, **22**, 35–56.

Godfrey-Smith, P. (1993) Functions: consensus without unity. *Pacific Philosophical Quarterly*, **74**, 196–208.

Hacking, I. (1998) *Mad Travelers. Reflections on the Reality of Transient Mental Illness*. University Press of Virginia, Charlottesville.

Hagen, E. (1999) The functions of postpartum depression. *Evolution and Human Behavior*, **20**, 325–59.

Ham, J. (1997) Taming the beast: Animality in Wedekind and Nietzsche. In J. Ham and M. Senior (eds), *Animal Acts: Configuring the Human in Western History*. Routledge, London, pp. 145–63.

Horrobin, D. (2001) *The Madness of Adam and Eve. How Schizophrenia Shaped Humanity*. Bantam Press, London.

Horwitz, A. and Wakefield, J. (2007) *The Loss of Sadness: How Psychiatry Transformed Normal Sorrow into Depressive Disorder*. Oxford University Press, Oxford.

Hull, D. (1998) On human nature. In D. Hull and M. Ruse (eds), *The Philosophy of Biology*. Oxford University Press, Oxford, pp 383–97.

Huxley, J., Mayr, E., Osmond, H., and Hoffer, A (1964) Schizophrenia as a genetic morphism. *Nature*, **204**, 220–1.

Irons, J. (1998) Adaptively relevant environments versus the environment of evolutionary adaptedness. *Evolutionary Anthropology*, **6** (6), 194–204.

Jamison, K. (1993) *Touched With Fire: Manic Depressive Illness and the Artistic Temperament*. Free Press, New York.

Keller, M. and Miller, G. (2006) Resolving the paradox of common, harmful, heritable mental disorders: Which evolutionary genetic models work best? *Behavioral and Brain Sciences*, **29**, 385–452.

Keller, M. and Nesse, R. (2005) Is low mood an adaptation? Evidence for subtypes that match precipitants. *Journal of Affective Disorders*, **86** (1), 27–35.

Keller, M. and Nesse, R. (2006) The evolutionary significance of depressive symptoms: Different adverse situations lead to different depressive symptom patterns. *Journal of Personality and Social Psychology*, **91** (2), 316–30.

Kevles, D. (1985) *In the Name of Eugenics: Genetics and the Uses of Human Heredity*. Alfred Knopf, New York.

Kitcher, P. (1987) *Vaulting Ambition: Sociobiology and the Quest for Human Nature*. MIT Press, Cambridge, MA.

Lewontin, R., Rose, S., and Kamin, L. (1984) *Not in Our Genes: Biology, Ideology and Human Nature*. Pantheon Books, New York.

Lloyd, G. (2007) *Cognitive Variations: Reflections on the Unity & Diversity of the Human Mind*. Oxford University Press, Oxford.

Lumsden, C. and Wilson, E. (1981) *Genes, Mind, and Culture*. Harvard University Press, Cambridge, MA.

Machery, E. (2008) The folk concept of intentional action: philosophical and experimental issues. *Mind & Language*, **23**, 165–89.

Marks, I. and Nesse, R. (1994) Fear and fitness: An evolutionary analysis of anxiety disorders. *Ethology and Sociobiology*, **15**, 247–61.

Maudsley, H. (1916) *Organic to Human: Psychological and Sociological*. Macmillan, London and New York.

McGuire, M. and Troisi, A. (1998) *Darwinian Psychiatry*. Oxford University Press, Oxford.

Moss, L. (2005) Darwinism, dualism and biological agency. In V. Hösle and C. Illies (eds), *Darwinism & Philosophy*. University of Notre Dame Press, Notre Dame, pp. 349–63.

Murphy, D. (2005) Can evolution explain insanity? *Biology & Philosophy*, **20**, 745–66.

Murphy, D. (2006) *Psychiatry in the Scientific Image*. MIT Press, Cambridge, MA.

Nagel, T. (1986) *The View From Nowhere*. Oxford University Press, Oxford.

Nesse, R. (1990) Evolutionary explanations of emotions. *Human Nature*, **1**, 261–89.

Nesse, R. (1999) Testing evolutionary hypotheses about mental disorders. In S. Stearns (ed.), *Evolution in Health and Disease*. Oxford University Press, New York, pp. 260–6.

Nesse, R. (2001) The smoke detector principle. Natural selection and the regulation of defensive responses. *Annals of the New York Academy of Sciences*, **935**, 75–85.

Nesse, R. (2005) Evolutionary psychology and mental health. In D. Buss (ed.), *The Handbook of Evolutionary Psychology*. John Wiley and Sons, New York, pp. 903–37.

Nettle, D. (2001) *Strong Imagination: Madness, Creativity and Human Nature*. Oxford University Press, Oxford.

Nettle, D. (2004) Evolutionary origins of depression: A review and reformulation. *Journal of Affective Disorders*, **81**, 91–102.

Nietzsche, F. (1984 [1878]) *Human, All Too Human*. M. Faber (ed.), University of Nebraska Press, Lincoln.

Nietzsche, F. (1994 [1887]) *On the Genealogy of Morality*. K. Ansell-Pearson (ed.), Cambridge University Press, Cambridge.

Ohman, A. and Mineka, S. (2001) Fears, phobias, and preparedness: Toward an evolved module of fear and fear learning. *Psychological Review*, **108**, 483–522.

Panksepp, J. and Panksepp, J. (2000) The seven sins of evolutionary psychology. *Evolution and Cognition*, **6** (2), 108–31.

Pearn, A. (2010) 'This excellent observer. . .': The correspondence between Charles Darwin and James Crichton Browne. *History of Psychiatry*, **21** (2), 160–75.

Pinker, S. (1997) *How the Mind Works*. Norton & Company, New York.

Potts, R. (1998) Environmental hypotheses of hominin evolution. *American Journal of Physical Anthropology*, **107**, 93–136.

Queller, D. (1995) The spaniels of St Marx and the Panglossian paradox: a critique of a rhetorical programme. *Quarterly Review of Biology*, **70**, 485–89.

Richardson, J. (2004) *Nietzsche's New Darwinism*. Oxford University Press, Oxford.

Richerson, P. and Boyd, R. (2005) *Not By Genes Alone: How Culture Transformed Human Evolution*. University of Chicago Press, Chicago.

Segerstrale, U. (2000) *Defenders of the Truth: The Sociobiology Debate*. Oxford University Press, Oxford.

Smith, E. (2000) Three styles in the evolutionary analysis of human behavior. In L. Cronk, N. Chagnon, and W. Irons (eds), *Adaptation and Human Behavior: An Anthropological Perspective*. Aldine De Gruyter, New York, pp. 27–46.

Smith, E., Borgerhoff-Mulder, M., and Hill, K. (2001) Controversies in the evolutionary social sciences: a guide for the perplexed. *Trends in Ecology & Evolution*, **16** (3), 128–35.

Sperber, D. (2005) Modularity and relevance: How can a massively modular mind be flexible and context-sensitive? In P. Carruthers, S. Laurence, and S. Stich (eds), *The Innate Mind: Structure and Contents*. Oxford University Press, Oxford, pp. 53–68.

Sterelny, K. (2003) *Thought in a Hostile World*. Blackwell, Oxford.

Sterelny, K. and Griffiths, P. (1999) *Sex and Death: An Introduction to Philosophy of Biology*. University of Chicago Press, Chicago.

Stevens, A. and Price, J. (2000) *Evolutionary Psychiatry: A New Beginning*. Routledge, London.

Sulloway, F. (1979) *Freud, Biologist of the Mind. Beyond the Psychoanalytic Legend*. Harvard University Press, Cambridge, MA.

Tinbergen, N. (1940) Die Ubersprungbewegung. *Zeitschrift für Tierpsychologie*, **4**, 1–40.

Tinbergen, N. and Tinbergen, E.A. (1972) *Early Childhood Autism—An Ethological Approach*. Parey, Berlin.

Tooby, J. and Cosmides, L. (1990) The past explains the present: Adaptations in the structure of ancestral environments. *Ethology and Sociobiology*, **11**, 375–424.

Tooby, J. and Cosmides, L. (1997) *Evolutionary psychology: a primer*. Online publication: http://www.psych.ucsb.edu/research/cep/primer.html.

Torrey, E. and Yolken, R. (2010) Psychiatric genocide: Nazi attempts to eradicate schizophrenia. *Schizophrenia Bulletin*, **36**, 26–32.

Vicedo, M. (2010) The evolution of Harry Harlow: from the nature to the nurture of love. *History of Psychiatry*, **21** (2), 190–205.

Wakefield, J. (1992) The concept of mental disorder: On the boundary between biological facts and social values. *American Psychologist*, 47, 373–88.

Wilson, E. (1975) *Sociobiology: The New Synthesis*. Harvard University Press, Cambridge, MA.

Wittgenstein, L. (1922) *Tractatus Logico-Philosophicus*. Routledge, London.

Part 1

Evolutionary psychiatry and its critics

Chapter 1

Fearing new dangers: phobias and the cognitive complexity of human emotions

Luc Faucher and Isabelle Blanchette

Philosophers of science have often been critical of ultimate explanations of mental illnesses. One of the few cases that managed, thus far, to keep up in this storm is the evolutionary account of phobias, of which Dominic Murphy once said that it is 'the best current candidate for an evolutionary explanation of mental disorder' (Murphy 2005, 746). The basic assumption behind this particular explanation is that our fear-regulating mechanisms are modular. This assumption would then explain why we tend to be selective in what we fear, i.e. why phobias are often about 'archaic' dangers, such as snakes and spiders, and why such phobic processes are mostly automatic, i.e. beyond cortical control. In this chapter, we challenge this theory on many counts. Thus we argue that the effects used to conclude to the existence of a prepared mechanism of fear are also shown by 'novel' objects, such as guns and syringes. Moreover, we also show that threat detection is the result of a larger process involving both limbic and cortical structures. In our view, these objections indicate that our fear mechanisms are probably much more flexible than evolutionary psychologists and psychiatrists seem to think.

In their seminal paper, Williams and Nesse (1991) announced that we were at the "dawn of a Darwinian medicine" that would revolutionize the way we think about medical disorders and the way we treat them. Psychiatry, Williams and Nesse proposed, would also profit from taking the Darwinian turn. According to Williams and Nesse, psychiatry would gain both better taxonomic tools (with categories like "evolved defense" and "environmental mismatch" added to its theoretical arsenal; see also Chapter 6) and a better understanding of particular mental disorders. Adopting a Darwinian perspective should result in theoretical as well as empirical progress in psychiatry.

In this chapter we will not question the capacity of an evolutionary perspective to produce theoretical progress in psychiatry. Instead, we will look at one of the paradigmatic explanations of a mental disorder: the explanation of phobias. Our goal is to challenge this explanation. In order to do that, we will first describe the explanation of specific phobias proposed by some evolutionary psychiatrists (section 1.1). Using empirical results from one of us (Blanchette 2006; Brown *et al.* submitted), we will argue that the effects used to conclude the existence of a prepared mechanism of fear (section 1.2), which plays a crucial role in the evolutionary explanation of phobias, are also shown by new (on an evolutionary scale) objects, like syringes and guns (section 1.3). This, together with other problems that we will present (section 1.4), makes us suspicious of the existence of the kind of fear mechanism postulated by some of the leading evolutionary psychiatrists. We will then discuss the mechanisms underlying fear behaviors, what they reveal about our cognitive/affective architecture, and the constraints placed on these mechanisms by evolution (section 1.5). This will lead us to conclude that the fear mechanism is probably much more flexible in humans than suspected by evolutionary psychiatrists. If we are right, we will have shown that the evolutionary perspective, at least in this paradigmatic case, has not produced the kind of empirical progress in psychiatry that Williams and Nesse predicted. We will also have shown that this perspective is compatible with a variety of different empirical theories, some of which might lead to much less change in the way we define and taxonomize mental disorders than is claimed by advocates of evolutionary psychiatry.

1.1 The case of phobia

1.1.1 Definition of phobia

If you define phobia as an unreasonable fear of a particular type of object or situation, then almost all of us are subject to phobia at some stage in our lives. However, the fear that most people experience is "sub-clinical" (Davey 2007, p. 247) in that it is not causing major distress or disrupting daily life and usually does not get labeled phobia. By contrast, when the fear experienced has severe and distressful effects, it is clinically diagnosable as a phobia. Indeed, interference with daily activities is what makes fear the object of clinical attention. As the *Diagnostic and Statistical Manual of Mental Disorders* (DSM-IV-TR) puts it,

> The essential feature of specific phobia is marked and persistent fear of clearly discernible, circumscribed objects or situations. Exposure to the phobic stimulus almost invariably provokes an immediate anxiety response [. . .]. Although adolescents and adults with this disorder recognize that their fear is excessive or unreasonable, this may not be the case with children. Most often, the phobic stimulus is avoided, although it is sometimes endured with dread. The diagnosis is appropriate *only if the avoidance fear, or anxious anticipation of encountering the phobic stimulus, interferes significantly with the person's daily routine, occupational functioning, or social life, or if the person is markedly distressed about having the phobia.*
>
> (DSM-IV-TR; our emphasis)[1]

As we mentioned before, being *disposed* to intense bouts of fear per se is not by itself enough to be the object of clinical attention. The person also has to be in a context where he or she is likely to be in contact with a token of the phobic stimulus, so the disposition he or she has gives rise to *actual* (debilitating) bouts of fear:

> For example, a person who is afraid of snakes to the point of expressing intense fear in the presence of snakes would not receive a diagnosis of Specific Phobia if he or she lives in an area devoid of snakes, is not restricted in activities by the fear of snakes, and is not distressed about having a fear of snakes.
>
> (DSM-IV-TR)

According to the DSM, although the objects of phobias can be exotic and strange, like chinophobia (fear of snow), dendrophobia (fear of trees), and cathisophobia (fear of sitting), phobias tend to focus around a few kinds of objects or situations, the main ones being:

- animal phobias (including snakes, spiders, rats, maggots and slugs)
- social phobias
- dental phobias
- water phobias
- height phobias
- claustrophobias
- blood–injury–injection phobias.[2]

---

[1] Note that the formulation of the conditions under which one is considered to be phobic leaves ample room for interpretation.

[2] Other features of phobias include an overall ratio of women to men of approximately 2 to 1, with a variation of the sex ratio across the different types of phobias (for instance, 70–90% of the individuals who fear animals and natural environments are female). The age at onset of many specific phobias (animal types, blood–injury–injection types) is usually in childhood, while, according to some, the incidence is highest during the

1.1.2 Evolutionary psychiatry and phobias

While theories in the past might have viewed anxiety and phobias in themselves as problematic and deserving clinical attention,[3] this is not the view taken by evolutionary psychiatrists. Much anxiety might seem, *prima facie*, pathological because it causes pain or is irrational, but in fact it could be a *normal defense* elicited in the presence of certain dangers. Indeed, evolutionary psychiatrists argue that both general anxiety and fear of particular objects[4] and situations can be normal and adaptive reactions. For instance, Nesse and Williams insist on this adaptive feature of anxiety:

> Everyone must realize that anxiety can be useful. We know what happens to the berry picker who does not flee a grizzly bear ... In the face of threat, anxiety alters our thinking, behavior, and physiology in advantageous ways.
>
> (Nesse and Williams 1997, p. 5)

According to this view, anxiety *per se* is not necessarily a disorder, but can be a normal response to certain objects or situations. But anxiety could also be the result of a dysregulation of these defense mechanisms. In order to understand how these mechanisms can *mal*function, one must first understand what their *function* is. Since these mechanisms are emotional in nature, we will first describe briefly what the "normal" functions of emotions are thought to be.

1.1.2.1 Evolutionary theory of emotion

According to proponents of the evolutionary view, emotions are "superordinate programs" (Tooby and Cosmides 2008, p. 116; for another recent

---

reproductive years and diminishes in the elderly (McGuire and Troisi 1998, p. 215). Genetic contributions are suggested by the fact that relatives of a person with a phobia are four to seven times more likely to develop phobias than relatives of persons without phobias (McGuire and Troisi 1998, p. 216). Finally, predisposing factors are of many kinds, including traumatic encounters with the phobic stimulus, observation of others undergoing trauma, unexpected panic attacks in the presence of the phobic stimulus, as well as social transmission (repeated warnings about the dangers of certain animals by parents). Evolutionary psychiatrists sometimes try to explain these features as well, but we won't look at their hypotheses in this chapter.

[3] "Psychiatric emphasis on anxiety as a classifiable "illness" has given rise to the erroneous belief, current through most of this century, that anxiety is "neurotic" and that no well-adjusted person should expect to suffer from it." (Stevens and Price 2000, p. 100)

[4] While fear is an intense negative emotion typically in reaction to an imminent threat, anxiety is a psychological state (more diffuse than fear) with a phenomenal component characterized by an unpleasant feeling that is typically associated with fear or worry. The DSM distinguishes between different kinds of anxiety disorders, phobias being one subgroup of anxiety disorders. Other subgroups include stress disorders, generalized anxiety disorders, and obsessive compulsive disorders.

statement of this position, see Nesse and Ellsworth 2009) whose function is to coordinate or orchestrate the different programs (or modules) that the mind comprises in order to produce an adaptive response to a recurrent challenge to survival. In other words, emotions are

> ... neurocomputational adaptations that have evolved in response to the adaptive problem of matching arrays of mechanism activation to the specific adaptive demands imposed by alternative situations [. . .] Thus each emotion evolved to deal with a particular, evolutionary recurrent situation type. The design features of the emotion program, when the emotion is activated, presume the presence of an ancestrally structured situation type (regardless of the actual structure of the modern world).
>
> (Tooby and Cosmides 2008, p. 117)

In a nutshell, emotions are coordinated patrons of reactions (cognitive, behavioral, physiological) adapted to meet the specific demands of particular adaptive problems that plagued our human ancestors. If emotions are solutions to enduring problems from our evolutionary past, we should expect a high degree of fit between their design features and the adaptive problems they correspond to.

This is what evolutionary psychiatrists claim is the case for non-clinical anxiety and fears: "Subtypes of anxiety probably evolved to give a selective advantage of better protection against a particular kind of danger" (Marks and Nesse 1997, p. 59). We should thus expect the design features of these particular fears to match the particular demands of the adaptive problems they have been selected to solve. This is what one finds according to Marks and Nesse, for instance "[h]eights induce freezing instead of wild flight, thus making one less liable to fall" (Marks and Nesse 1997, p. 61). In this case, as in many others, evolution (through natural selection) would have provided us with a built-in (i.e. innate) solution to the problem of generating the most adaptive behavior in this kind of situation.[5]

Because fear responses are relatively inexpensive and because not responding with fear might be costly or fatal, another feature of fear responses is that they will be triggered more often than not. In other words, false alarms are to be expected. This is what Nesse (2005) calls the "smoke detector principle". According to this principle, evolution has shaped emotions to maximize our reproductive benefits, not our pleasure or our satisfaction. Therefore, even if it would be better from our point of view to experience anxiety or fear less often, from an evolutionary point of view it is better to be safe than sorry.

---

[5] The solution is thought to be built-in because either the most adaptive solution to the problem is somewhat invisible to the individual or it is too costly to make an error or to produce a response too late.

Finally, because the sources of danger that have shaped our fear system have been relatively similar through our history, evolution has prepared us to associate fear more easily with some kinds of stimuli than with others. The stimuli that are more readily associated with fear would be stimuli that were cues for danger in our environment of evolutionary adaptedness (EEA). As Marks puts it:

> It is easy to think of fears that enhanced survival in the past (say, fear of animals) or continue to do so in the present (fear of heights, separation, and perhaps strangers). This idea is intrinsic to more recent and related concepts of prepotency... and preparedness... prepotency indicates that particular stimuli are salient for a given species, which attends selectively to them rather than to others even at their first encounter. Preparedness is the idea that certain stimuli associate selectively with one another and with particular responses, some connections being more available than others.
>
> (Marks 1987, p. 230)

1.1.2.2 Theory of phobias

As we mentioned earlier, anxiety disorders are thought to be mainly disorders of regulation of functional defensive responses to threats. Disorders of regulation could relate to excessive or deficient responses. However, most clinical attention has focused on excessive fear responses, with less concern over lack of fear responses (although this is sometimes mentioned as a feature of mental disorders that are not conceived as the opposite of phobias, like in certain cases of "acquired psychopathy"; e.g., see Anderson *et al.* 1999). Marks and Nesse suggest that cases of phobias are to be understood as cases of overactivity of the specific anxiety mechanisms designed to defend the organism against particular threats.

Murphy (2004, 2005, 2006) claims that Marks and Nesse's suggestion can be given two different readings:

> Sometimes the argument seems to be that phobics suffer from broken anxiety-producing mechanisms, with the result that they become unduly anxious. That makes the theory a breakdown explanation. The mismatch interpretation starts from the claim that anxiety, like most traits, shows phenotypic variation. Individuals toward the sensitive end of the distribution, who become anxious more readily, might have functioned quite normally in ancestral environments. But they suffer in contemporary environments where the stimuli are unreliable guides to danger...
>
> (Murphy 2005, p. 748–9)

Murphy is right to point out that Marks and Nesse seem to favor the second reading, where an intact anxiety-producing mechanism is not suited to modern environments, arguing that it is possible that clinical phobias might just be "extremes of normal forms of anxiety" (Marks and Nesse 1997, p. 61; see also Nesse 1999). They indeed argue that this variation between individuals would be a way to adapt to the variation in the dangerousness of some threats that different generations had to face (Nesse and Williams 1997, p. 21). Many cases of phobias seen in clinic would then be due to the fact that certain individuals exhibit a level of anxiety for objects or situations that might have been required in the past in certain environments, but that is not required in our present-day environments.[6] So in new environments (Murphy 2005, p. 748):

> ... people toward the sensitive end of the distribution of phenotypic variation may be incapable of coping with many ordinary situations despite the fact that all of their mental mechanisms are functioning just as natural selection designed them to function.

If this is the case, the central problem for patients suffering from phobias would be an oversensitive fear and anxiety mechanism. Murphy has suggested that Marks and Nesse's hypothesis should be supported by more work on the underlying (proximal) mechanisms responsible for anxiety. He cites Clark's theory of panic disorder (1997) as a theory that might inspire evolutionary psychiatrists. In Clark's theory, people's panic attacks are caused by *catastrophic misinterpretations of bodily signals*. Murphy thinks that similar misinterpretations caused by oversensitivity to certain kinds of environmental stimuli might explain phobias. For instance, Davey explains that height phobia could be the result not only of heightened discrimination of bodily signals, but also by a "bias toward interpreting ambiguous bodily sensations as threatening" (Davey 2007, p. 249).

If Murphy thinks that this reading is a better interpretation of what Marks and Nesse have in mind than the breakdown hypothesis (phobias being caused by some defective psychological mechanisms), the latter hypothesis should not be dismissed too quickly. Indeed, until they are tested, either hypothesis could be right. In fact some evolutionary psychiatrists seem to prefer the breakdown hypothesis. For instance, McGuire and Troisi write that:

> ...[v]iewing phobias as evolved defensive responses to specific contingencies implies that they can be adaptive. While this is an acceptable interpretation of transient phobias, *it is not easily reconciled with enduring and debilitating phobias*.
>
> (McGuire and Troisi 1998, p. 217; our emphasis)

It is plausible to assume that some cases of phobias might have been as counteradaptive in the past as in the present in that they would have compromised the attainment of biological goals. Being afraid of the dark or of snakes might

---

[6] So if snakes are very dangerous in your current environment, people who are very afraid of them might have an advantage over others who are not or are only mildly afraid of them. In other environments where snakes are not dangerous at all, being extremely afraid of them would be regarded as more of a problem.

be adaptive, but if it does not lead the individuals to take actions to protect themselves against the feared dangers and thus reduce the anxiety, it is not adaptive. It seems that while some cases of phobia might be normal and adaptive, some can be maladaptive if they are extended in time and/or if they are debilitating. As McGuire and Troisi pointed out, the source of maladaptive phobia can be "reduced capacities to process information accurately, to correct recognition distortions, and to discontinue emotional responses once they are initiated" (McGuire and Troisi 1998, p. 219). If such is the case, the problems of phobic individuals would be related to the capacity to accurately evaluate the danger represented by a stimulus, to modify the interpretation of the stimulus in light of further evidence, and to control both the type and the intensity of the response to it.

Whichever type of evolutionary explanation of phobias one favors, it should be clear from the preceding paragraphs that more details about the psychological mechanism(s) involved in phobias will have to be provided to allow an understanding of the condition. Both readings propose that certain characteristics of phobias (like their distribution, but also the fact that they seem irrational and disproportionate—even to those who experience them—as well as the fact that they are experienced as uncontrollable) are to be explained by a particular kind of psychological mechanism, described by some as a "fear module". In the next section, we examine one model of that mechanism adopted by many evolutionary researchers, including Marks and Nesse.[7]

1.2 A module for fear

In their discussion of the distribution of phobias, Nesse and Williams remark that:

> Most of our excessive fears are related to prepared fears of ancient dangers . . . Do anxiety disorders, . . ., result from novel stimuli not found in our ancestral environment? Not often. New dangers such as guns, drugs, radioactivity, and high-fat meals cause too little fear, not too much.
>
> (Nesse and Williams 1997, p. 7)

As we saw earlier, the evidence concerning the existence of prepared fears rests on the mesh between presumed dangers in our EEA and different features of the fear responses that are adapted to the nature of these dangers. It also rests on work in psychology on learning and attention by scientists proposing

---

[7] It should be noted that even if this view is popular among evolutionary psychiatrists, some evolutionary social scientists (e.g. Smith *et al.* 2001) do not endorse this view and are more sympathetic with the view we defend in the last section of this chapter.

the "fear module theory". We will present some of this work in the present section. In the next section, we will review some empirical data that make us skeptical of the conclusions reached by proponents of that theory (section 3). We will also put forth some theoretical reasons that further motivate us to doubt the explanation in terms of prepared fears (section 4).

The "fear module theory" has been proposed by Öhman and Mineka (2001), who say they were inspired to use the concept of a fear module by Fodor (1983), Griffiths (1990), and Tooby and Cosmides (1992), to summarize a large amount of empirical work on the subject of fear acquisition and threat processing. It is a particularly well worked out and well-specified example of an evolutionary theory of emotional processes. The central assertion of this theory is that recurrent threats to survival have shaped the evolution of a fear module, a specialized mental structure that accomplishes the task of constantly monitoring the environment to detect possible threats. Promptness in detecting threats to survival would obviously be adaptive, and thus organisms that possessed this trait would have a greater chance of survival and pass on the genes responsible for the fear module.

The fear module is proposed to have four central characteristics: selectivity, automaticity, encapsulation, and a dedicated neural circuit. Selectivity is the idea that certain classes of stimuli, recurrent threats to survival, preferentially trigger the fear module. Because this is one of the strongest and most debated claims of the fear module, we will focus much of the following discussion on it. Automaticity is the hypothesis that the fear module may be activated without conscious awareness or without the engagement of cognitive resources. Studies showing that it is possible to develop fear reactions to stimuli that are not perceived consciously, through subliminal conditioning for instance, have been presented to supported this claim (Esteves *et al.* 1994; Öhman and Soares 1998). Encapsulation suggests that once the fear module is triggered, processing of fearful stimuli is immune from the influence of other cognitive processes. The fact that most phobics readily recognize that their fear is irrational (or at least disproportionate) is consistent with encapsulation. Finally, recent data from neuroscience suggest a key role for the amygdala in fear processing and fear acquisition, consistent with the proposal of a dedicated neural circuit for the fear module (Öhman 2005).

We return to selectivity, which is one of the central claims of the fear module theory, and the claim that sets it apart from non-evolutionary accounts of fear processing and fear acquisition. According to the fear module, recurrent threats to survival should be particularly potent triggers of the fear module. The theory does not propose that fear of specific stimuli is necessarily hardwired, but that organisms are *prepared to learn* the association between certain stimuli and negative outcomes more readily than others.[8] Thus, the model proposes an essential role for learning, but learning that operates on a system with a readiness to selectively associate certain stimuli (evolutionary relevant threats) with negative outcomes. For instance, people should more easily detect the association between snakes and negative outcomes than that between flowers and negative outcomes.

Among the strongest evidence for selectivity is the work on fear conditioning (Öhman *et al.* 1975; Cook *et al.* 1986). In this paradigm, pictures of different types of stimuli are presented, some of which will be paired with a negative outcome (for instance an electric shock). Over pairings, participants develop a fear response to the stimulus associated with the electric shock even if it is presented independently of it. Fear responses can be measured in different ways, but increased skin conductance responses (SCRs) are often used. The fear-conditioning paradigm has shown differences in the development of fear responses to evolutionary relevant (snakes, spiders) and irrelevant fear stimuli (flowers, mushrooms). Participants can acquire fear responses (as measured by SCRs) to both types of stimuli. However, when the stimulus is no longer associated with the negative outcome, the fear response persists longer in the case of fear-relevant stimuli compared to evolutionary neutral stimuli.

Data on observational fear learning in rhesus monkeys raised in captivity (Cook and Mineka 1989) has also been used to support selectivity. The monkeys will readily develop a fear of snakes if they observe a confederate behaving fearfully while manipulating a snake. However, monkeys will not develop the same fear response if the confederate is manipulating a rabbit or a flower, an evolutionary-irrelevant fear object.

Work on illusory correlations in humans has also been used to further support the claim that threat processing is selectively primed for evolutionary-relevant stimuli (Tomarken *et al.* 1989). In this paradigm, images are presented followed by negative or neutral outcomes (electric shock vs tone or nothing), similar to the fear-conditioning paradigm. In this case, however, the measure is cognitive rather than physiological. Participants are asked to estimate how often each type of stimuli is followed by a shock. Participants systematically

---

[8] The phrase "prepared learning" was initially proposed by Seligman (1970). According to him, evolution has predisposed humans and many animal species to easily and quickly learn, as well as persistently retain, associations or responses that foster survival when certain objects or situations that have been evolutionary significant are encountered. Seligman's idea was inspired by Garcia and Koelling's (1966) work on the selectivity and robustness of aversion learning following nausea.

overestimate the likelihood of a negative event following the threatening stimuli, especially when the stimuli are evolutionary relevant. For instance, participants may perceive an association between snakes and electric shocks and not between mushrooms and shocks, even when the probabilities are equal in both cases.

Evidence also comes from the visual search paradigm (Öhman *et al.* 2001a,b). In this task, arrays of pictures are presented and participants must detect discrepant targets amongst distracters. This task has been used to assess both the selectivity and the automaticity features of the fear module. We go into more detail on the methodological aspects in the next section, but influential studies by Öhman and colleagues (2001) have been used to argue that fear-relevant stimuli are detected more rapidly and more efficiently than neutral stimuli. For instance, participants detect a discrepant snake in an array of flowers more rapidly than they detect a discrepant flower in an array of snakes.

In all these paradigms, most of the claims on selectivity come from a comparison of fear-relevant stimuli and fear-irrelevant stimuli, for instance comparing snakes with flowers, or spiders with mushrooms. Some studies have also performed a more stringent test of evolutionary claims by comparing phylogenetic fear stimuli and ontogenetic fear stimuli (Flykt *et al.* 2007). The former are stimuli that would evoke fear in the EEA. The latter are stimuli that evoke fear in contemporary situations but are not recurrent threats to survival in the EEA. This comparison is important to delineate the effects that result from fear from the effects that are specific to evolutionary relevant fears. For instance, studies have compared the extinction of conditioning to snakes versus guns. Illusory correlations have also been examined for snakes versus broken electrical equipment. In both cases it has been found that the evolutionary-relevant fear held special status; evolutionary-relevant fear produced less extinction (McNally, 1989) and a stronger covariation bias (Amin and Lovibond 1997).

The fear module theory does allow for the possibility that fear responses can develop to novel fear stimuli. Öhman and Mineka (2001) explicitly acknowledge that ontogenetic and phylogenetic fears are not mutually exclusive. However, the idea of selective association implies that phylogenetic fear stimuli should have a special status and be particularly good triggers of the fear module. For instance, Öhman and colleagues state that there is a "general bias preferentially to direct attention toward evolutionarily fear-relevant stimuli among humans" (Öhman and Mineka 2001, p. 475). Thus the fear module theory implies that the two types of fears may be based on different mechanisms, or at least that less extensive processing would be necessary for evolutionary-relevant fear stimuli (Mineka and Öhman 2002). Furthermore, features such as automatic processing and encapsulation should be particularly evident for phylogenetic fear stimuli. This is what we explore in more detail in the next section.

1.3 Snakes and spiders vs syringes and guns

Having to detect a particular stimulus in a complex environment where one is receiving a large amount of visual information simultaneously is a complex task. Imagine walking in a savannah, with all the colors, shapes, and textures there that stimulate the visual system. How do people single out and identify one particular stimulus from the whole visual environment? Obviously, being able to detect threatening objects quickly and efficiently gives individuals additional time to react to the threat, which increases chances of survival.

The features of this type of situation are encapsulated in the laboratory paradigm of the visual search task. Arrays of items are presented simultaneously and participants are asked to determine whether all items are taken from the same category (e.g., all oranges) or whether there is a discrepant item (e.g., eight oranges and one apple). Reaction times are an indication of the efficiency with which particular targets can be detected against a background of distracters. Numerous experiments have demonstrated that threatening targets are detected particularly efficiently. Participants are quicker to make a response when the discrepant item is a threatening target against non-threatening distracters (e.g., a snake amongst flowers) than when the discrepant target is non-threatening against a background of threatening distracters (e.g., a flower amongst snakes). This threat superiority effect has been shown to be quite robust (Brosch and Sharma 2005; Flykt 2005; Blanchette 2006; Fox et al. 2007) and can be observed with different types of stimuli, such as angry facial expressions, snakes, and spiders. These data suggest a cognitive system that is geared towards the efficient processing and detection of threatening stimuli, in line with the predictions of the evolutionary models of fear and anxiety.

A number of features of the visual search paradigm have shed light on the cognitive mechanisms that underpin threat detection, in particular the issue of automaticity. By varying the number of distracters, experimenters can draw conclusions about the type of search that is employed in detecting targets. If participants rely on a serial search, examining items sequentially, reaction times should increase linearly with the number of items. By contrast, if participants rely on parallel searches, processing all items at once, the number of distracters will not affect reaction times. Serial searches are typically associated with controlled, effortful processing, whereas parallel searches are thought to reflect automatic processing. Target position can also be used to determine whether the search is serial or parallel. If participants are using a parallel search, whether the target is presented centrally or in the periphery, on the top or the bottom row, will not have an effect on reaction times.

According to evolutionary models, because threat detection is an ability shared by all mammals, it is likely to be based on automatic rather than controlled processes, which appeared later in evolution. Thus, one would expect a differential impact of the number of distracters and of location of the target for threatening and neutral cues. Using these two indices, there is indeed some evidence that threat detection may be particularly efficient, although the data do not support a purely automatic detection of threat (Fox *et al.* 2000; Öhman *et al.* 2001). Searches for threatening targets are less affected by the number of items and less affected by the location of the target than are searches for non-threatening targets (although they may not be entirely automatic).

These conclusions about the efficiency of threat detection fit in nicely with an evolutionary model. The threat superiority effect would be particularly adaptive if it were automatic, independent from higher level goals and volitional control. Furthermore, a fear detection system shared by all mammals would likely be based largely on subcortical structures and operate largely outside conscious awareness, thus acting automatically. In addition to the work on attention, there are data from human and animal studies concerning neural circuitry of fear, centered on the amygdala, that also support this conclusion. LeDoux's work shows that fear conditioning may be based, in animals at least, on subcortical structures. Evidence for this comes from animal lesion studies blocking the cortical pathways carrying the sensory information to the amygdala. In these cases, sensory information that has been processed entirely in subcortical structures (auditory thalamus → lateral nucleus → central nucleus of the amygdala) without cortical processing can suffice to establish fear conditioning (LeDoux 2000). Thus, different lines of evidence converge to support the idea of an automatic processing of threat.

While these data are consistent with an evolutionary model of fear processing, they are also consistent with the view that fear is much less defined by evolutionary constraints. One view is that because phylogenetically relevant fear stimuli such as snakes, spiders, and angry faces are detected more efficiently than neutral stimuli, there must be important phylogenetic constraints on threat processing. Another view is that it is fear, and not evolutionary relevance, that is primarily responsible for the attentional effects. The crucial comparison must be between phylogenetically relevant fear stimuli and ontogenetically relevant fear stimuli. If fear-relevant but biologically irrelevant stimuli (stimuli that could only have acquired their threatening nature through ontogenetic contingencies) elicit the same detection advantage, the same data actually support the view that fear processing is not phylogenetically constrained. As we mentioned previously, specificity is one of the core features of the fear module.

In fact, recent experiments fail to confirm such specificity. Recent work in different laboratories, including ours, has compared reaction times to phylogenetic and ontogenetic fear stimuli (e.g., snakes and spiders vs guns and syringes, see Fig. 1.1) (Brosch and Sharma 2005; Blanchette 2006; Fox *et al.* 2007). In a number of studies, ontogenetic threatening stimuli produced the same threat superiority effects as phylogenetic stimuli. For instance, in one study, guns and knives led to faster detection amongst clocks or toasters just as snakes and spiders led to faster detection amongst flowers and mushrooms. We also found that the increase in reaction time as a function of the number of distracters was less pronounced for threatening targets, suggesting more automatic processing. This was the case for both evolutionary relevant and irrelevant threats, as can be seen in Fig. 1.2.

This threat superiority effect has been shown with guns, syringes, hand-grenades, and knives amongst others. It has also been shown with symbolic representation of threatening stimuli, using cartoon depictions of snakes,

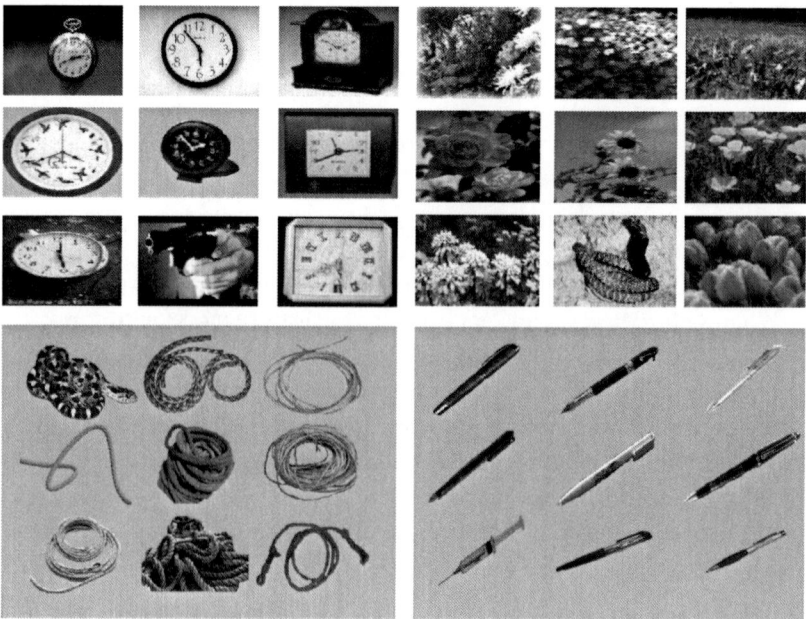

Fig. 1.1 Examples of stimuli in our experiments using the visual search task, comparing evolutionary-relevant and evolutionary-irrelevant stimuli.

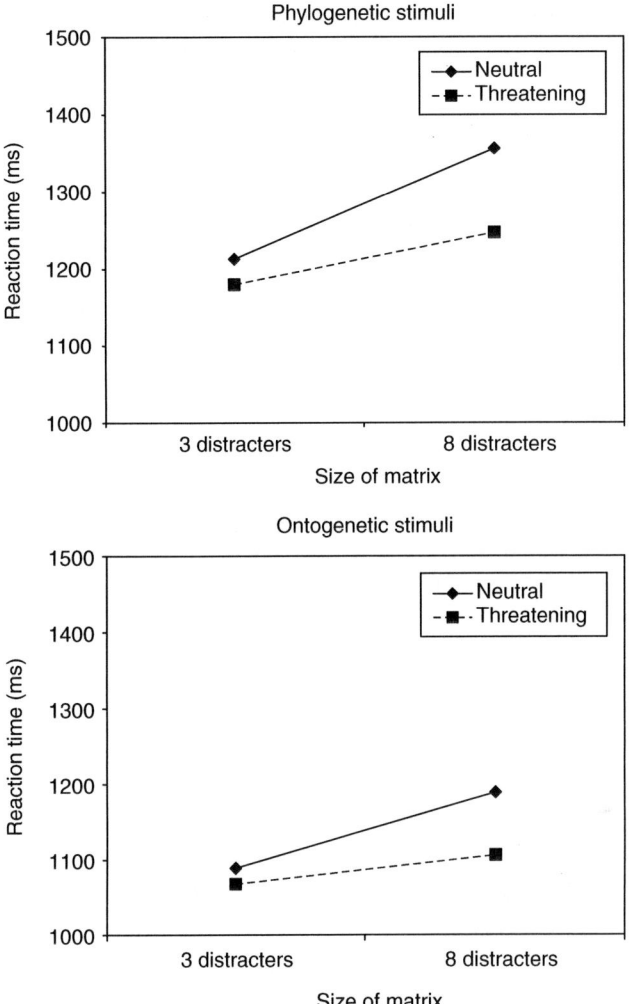

Fig. 1.2 Aggregate data from three studies showing reaction times to detect phylogenetic and ontogenetic threatening and neutral targets amongst small (three distracters) and large (eight distracters) matrices.

guns, and knives. To control for the inevitable confounds present when using complex realistic images, in different studies, we have equated the stimuli as closely as possible for features such as shape, color, and background across phylogenetic and ontogenetic conditions. For instance, we used palm trees as a distracter for spiders and pens as distracters for syringes. In all cases, the robust threat superiority effect was evident for ontogenetic as well as phylogenetic stimuli.

There is also developmental evidence that supports a role for experience in shaping the modulation of attentional processes. In two studies, LoBue and DeLoache (LoBue and DeLoache 2008; DeLoache and LoBue 2009) found a threat superiority effect in children of 3–5 years old for snakes and spiders, similar to that found in adults. In a further study, LoBue (in press) also found a threat superiority effect in 3-year-old children for syringes, but not for knives. She hypothesized that children were more likely to have had direct experience with syringes (through vaccination), but not knives, a pattern that was confirmed by the parents. These data suggest a direct role for ontogenetic contingencies in the attentional effects observed with fear-related stimuli.

Thus, overall, there is mounting evidence that phylogenetic and ontogenetic threats produce similar effects in terms of attentional capture. Similar behavioral outcomes may, however, be produced by different underlying mechanisms. Faster detection of ontogenetic threats could result from different cognitive or neural mechanisms. This possibility cannot be explored with measures of reaction time. In the face of similar outcomes it is more parsimonious to assume similar mechanisms. Nevertheless, selectivity could mean different mechanisms or differential activation of similar mechanisms for special classes of stimuli. We have recently started to examine the neural correlates of threat processing to specifically compare phylogenetic and ontogenetic threats using event-related potentials (ERPs).[9]

ERPs represent a perfect methodology to examine whether the faster detection of threatening targets is produced by similar or different neural mechanisms for phylogenetic and ontogenetic threats. We examined this question in a recent experiment, using a dot-probe paradigm. In this task, two cues are presented in two locations. One of these cues is neutral and one is threatening. A neutral target then follows in the location of either the threatening cue or the neutral cue. Because threat captures attention, in this paradigm, reaction times are generally faster when the target follows in the location of the threatening cue. Research has shown that a positive peak occurring approximately 100 ms

---

[9] ERPs allow us to observe the electrical potentials that are evoked as a result of processing specific stimuli. Surface electrodes are applied to the scalp and a continuous electroencephalogram (EEG) is recorded. This electrophysiological trace is then averaged over repeated presentations of the same category of stimuli. The resulting peaks, their amplitude, and timing, can be extracted and related to specific cognitive mechanisms. Because they provide very good time resolution, ERPs have been used extensively to study the effect of emotion on attention. Studies have shown that the threatening value of stimuli modulates early ERP components such as the P1, a positive deflection occurring around 100 ms post stimulus, thought to reflect early perceptual processing of stimuli by the primary visual cortex (Luck 2005).

after target presentation, the P1, is amplified when the target follows in the location of the threatening cue. Crucially, we wanted to compare whether this P1 modulation by threat would be different for phylogenetic and ontogenetic threatening cues. The answer is no. Whether the cues were snakes, spiders, knives, or syringes, the subsequent neutral targets evoked larger P1s than when the cues were neutral: ropes, trees, spoons, and pens. Thus, from these data, there is no reason to believe that the same behavioral outcomes are produced by different neural mechanisms. The electrophysiological signatures of phylogenetically relevant and irrelevant threats were, as far as we can tell, the same. Thus, not only do all types of threats produce the same effects on attention, they seem to produce this effect through the same neural mechanisms. This pattern is inconsistent with a specific and encapsulated fear module, with dedicated neural circuitry, which would preferentially process phylogenetically relevant stimuli.

These conclusions are consistent with recent findings resulting from other paradigms. For instance, using a conditioning paradigm, Flykt *et al.* (2007) hypothesized that although fear reactions could be acquired equally for ontogenetic (guns) and phylogenetic (snakes) stimuli, only phylogenetic stimuli should show persistence of the fear response in an extinction phase if the stimuli are presented masked (subliminally). In fact, they found that participants continued to show fear responses (as indexed by skin conductance) to masked guns as well as snakes. These data point to a much less constrained fear-learning mechanism than the one previously proposed by evolutionary-oriented psychologists.

As one of us argued (Blanchette 2006, p. 1499), the results concerning the generality of the threat-superiority effect are inconsistent with a strict interpretation of the evolutionary fear module theory. There are two possible versions of the evolutionary thesis concerning the effect of threat on attention. One version suggests that not only are the basic processes biologically determined, but that some of the triggers are hard-wired as well. In this case, threats that were relevant at the time that the fear module evolved, for instance, would be more effective triggers and would lead to more efficient, automatic detection. Another version of the evolutionary thesis might suggest that stimuli that elicit fear are automatically detected and that this is based on hard-wired neural circuitry. However, there could be flexibility in the specific stimuli that will trigger these processes. Associative, observational, or verbal learning could lead novel stimuli to become effective triggers of these mechanisms. In such a case, recent threats could be detected as efficiently, through the same neural mechanisms. This would allow for more influence of ontogenetic contingencies in the relation between fear and attention. The data presented in this section are inconsistent with the first version of the evolutionary thesis, but could be consistent with the second.

1.4 **Further problems with the evolutionist's explanation of phobias**

In the next section, we will review evidence from the neurosciences that gives further support to the picture of a flexible fear-processing mechanism that we just suggested. Before we present this revised conception of the fear-processing mechanism, we review additional findings that, we argue, represent problems for the evolutionary explanation of phobias that makes reference to a fear module. These findings are related to the prevalence of ontogenetic fears, the biological relevance of prepared fears, and the role of fear in phobias.

First, Öhman and his colleagues argue that particular fears (like the fear of snakes) are phylogenetically shared with some of our primate ancestors. They rest their claim on experiments done by Mineka and Cook in the 1980s involving captive monkeys that had never been in contact with snakes. These monkeys spontaneously exhibited fear when shown real or artificial snakes, but not when shown a flower. On closer examination, however, Kagan and Fox observe:

> . . .the behavioral reactions of most monkeys, chimpanzees, and human infants to a snake are no different from their reaction to discrepant events that are harmless, like a tortoise or seaweed. Monkeys born and raised in a laboratory, and therefore protected from contact with live snakes, showed a longer period of motor inhibition to the presentation of a snake, whether alive or an artefact, than to blue masking tape. However, that restraint only occurred on the first testing session. During later sessions, the animals showed no more restraint to the snake than to the masking tape. Moreover, a majority of animals failed to show any difference in withdrawal behavior to the snake compared to the harmless masking tape. If snakes were a biological potent incentive for a fear state, motor restraint should have not habituated so quickly and the majority of monkeys, rather than just 30%, should have shown a withdrawal reaction.
>
> (Kagan and Fox 2006, p. 202)

Secondly, evolutionists suggest that fear of phylogenetically older objects will be learned faster and more effectively than phylogenetically recent objects. Because phobias are explained by a breakdown or a lower threshold of a module built to be more sensitive to snakes than cars, we should expect to find more cases of snake phobia than car phobia. Marks and Nesse (1997) claim that few people fear cars. But as Merkelbach and de Jong remark:

> . . .the claim is problematic. A substantial proportion (38%) of survivors of road vehicle accidents develop a phobia related to driving a car (accident phobia; Kuch *et al.*

1994). Thus phobias directed at evolutionarily recent objects do occur and it is far from clear whether they are rare.

(Merckelbach and de Jong 1997, p. 334)

Marks and Nesse's claim is comparative and they are not denying that car phobias can happen, but they claim that these phobias are rare. If Merckelbach and de Jong's figure is accurate and 38% of road vehicle accident survivors do indeed develop a phobia, that is a very large number of people (given that the number of car accidents in the USA in 2005 was about 5 000 000. Even if we take only head-on accidents, collisions with pedestrians and bicyclists, and rollovers, which are probably more traumatic, we get a figure of around 400 000 accidents).

Thirdly, even if specific fears are prepared, it does not necessarily entail that they are evolutionary relevant. For instance, the list of typical phobic objects in section 1 features slugs. As Davey once put it: "It is unlikely that our ancestors ever had to avoid packs of predatory slugs or snails!" (Davey 1994). As for spiders, it seems that a very small percentage of them are dangerous (i.e. 0.1% of the 35 000 spiders varieties according to Renner (1990), quoted by Merckelbach and de Jong (1997, p. 336[10])). At the same time, the mushrooms that Öhman and his colleagues considered to be phylogenetically non-threatening are at least as dangerous if not more dangerous than snakes (as there are at least 100 species of dangerous mushrooms just on US soil).[11] This is to say nothing, as Murphy (2005) observed, about the fact that phobias of large predators (like bears or tigers), which our ancestors surely lived around for a long time, are fairly uncommon.[12]

---

[10] Gerdes *et al.* (2009) add to this that "[s]ince spiders generally prey upon insects or other spiders, their venom has not evolved to harm large vertebrates such as humans. Spiders rarely use venom in response to vertebrates for defense and generally do so in last resort ... Moreover, most studies of spider bites have been retrospective and bites have not been confirmed by eyewitnesses ... For example, 80% of suspected cases of spider bites in Southern California were caused by other arthropods, mostly ticks and reduvid bugs" (p. 66). Compared to spider bites, bee and wasp stings are more likely to be lethal.

[11] We are not suggesting that one would be better to fear mushrooms, but only that we should not rely on our intuition concerning what has been a source of danger for our ancestors and what has not.

[12] Jesse Prinz (personal communication) told one of us that snake phobia is rarer in areas where people are in frequent contact with snakes than where the contact is infrequent, which seems to go against the evolutionary hypothesis of Öhman and his colleagues. Indeed, when school-aged children from a Dakota Indian tribe in Manitoba were asked to say what they were most afraid of, most of them mentioned bulls or horses, some mentioned witches or ghosts, but almost none mentioned snakes, even though snakes are very common in that area and, we suspect, ghosts are not (see Kagan and Fox 2006, p. 202).

A fourth problem has been noted by Davey (1995), who showed that fear of spiders co-varies with fear of other animals with disgust-evoking status (like bats, lizards, slugs, and snails). As Davey puts it:

> Many animal phobias, especially small-animal phobias, appear to be related to the disgust emotion rather than fear alone. Thus, it seems that people who have a highly developed disgust reaction are also likely to be more fearful of a whole range of disgust-relevant animals, and these include snakes, spiders, rats, mice, insects and [other] invertebrates.
>
> (Davey 2007, p. 249)

Patients suffering from blood–injection–injury fears also have a higher disgust sensitivity (Page 1994).

Finally, it seems that the effects attributed to prepared learning could instead be due to what Davey (1992, 1995) has called "general expectancy bias" rather than pre-wired mechanisms. For certain objects, participants may have *a priori* expectancy biases (expectation that the stimuli will be followed by an aversive consequence) that influence the *a posteriori* perception of covariation biases and thus increase illusory correlations between a stimuli and negative outcomes (like a spider and pain). This readiness to associate certain stimuli with negative outcomes could also explain differential resistance to extinction.[13] The fact that fear responses to phylogenetic stimuli persist for a longer time may depend on the more or less articulated expectations concerning the dangerousness of these objects. The sources of those expectations might be personal experiences of encounter with the object, but social or cultural transmission might also be involved. For instance, Davey (1994) suggests that fear of spiders might be the result of cultural ideas that appeared in Europe in the Middle Ages concerning their dangerousness (these ideas would be cultural vestiges in presnet-day societies; see also Bartholomew 1994). This particular explanation might not be true (someone we know called it a "just-so cultural story"), but what seems to be true is, as de Jong and Merckelbach noted, that:

> ... small-animal phobias *are not at all noncognitive* anxiety disorders. For instance, spider phobics seem to have highly developed sets of negative ideas about spiders and about their own reactions during confrontation with spiders.
>
> (de Jong and Merckelbach 1997, p. 362; our emphasis)

---

[13] And indeed, as Davey remarks, "...compared to nonphobics, phobics endorse a significantly higher expectancy of aversive outcomes following presentation of their phobic stimulus...," while they also differ in terms of "the range of specific articulated beliefs about how harmful and dangerous contact with their phobic situation might be (phobic threat beliefs)" (Davey 2002, p. 152). The level of threat beliefs returns to normal levels after successful therapy.

It is thought that these negative ideas are playing a role in either or both the aetiology and maintenance of phobias (Thorpe and Salkovskis 1995).

These represent important problems for the standard evolutionary psychiatry view of phobia. Firstly, the original evidence for the phylogenetic origin of some fears—the fact that we would share these fears with monkeys—is questionable. Secondly, phobias of evolutionarily recent stimuli may not be as rare as predicted. Thus, one of the reasons to invoke the existence of fear modules of the sort suggested in section 2 is not supported. Furthermore, we have suggested that our knowledge of what is and what isn't evolutionary relevant relies largely on intuitions about the evolutionary pressures that would have shaped the fear-learning modules. More work needs to be done in order to identify the factors that really played a role in shaping fear mechanism. Also, as we saw, different phenomena cast doubt over the role of a fear module in all cases of phobia. For instance, some data suggest the possible importance of disgust in certain types of phobia. The elicitors of disgust are usually thought (even by evolutionists) to be much less constrained than the triggers for fear (which is not to say that the class of elicitors is not constrained at all; cf. Fessler and Navarrete 2003). If such is the case, this suggests that the elicitors of certain types of phobias will be more varied than suggested by the prepared learning theory. Finally, the effects found in the literature on the fear module could be explained by another, more general, mechanism (what Davey has called the "cognitive associative mechanism") that would be fed by our subjective estimate of dangerousness (which, in certain cases, as with spiders, do not align with objective dangerousness; see Gerdes *et al.* 2009).

All of the objections we raised point in the same direction. The fear mechanism involved in phobia may be more flexible than the one posited by the standard evolutionary account. In the next section we will take a look at recent evidence from neuroscience that confirms this intuition. This evidence is also suggestive of an alternative theory of phobia, which we will propose at the end of the next section.

1.5 **An alternative conception of emotion**

As we mentioned in section 1, the explanation of phobias by Marks and Nesse involves a fear module geared to solve specific adaptive problems, for example to learn to fear, and pay attention to certain kinds of objects and organisms (snakes, spiders, angry strangers, etc.) faster than to other, evolutionary-irrelevant stimuli (mushrooms, guns, syringes, etc.). It is also postulated that the fear module is responsible for the rapidity and efficiency with which evolutionary-relevant stimuli are detected. As we mentioned before, for Öhman and colleagues (Mineka and Öhman 2002), the fear module is assumed to have the following characteristics: it is (1) selective, (2) automatic, (3) encapsulated, and (4) has specific neural circuitry. About this specific neural circuitry, Öhman says, following Ledoux (1996, 2000), that:

> ... fear is controlled from ancient systems in the brain, primarily the amygdala, that may act relatively independently of the later emerging higher cognition .... A central point in [this] model of fear activation is that the amygdala can be rapidly activated by a "low road", via the thalamus, *that does not require the cortex.*
>
> (2005, p. 954–5; our emphasis)

It is thought that this low road operates on lower grained information than the higher "cortical" road. It would also work faster and activate the amygdala before a "full cognitive" evaluation of the stimuli is done (or even independently of this evaluation). Some innate triggers are thought to be somehow recognizable by the amygdala through this low road, while other stimuli would need more analysis. This is presented as a proximate explanation of the presumed differences between evolutionary relevant and irrelevant stimuli in terms of learning and attention.

We have seen in section 3 that non-evolutionary-relevant stimuli can behave exactly like evolutionary-relevant stimuli when it comes to allocation of attention and that we have reason to think that they rely on the same neurophysiologic mechanisms in both cases. We have also suggested at the end of section 4 that complex cognitive processes and associative learning might be involved in phobias. Both ideas would go against positing the kind of sharp divide between emotion and cognition that is presently at the basis of the fear module theory. In this section we want to look at some evidence from neuroscience that questions this divide. We concur here with the diagnostic of a commentator of the recent emotion literature for whom in the last few years researchers have been a bit too much "amygdaloid-centric" (Hardcastle 1999, p. 239).[14] According to her, the results of Ledoux's (and others') neuroscientific investigations has made us focus almost exclusively on limbic structures to understand emotions. In doing so, we neglected the fact that in humans[15] "subcortical and cortical areas working together as a complex dynamical system produce our emotions" (idem).[16] Indeed, to understand how the

---

[14] Indeed, not everybody who has studied emotion has been suffering from "amygadaloid-centricity", but in the literature on the fear module the diagnostic seems to be adequate.

[15] We specify "in humans" because as Berridge (2003, p. 41) puts it "[a h]uman can be devastated, rendered into vegetative states, by large neocortical lesions, whereas a rat can lose its entire neocortex and continue on remarkably normal."

[16] This diagnostic is now shared by many (see Groenewegen and Uylings 2000; Adolphs

amygdala works in humans to produce emotions we should take into account the fact that there are many top-down projections coming from the cortex. We propose that these projections are central to understand the behavior of the amygdala (in the normal as well as in the phobia state).[17]

The necessity of these top-down projections for fear is made clear when one looks at the properties of the cells that form the visual pathway that goes from the thalamus to the amygdala. As Johnson puts it:

> The cells in the areas of the visual thalamus that project directly to the amygdala only respond to some of the cues that are encoded at the periphery. That is, cues that are indications that the stimulus is moving quickly and it is "looming", but do not include any indications of the form of the stimulus [. . .] There is no point prior to cortical processing when the form of an object is represented. [. . .] But there is no place in the thalamus, or prior to reacting it, when the information that all the cells are responding to is integrated or combined. Therefore, there is no way for these areas to respond to a snake-like form.
>
> (Johnson 2008, p. 748–9)

What this implies is that the quick, modular system using the low road does exist, but what it processes is not snake shapes or spider shapes, but much lower levels properties that are not specific to snakes or spiders. To react to a snake or a spider, per se, cortical processing is needed.

Recent work on attention and fear has also painted a different picture from the one proposed by evolutionary psychiatrists (and the evolutionary psychologists' work on which they based their theory). For instance, Pessoa argues that a number of studies (including his own; Pessoa 2005; Pessoa and Ungerleider 2005), have shown that:

> . . . *the amygdala functions in a manner that is closely tied to top-down factors*. For instance, amygdala responses are strongly dependent on attention, even for stimuli that are affectively significant (owing to previous pairing with mild shocks). Amygdala responses appear to be closely linked to perception, and are not simply predicted by the physical characteristics of the stimulus — for example, responses to a briefly presented fearful face differ greatly depending on whether participants actually report perceiving a fearful face. *In general, controlling attention to, and 'cognitively' changing the meaning of, emotionally evocative stimuli greatly affects amygdala responses.*
>
> (Pessoa 2008, p. 149–50; our emphasis)

2008; Pessoa 2008). For instance, Pessoa (2008, p. 150) writes that ". . .concluding evidence is starting to paint a dynamic and context-dependent picture of amygdala function."

[17] Work on the connectivity between the amygdala and more "cognitive" areas confirms this picture; see, for instance, Groenewegen and Uylings (2000).

A number of other recent studies have highlighted the influence of strategic, or top-down factors on amygdala function specifically and automatic threat processing generally. For instance, the fMRI findings of Pessoa are consistent with findings of ERP studies run by Holmes, Eimer, and colleagues (Eimer *et al.* 2003; Holmes *et al.* 2003) who show an "automatic" modulation of electrophysiological traces by threat only when threat is voluntarily attended to. Consistent findings have also been observed in behavioural tasks where automatic effects of fear are observed only when it is strategically relevant to process emotions (Huang *et al.* 2008). These findings show that automatic threat detection and amygdala function is permeable to prefrontal influences (which host complex cognitive processes). If this is the case, it might explain why guns and syringes can behave like snakes and spiders. The question is, though, does the effect of snakes and spiders involve the same kind of top down influence on the amygdala? Given what we know about the property of the cells of the low-road as well as the role of beliefs in the etiology of phobia, we have no reason to think otherwise.

Other work in neuroscience is also giving us reasons to think of the function of the amygdala differently than as a simple threat detector. It appears that unpredictability is the factor that is responsible for first engaging the amygdala. This is not to say that the amygdala has no role in fear (it does, obviously), but it sheds light on the aspect of stimuli the amygdala is particularly sensitive to. Adolphs has gone as far as saying that in the case of emotional face recognition, the role of the amygdala "does not appear to be specialized for processing threat or fear as such, or perhaps even emotion or reward . . ." (Adolphs 2008, p. 169), but rather unpredictability (see Whalen 2008 for a summary of experiments backing this position). For instance, experiments show amygdala activity for ambiguous facial expressions. The amygdala is active in response to fearful, but also to surprised, faces.

These results suggest the following picture of amygdala function: it detects unpredictable events, directs attention towards them for further analysis, and then, depending on the result of the analysis, a "decision" is taken concerning their dangerousness. This decision may be based on previous "direct" experience of encounter with the stimulus, but could also be based on the individual "indirect", social, or cultural learning history concerning the stimulus.[18] In the case where the events are not considered dangerous, amygdala activity is dampened; in the case that the events are considered dangerous, the amygdala

---

[18] Genetics/temperamental traits might bias individuals towards conclusions of vulnerability or safety (see Armfield 2006; Mineka and Zinbarg 2006).

is maintained in preparation for action. According to this account, subsequent activity of the amygdala is modulated by areas of the brain in charge of control (the prefrontal cortex; Ochsner 2007). Some accounts of anxiety disorders close to ours also highlight the importance of the links between the amygdala and prefrontal areas. For instance, Bishop (2007) reviews evidence showing that the amygdalo-prefrontal network is important in the acquisition and extinction of fear as well as the attentional effects of anxiety and its effects on the interpretation of ambiguous stimuli. According to her view, anxiety disorders are related to an imbalance in the regulation of amygdala activity by prefrontal areas, consistent with our suggestion of the intrinsic cognitive elements of threat processing.

The problem of phobics could thus be with the way the prefrontal cortex handles the unpredictability. It might come from the fact, as suggested by Larson and her colleagues (2006), that the amygdala of phobics is characterized by a rapid onset when presented with phobia-related stimuli, followed by a rapid decreased firing that contrasts with the sustained pattern of activity in controls. This is consistent with studies that show that phobics disengage from phobic stimuli more rapidly than normal. This rapid disengagement might lead the prefrontal to "play safe" and assume that the stimulus is dangerous. If engagement is necessary to extinguish fear responses, it might also explain how phobias can be maintained.

These novel conceptions of the role of the amygdala in fear and anxiety point in the direction of a more flexible, adaptable, and cognitive fear-processing mechanism. Although subcortical and automatic processes clearly play a role in threat processing, research increasingly documents how these threats are not cognitively elaborated (at least, not enough to represent evolutionary-relevant stimuli like snakes or spiders) and how strategic factors and higher-level skills form an integral part of threat processing. The fact that the amygdala is sensitive to unpredictability, a notion that can admit important cognitive elaboration (the fact that you consider a dog as unpredictable depends on previous experience with dogs, on what people tell you about dogs, etc.), gives us clues as to why guns and syringes as well as snakes and spiders elicit the same patterns of responses. All of them might be represented as sources of "unpredictable" events.

1.6 Conclusion

Williams and Nesse have proposed that evolutionary biology should be considered as one of the "essential basic sciences" of psychiatry, that is, on the same footing as other sciences that are claimed by others to be essential (like cognitive science or neuroscience). Among the reasons for giving such a status to evolutionary biology is the fact that it is thought that it could be conducive of empirical progress and help us get a better understanding of particular mental disorders. We think that in the case of phobia (one of the paradigmatic cases of evolutionary psychiatry), the explanation given by Marks and Nesse is wrong. The work we presented in section 1.3 (and the objections we raised in section 1.4) challenges some of the central assumptions of the evolutionary model and leads us to propose a different picture of the mechanisms involved in fear and phobia. As we have shown (section 1.5), our picture is consistent with most recent work in neuroscience concerning the role of the amygdala, a neural structure thought to be central to the explanation of phobias. Evolutionary psychiatry has made a point of respecting one of the constraints of interdisciplinary work by looking for support and validation of its main assumptions in other disciplines (like experimental psychology and neurosciences in this case), but we have shown that it has failed to integrate some of the more recent changes in these disciplines. This is not to say that evolutionary psychiatry should be abandoned. It is rather our opinion that at this point we should restrain our enthusiasm concerning the potential of evolutionary psychiatry (at least concerning its potential to generate empirical progress in psychiatry) and postpone the advent of a Darwininan revolution until we have results that speak for it.

Acknowledgments

We would like to thank Sylvain Sirois, Edouard Machery, and the editors of this collection who made numerous comments on previous drafts of this chapter. LF wishes to thank the Centre de Recherche en Ethique de l'Université de Montréal (CREUM) where he was fellow researcher while writing this chapter. IB received funding from the University of Manchester, Research Support Fund.

References

Adolphs, R. (2008) Fear, faces and the human amygdala. *Current Opinion in Neurobiology*, **18**, 166–72.

Anderson, S.W., Damasio, H., Tranel, D., and Damasio, A.R. (1999) Impairment of social and moral behavior related to early damage in the human prefrontal cortex. *Nature Neuroscience*, **2**, 1032–37.

Amin, J.M. and Lovibond, P.F. (1997) Dissociations between covariation bias and expectancy bias for fear-relevant stimuli. *Cognition & Emotion*, **11** (3), 273–89.

Armfield, J.M. (2006) Cognitive vulnerability: a model of the etiology of fear. *Clinical Psychology Review*, **26** (6), 746–68.

Bartholomew, R.E. (1994) Tarantism, dancing mania, and demonopathy: the anthropolitical aspects of mass psychogenic illness. *Psychological Medicine*, 24, 281–306.

Bishop, S.J. (2007) Neurocognitive mechanisms of anxiety: An integrative account. *Trends in Cognitive Sciences*, 11 (7), 307–16.

Blanchette, I. (2006) Snakes, spiders, guns, and syringes: How specific are evolutionary constraints on the detection of threatening stimuli? *The Quarterly Journal of Experimental Psychology*, 59 (8), 1484–504.

Brosch, T. and Sharma, D. (2005) The role of fear-relevant stimuli in visual search: A comparison of phylogenetic and ontogenetic stimuli. *Emotion*, 5 (3), 360–64.

Brown, C., El-Deredy, W., and Blanchette, I. (in press) Attentional modulation of visual-evoked potentials by threat: Investigating the effect of evolutionary relevance. *Brain & Cognition*.

Cook, M. and Mineka, S. (1989) Observational conditioning of fear to fear-relevant versus fear-irrelevant stimuli in rhesus monkeys. *Journal of Abnormal Psychology*, 98, 448–59.

Cook, E.W., Hodes, R.L., and Lang, P.J. (1986) Preparedness and phobia: Effects of stimulus content on human visceral conditioning. *Journal of Abnormal Psychology*, 95 (3), 195–207.

Davey, G. (1992) Classical conditioning and the acquisition of human fears and phobias: a review and synthesis of the literature. *Advances in Behavioural Research and Therapy*, 14, 29–66.

Davey, G.C.L. (1994) The disgusting spider: the role of disease and illness in the perpetuation of fear of spider. *Society and Animals*, 2, (1), 17–25.

Davey, G.C.L. (1995) Preparedness and phobias: specific evolved associations or a generalized expectancy bias? *Behavioral and Brain Sciences*, 18, 289–97.

Davey, G.C.L. (2007) Psychopathology and treatment of specific phobias. *Psychiatry*, 6 (8), 247–53.

DeLoache, J. S. and LoBue, V. (2009) The narrow fellow in the grass: Human infants associate snakes and fear. *Developmental Science*, 12 (1), 201–7.

Eimer, M., A. Holmes, and McGlone, F.P. (2003) The role of spatial attention in the processing of facial expression: an ERP study of rapid brain responses to six basic emotions. *Cognitive, Affective & Behavioral Neuroscience*, 3 (2), 97–110.

Esteves, F., Parra, C., Dimberg, U., and Öhman, A. (1994) Nonconscious associative learning: Pavlovian conditioning of skin conductance responses to masked fear relevant facial stimuli. *Psychophysiology*, 31 (4), 375–85.

Fessler, D. and Navarrete, C.D. (2003) Meat is good to taboo: Dietary proscriptions as a product of the interaction of psychological mechanisms and social processes. *Journal of Cognition and Culture*, 3 (1), 1–40.

Flykt, A. (2005) Visual search with biological threat stimuli: Accuracy, reaction times, and heart rate changes. *Emotion*, 5 (3), 349–53.

Flykt, A., Esteves, F., and Öhman, A. (2007) Skin conductance responses to masked conditioned stimuli: Phylogenetic/ontogenetic factors versus direction of threat? *Biological Psychology*, 74 (3), 328–36.

Fodor, J. (1983) *The Modularity of Mind*. MIT Press, Cambridge.

Fox, E., Lester, V., Russo, R., Bowles, R. J., Pichler, A., and Dutton, K. (2000) Facial expressions of emotion: Are angry faces detected more efficiently? *Cognition & Emotion*, **14** (1), 61–92.

Fox, E., Griggs, L., and Mouchlianitis, E. (2007) The detection of fear-relevant stimuli: Are guns noticed as quickly as snakes? *Emotion*, **7** (4), 691–6.

Garcia, J. and Koelling, R. (1966) Relation of cue to consequence in avoidance learning. *Psychonomic Science*, **4**, 123–4.

Gerdes, A.B.M., Uhl, G., and Alpers, G.W. (2009) Spiders are special: fear and disgust evoked by pictures of arthropods. *Evolution and Human Behavior*, **30**, 66–73.

Griffiths, P.E. (1990) Modularity and the psychoevolutionary theory of emotion. *Biology and Philosophy*, **5**, 175–96.

Groenewegen, H.J. and Uylings, H.B. (2000) The prefrontal cortex and the integration of sensory, limbic and autonomic information. *Progress in Brain Research*, **126**, 3–28.

Hardcastle, V.G. (1999) It's O.K. to be complicated: the case of emotion. *Journal of Consciousness Studies*, **6**, 237–49.

Holmes, A., Vuilleumier, P., and Eimer, M. (2003) The processing of emotional facial expression is gated by spatial attention: evidence from event-related brain potentials. *Cognitive Brain Research*, **16**, 174–84.

Huang, Y.M., Baddeley, A., and Young, A.W. (2008) Attentional capture by emotional stimuli is modulated by semantic processing. *Journal of Experimental Psychology: Human Perception and Performance*, **34** (2), 328–39.

Johnson, G. (2008) Ledoux's fear circuit and the status of emotion as a non-cognitive process. *Philosophical Psychology*, **21** (6), 739–57.

Kagan, J. and Fox, N. (2006) Biology, culture, and temperamental biases. In W. Damon, R. M. Lerner and N. Eisenberg (eds), *Handbook of Child Psychology: Social, emotional, and personality development*. John Wiley and Sons, Hoboken, NJ, pp. 167–225.

Kuch, K., Cox, B.J., Evans, R. and Shulman, I. (1994) Phobias, panic and pain in 55 survivors of road vehicle accidents. *Journal of Anxiety Disorders*, **8**, 181–7.

Larson, C.L., Schaefer, H.S., Siegle, G.J., Jackson, C.A.B., Anderle, M.J., and Davidson, R.J. (2006) Fear is fast in phobic individuals: Amygdala activation in response to fear-relevant stimuli. *Biological Psychiatry*, **60**, 410–7.

LoBue, V. (in press) What's so scary about needles and knives? Examining the role of experience in threat detection. *Cognition & Emotion*.

Ledoux, J. (1996) *The Emotional Brain*. Simon & Schuster, New York.

LeDoux, J.E. (2000) Emotion circuits in the brain. *Annual Review of Neuroscience*, **23**, 155–84.

LoBue, V. and DeLoache, J.S. (2008) Detecting the snake in the grass: Attention to fear-relevant stimuli by adults and young children. *Psychological Science*, **19** (3), 284–89.

Luck, S.J. (2005) *An Introduction to the Event-Related Potential Technique*. MIT Press, Cambridge, MA.

Marks, I. (1987) *Fears, Phobias, and Rituals: Panic, Anxiety, and their Disorders*. Oxford University Press, Oxford.

Marks, I. and Nesse, R. (1997) Fear and fitness: An evolutionary analysis of anxiety disorders. Reprinted in S. Baron-Cohen (ed.), *The Maladapted Mind*, Psychology Press, East Sussex, pp. 57–72.

McGuire, M. and Troisi, A. (1998) *Darwinian Psychiatry*. Oxford University Press, Oxford.

McNally, R.J. (1987) Preparedness and phobias: A review. *Psychological Bulletin*, **101**, 283–303.

McNally, R.J. (1989) On nonassociative fear emergence. *Behaviour Research and Therapy*, **40** (2), 169–72.

Merkelbach, H. and de Jong, P.J. (1997) Evolutionary models of phobias. In G.C. Davey (ed.), *Phobias: A Handbook of Theory, Research and Treatment*. Wiley, Chichester.

Mineka, S., and Öhman, A. (2002) Phobias and preparedness: The selective, automatic, and encapsulated nature of fear. *Biological Psychiatry*, **52** (10), 927–37.

Murphy, D. (2004) Darwinian models of psychopathology. In J. Radden (ed.), *The Philosophy of Psychiatry: A Companion*. Oxford University Press, New York, pp. 329–37.

Murphy, D. (2005) Can evolution explain insanity?. *Biology and Philosophy*, **20**, 745–66.

Murphy, D. (2006) *Psychiatry in the Scientific Image*. MIT Press, Cambridge, MA.

Nesse, R. (1999) Testing evolutionary hypotheses about mental disorders. In S. Stearns (ed.), *Evolutionary Medicine*. Oxford University Press, New York, pp. 260–66.

Nesse, R. (2005) Evolutionary psychology and mental health. In D. Buss (ed.), *The Evolutionary Psychology Handbook*. John Wiley and Sons, Hoboken, NJ, pp. 903–27.

Nesse, R. and Berrige, K.C. (1997) Psychoactive drug use in evolutionary perspective. *Science*, **278**, 63–66.

Nesse, R. and Ellsworth, P.C. (2009) Evolution, emotions, and emotional disorders. *American Psychologist*, **64** (2),129–39.

Nesse, R. and Williams, G. (1997) Are mental disorders diseases? In S. Baron Cohen (ed.), *The Maladapted Mind: Classic Readings in Evolutionary Psychopathology*. Psychology Press, Hove, pp. 1–22.

Ochsner, K. (2007) How thinking controls feeling: a social cognitive neuroscience approach. In E. H. Jones and P. Winkielman (eds), *Social Neuroscience: Integrating Biological and Psychological Explanations of Behavior*. Guilford Press, New York, pp. 106–36.

Öhman, A. (2005) The role of the amygdala in human fear: Automatic detection of threat. *Psychoneuroendocrinology*, **30** (10), 953–8.

Öhman, A. and Mineka, S. (2001) Fears, phobias, and preparedness: Toward an evolved module of fear and fear learning. *Psychological Review*, **108** (3), 483–522.

Öhman, A. and Soares, J.J.F. (1998) Emotional conditioning to masked stimuli: Expectancies for aversive outcomes following nonrecognized fear-relevant stimuli. *Journal of Experimental Psychology: General*, **127** (1), 69–82.

Öhman, A., Eriksson, A., and Olofsson, C. (1975) One trial learning and superior resistance to extinction of autonomic responses conditioned to potentially phobic stimuli. *Journal of Comparative & Physiological Psychology*, **88**, 619–27.

Öhman, A., Flykt, A., and Esteves, F. (2001a) Emotion drives attention: Detecting the snake in the grass. *Journal of Experimental Psychology: General*, **130** (3), 466–78.

Öhman, A., Lundqvist, D., and Esteves, F. (2001b) The face in the crowd revisited: A threat advantage with schematic stimuli. *Journal of Personality and Social Psychology*, **80** (3), 381–96.

Page, A.C. (1994) Blood-injury phobia. *Clinical Psychology Review*, 14, 443–61.

Pessoa, L. (2005) To what extent are emotional stimuli processed without attention and awareness? *Current Opinion in Neurobiology*, 15, 188–96.

Pessoa, L. (2008) On the relation between emotion and cognition. *Nature Review of Neuroscience*, 2, 148–58.

Pessoa, L. and Ungerleider, L.G. (2005) Visual attention and emotional perception. In L. Itti, G. Rees, and J.K. Tsotsos (eds), *Neurobiology of Attention*. Elsevier, San Diego, pp. 160–6.

Renner, F. (1990) *Spinner: Ungeheuer Sympathisch*. Nitzsche Verlag, Kaiserlslautern.

Seligman, M.E.P. (1970) On the generality of the laws of learning. *Psychological Review*, 77, 406–18.

Smith, E.A., Borgheroff Mulder, M., and Hill, K. (2001) Controversies in evolutionary social sciences: a guide to the perplexed. *Trends in Ecology and Evolution*, 16 (3), 128–33.

Stevens, A.S. and Price, J. (2000) *Evolutionary Psychiatry: A New Beginning*, 2nd edition. Routledge.

Thorpe, S.J. and Salkovskis, P.M. (1995) Phobic beliefs: Do cognitive factors play a role in specific phobias? *Behavior Research and Therapy*, 33, 805–16.

Tomarken, A.J., Mineka, S., and Cook, M. (1989) Fear-relevant selective associations and covariation bias. *Journal of Abnormal Psychology*, 98 (4), 381–94.

Tooby, J. and Cosmides, L. (1992) The psychological foundations of culture. In J. Barkow, L. Cosmides, and J. Tooby (eds), *The Adapted Mind: Evolutionary Psychology and the Generation of Culture*. Oxford University Press, New York, pp. 17–67.

Tooby, J. and Cosmides, L. (2008) The evolutionary psychology of emotions and their relationship to internal regulatory variables. In M. Lewis, J.M. Haviland-Jones, and L.F. Barrett (eds), *Handbook of Emotions*, 3rd edition. Guilford, pp. 114–37.

Whalen, P.J. (2008) The uncertainty of it all. *Trends in Cognitive Sciences*, 11 (12), 499–500.

Williams, G. and Nesse, R. (1991) The dawn of Darwinian medicine. *Review of Biology*, 66, 1–22.

Chapter 2

Sexual imprinting and fetishism: an evolutionary hypothesis

Hanna Aronsson

Traditionally, evolutionary psychology has conceptualized sexual preferences as genetically determined adaptations, enabling organisms to single out high quality partners. In this chapter, I argue that the existence of paraphilias, such as fetishism, poses a serious problem for such traditional evolutionary accounts. My own proposal revives the ethological notion of sexual imprinting – a process observed in animals where sexual preferences are acquired through experience with parents and siblings during a sensitive period in early life. Although this process usually generates biologically functional preferences for conspecifics, in certain situations another species or even artefacts can be imprinted on. Acknowledging that it is difficult to provide evidence for the existence of sexual imprinting in humans (and to design studies that would generate such evidence), I suggest that sexual imprinting may provide an explanation for both common and uncommon human sexual preferences.

This chapter reviews the evidence in favor of a theory about the development of human sexual preferences that provides an explanation for both common and uncommon sexual preferences within an evolutionary framework. The theory I have in mind is the ethological theory of sexual imprinting, which has been extensively studied in birds and a few mammals. Essentially, these studies, which are reviewed below, show that sexual imprinting is characterized by a sensitive period early in life during which adult sexual preferences are shaped through social experience. Reasons for believing that a similar mechanism exists in humans are presented. Finally, the explanatory value of sexual imprinting is compared to prevailing evolutionary views presupposing genetically determined sexual preferences. The conclusion is that sexual imprinting has higher explanatory power since it provides an explanation not only for typical preferences, but also for rare fetishistic preferences, as well as individual and cultural differences in sexual preferences.

2.1 The science of fetishism: a history

"Paraphilia" is Greek for love (philia) beyond the usual (para). Paraphilias include a range of behaviors, all having in common sexual arousal in response to an unusual personally or socially unacceptable stimulus (Bullough 1988). Thus, what is regarded as a paraphilia is largely culturally determined. Bullough, for instance, claims that the labeling of sexual practices as deviant or bizarre is a legacy from the Western Christian tradition of regarding everything but procreative sex as sinful. Moreover, many sexual practices once regarded as sinful, and thus paraphilic, are no longer regarded as deviant. Examples include homosexuality,[1] oral-genital sex, and masturbation (Bullough 1988; Steele 1996).[2]

According to the *Diagnostic and Statistical Manual of Mental Disorders*, paraphilias include such conditions as sexual sadism and masochism, exhibitionism, and fetishism. This chapter mainly focuses on a possible explanation for fetishism. Although, according to Valerie Steele, fetishism as we see it today developed in eighteenth century Europe and crystallized as a distinct sexual phenomenon in the second half of the nineteenth century, the sexualizing of objects has existed in many cultures for thousands of years, for example Chinese foot binding.

Fetishism was first used in the modern psychological sense by Alfred Binet in an essay published in the late nineteenth century. In *Psychopatia Sexualis*, contemporary sexologist Richard von Krafft-Ebing defined fetishism as "The association of lust with the idea of certain portions of the female person, or with certain articles of female attire" (quoted in Steele 1996, p. 11). As this quote reflects, it is a common impression that fetishes are more common in men than in women (Bullough 1988; Steele 1996). In the *Diagnostic and Statistical Manual of Mental Disorders* (DSM) fetishism is defined as "recurrent, intense sexually arousing fantasies, sexual urges or behaviours involving the use of nonliving objects (e.g., female undergarments)" (APA 1994, p. 522–3). To be classified as pathological, the fetish is required to cause significant distress or impairment in social functioning. Krafft-Ebing recognized that there is a spectrum from "normal" fetishists who are attracted to such things as hair color and body shape, to "pathological" fetishists where

---

[1] Homosexuality was removed from the American Psychiatric Association's *Diagnostic and Statistical Manual of Mental Disorders* in 1973 (Dailey 1988).
[2] This section is based on Steele (1996) unless otherwise noted.

the attraction to the object overshadows the person possessing it and the fetishist is unable to appreciate the whole (Steele 1996; Mason 1997). The fetish then becomes the exclusive object of sexual desire. Although the exact incidence and prevalence of fetishism is unknown (Penix and Picket 2006), it is clear that extreme pathological fetishism is rare (Mason 1997) while "normal" fetishism is probably more common. Psychiatrist Robert Stoller, for instance, claimed that most males of most cultures are minifetishists. Moreover, certain phenomena, such as women wearing lingerie, have become normative sexual imagery.

What causes fetishism is poorly known. Proposed explanations include psychoanalytic theories, brain injuries (e.g., Epstein 1960, 1961), and different learning mechanisms. In the late nineteenth century Alfred Binet suggested that fetishes could be the result of associative learning. Sigmund Freud also believed in the importance of experience. In his 1905 *Three essays on the theory of sexuality* he suggested that fetishism was the adult consequence of childhood trauma and later, in his 1927 work *Fetishism*, viewed it as an expression of castration anxiety, where the fetish served as the symbolic meaning of a penis substitute (Mason 1997). Today, such ideas are believed to hold little scientific validity (Steele 1996). Association learning—or, in this context, sexual conditioning—have continued to be discussed as a possible causal factor in fetishism (O'Donohue and Plaud 1994; Mason 1997; Pfaus *et al.* 2001). Sexual conditioning occurs when a certain stimulus becomes sexually arousing after having been experienced in conjunction with a sexual reward, such as an orgasm. However, this explanatory model has been criticized for not being able to explain why the preference remains stable throughout a lifetime (Wilson 1987; Mason 1997). Conditioned responses will normally disappear in the absence of reinforcement (known as extinction) (Wilson 1987). It is therefore interesting that ethologists observed a phenomenon in birds (Lorenz 1935; Immelmann 1972) where sexual preferences were learnt from parents and siblings during a sensitive period early in life and thereafter remained remarkably stable. This learning took place long before sexual maturation, in the absence of sexual reward, and seemed to be the result of mere exposure. Sexual imprinting as an explanatory model for fetishism has been advocated by psychologist Glenn Wilson (1987). Before that, Desmond Morris, in his popular scientific 1969 book *The Human Zoo*, speculated that sexual fetishism might be the consequence of "sexual imprinting" in humans. He cites a number of cases in support of this theory, all having in common a first sexual experience in the presence of an object that subsequently becomes the object of sexual arousal. For instance, a boy becomes a glove fetishist as a result of having experienced his first ejaculation while playing with a glove and rubbing it against his penis. However, although Morris uses the term "sexual imprinting", his examples are in fact more consistent with sexual conditioning since learning takes place in a sexual context and with a sexual reward.[3] An imprinting-like learning mechanism as an explanation to paraphilia was also proposed by John Money (1984). His idea was that sexual preferences exist as so-called "lovemaps". Lovemaps are templates or schemas in the brain that are not complete at birth but need input from the social environment (Mason 1997). The development of lovemaps is reminiscent of imprinting in that they depend on experience in early childhood and once formed they are resistant to change. However, they differ from imprinting in that the kind of experience that influences them is sexual in nature. The normal lovemap develops from "sexuoerotic" play in early childhood, while paraphilias result from experience "vandalizing" the lovemap, for instance corporal punishment inducing genital arousal (Money 1986).

The rise of the evolutionary study of human behavior in the 1970s brought with it a few attempts to explain preferences that can be regarded as "fetishistic" (see Wilson 1975). In *The Human Zoo* (1969), Morris speculated that certain clothes and make-up exaggerate "natural", "biological", sexual signals. For instance, a corset exaggerates the female hourglass shape that attracts many heterosexual males. Furthermore, a connection between the evolution of sexual arousal patterns and the shininess, smell, and shape of objects and materials has been proposed (see Steele 1996). Evolutionary considerations have also been invoked to explain why more men than women have been observed to demonstrate fetishistic tendencies. The reason is hypothesized to be that males are often less selective than females in their sexual choices (Wilson 1987). However, as this is an evolutionary account of paraphilias, we will start by looking at the evolutionary function of sexual preferences generally.

2.2 The evolution of human sexual preferences

To start with, it might be helpful to discuss what is meant by sexual preferences. In biology, for instance, it is more common to talk about partner preferences, mate choice, or mating preferences. Such preferences are investigated by observing to which of a number of potential mates an animal prefers to direct courtship. In animals it is fair to assume that partner preferences and sexual preferences are the same. In humans, factors other than sexual attraction affect partner preferences. Using the term *sexual* preferences emphasizes that it is the sexual arousability of a potential partner that is in focus. In humans, sexual

---

[3] But see my discussion of a second sensitive period in sections 2.3.2 and 2.3.5.

preferences can be measured by having a person judge the sexual attractiveness of a stimulus or directly measuring physical arousal in response to a stimulus. People can, of course, also prefer certain sexual acts, e.g., oral sex, anal sex, and bonding. For the purposes of this text, however, an individual is said to have a sexual preference for a trait, such as a certain body type, if this trait works as a stimulus that does better than other stimuli in eliciting a sexual response.

From an evolutionary viewpoint, the function of sexual preferences is to enable successful reproduction. Successful reproduction requires that the sexual partner is a sexually mature conspecific of the opposite sex. We can thus expect sexually reproducing organisms to have evolved mechanisms for recognition of age, sex, and species identity in a potential partner. Assessment of these things is not necessarily a trivial task, and mistakes sometimes occur (e.g., see Gray 1958). Nevertheless, the evolutionary analysis of sexual preferences often assumes that sexual preferences have been fine-tuned by genetic evolution to allow an individual to choose a mate of high genetic quality (see Enquist *et al.* 2002 for a critical discussion of this analysis).

Human mating behavior has attracted a lot of attention from researchers seeking an evolutionary understanding of human behavior (Laland and Brown 2002). Such research mostly takes place within an adaptationist framework (Futuyma 1998) and is based on the idea that psychological adaptations underlie behavior. Defining "adaptation", Futuyma (1998, p. 355) explains that "a feature is an adaptation for some function if it has become prevalent or is maintained in a population (or species, or clade) because of natural selection for that function". As such, adaptations must have a genetic basis, since selection has no evolutionary effect unless there is inheritance. In evolutionary psychology, where adaptationist reasoning is readily applied, human psychological mechanisms are thought to be adaptations to our ancestral environment, or environment of evolutionary adaptiveness (EEA) (Laland and Brown 2002).

One particularly popular solution to the partner recognition problem spelled out above is that sexual preferences are such adaptations. For instance, males are thought to have evolved preferences for female traits that are associated with youth and high reproductive potential, such as smooth skin, good muscle tone, and an optimal waist-to-hip ratio (Futuyma 1998; Laland and Brown 2002). Furthermore, evolutionary psychologists generally believe that preferences have evolved to detect signs of a partner's genetic quality (Trivers 1972; Andersson 1994; Cartwright 2000). This is because choosing a mate of high genetic quality confers the advantage of having offspring that inherit high-quality genes and therefore increase one's inclusive fitness (e.g., see Andersson, 1994; Bradbury and Verhencamp 1998). Body symmetry has been suggested to be a sign of genetic quality (Cartwright 2000), and studies show that women find men with symmetrical features attractive. Moreover, empirical studies showing that individuals agree to a considerable extent (but not completely) on sexual preferences, both within and between populations (e.g., see Buss 1989; Cunningham *et al.* 1995; Rhodes *et al.* 2002) have been proposed as support for genetically evolved preferences (e.g., Buss 1989; see Grammer *et al.* (2003) for discussion).

However, it is easy to find examples of sexual preferences that differ between individuals (Little *et al.* 2003), cultures, and periods (Grammer *et al.* 2003). For instance, there are cultural differences in attraction to such things as preferred build and skin color (Laland and Brown 2002) and further cultural differences in adornment and the extent to which these adornments are perceived as sexually attractive. For example, lip enlargement, which occurs in some African and South American tribes (Zebrowitz 1997), is not perceived as particularly attractive by people from other cultures (see Enquist *et al.* 2002 for discussion). Standards of beauty have also varied considerably over time with respect to things like ideal body type and clothing.

It is not a logical impossibility that variations in preferences that concern "natural" features of the human body are due to variation in the genes determining such preferences. However, as already touched on above, artifacts such as clothing and other adornment can also be perceived as attractive (e.g., see

Table 2.1 Number of members subscribed to Yahoo! discussion groups concerned with sexual preferences for various objects associated with the body

Preferred object	**Group members**
Footwear	29 022
Objects worn on legs and buttocks (stockings, skirts, etc.)	27 490
Underwear	17 951
Whole-body wear (costumes, coats, etc.)	9306
Objects worn on trunk (jacket, waistcoat, etc.)	6886
Objects worn on head and neck (hats, necklaces, etc.)	2210
Stethoscopes	933
Wristwatches, bracelets, etc.	844
Diapers	483
Hearing aids	480
Catheters	28
Pace-makers	2

Modified from Scorolli *et al.* (2007). Reprinted by permission from Macmilla Publshers Ltd: International Journal of Impotence Research, Scorolli C, Ghirlanda S, Enquist M, Zattoni S & Jannini EA, Relative prevalence of different fetishes, copyright 2007

Enquist *et al.* 2002). As noted above, there is actually a striking diversity of sexual preferences involving objects, most of them somehow associated with the human body (Scorolli 2007). These objects are sometimes referred to as sexual fetishes, even though they are not necessarily part of a condition that meets the diagnostic criteria of fetishism. This is evidenced by the plethora of discussion groups on the internet devoted to sexual interest in a diversity of objects (Table 2.1; see Scorolli *et al.* 2007). Although the existence of both homosexuality and incest has attracted considerable attention from evolutionary psychologists, there is a lack of adaptationist theories of fetishism and other paraphilias. Obviously, fetishistic preferences are difficult to reconcile with the adaptationist idea that preferences evolved to assess mate quality in the EEA, especially since most of them involve recent human inventions, such as glasses and cigarettes. Apart from the fact that there were no glasses and cigarettes in the Pleistocene, it is also highly unlikely that wearing glasses, for instance, is a sign of good mate quality. In fact, glasses could even be considered an indicator of poor mate quality.

Is there another way of understanding sexual preferences from an evolutionary perspective that accommodates the above observations? Insight may be gained by considering how sexual preferences develop in animals. Ethologists have studied how animals learn sexual preferences through sexual imprinting (see reviews by Kruijt 1985; Bolhuis 1991; Clayton 1994; ten Cate 1994). This usually results in biologically functional preferences, but, in unusual circumstances, might result in atypical preferences, such as a preference for another species or for an artifact. The following sections will deal with the phenomenon of sexual imprinting in depth, starting with what is known from animal studies and followed by a discussion of the reasons for believing that a similar mechanism might be at work in humans.

2.3 **Sexual imprinting in animals**

The basic principle of imprinting is often illustrated by the picture of a trail of goslings following Konrad Lorenz as if they believed him to be their mother. Lorenz showed that this following response, normally triggered by the goslings' mother, could also be triggered by a diversity of moving objects, such as a human being or even a moving box. According to the terminology used by Lorenz, the following response of the goslings became *imprinted* to the object in question (Eibl-Eibesfeldt 1975). This instance of imprinting, whereby a newly hatched gosling forms an attachment to the first moving object it sees and follows it around, is called *filial* imprinting (Avital and Jablonka 2000).

However, a range of other reactions involved in different kinds of behaviors was also found to be imprinted to specific objects. Of crucial importance to this chapter, imprinting was also observed to occur with respect to sexual reactions (Eibl-Eibesfeldt 1975; Clayton 1994). Lorenz (1931), for instance, studied jackdaws that were raised by humans from the nestling stage. These birds were later observed to court humans. They had been *sexually imprinted* to humans. Interestingly, even if a hand-raised jackdaw joined a flock of conspecifics when fledged, it would prefer to court humans once sexually mature. The imprinting must thus have taken place at a point in the young bird's development when it was still too young to actually display any sexual behavior (Eibl-Eibesfeldt 1975).

Lorenz (1935, as cited in Hogan 2001, p. 263) originally defined imprinting as "the acquisition of the object of instinctive behavior patterns oriented towards conspecifics". According to Hogan's more recent reformulation of Lorenz' definition, imprinting would be "the development of a perceptual mechanism (or schema) that is responsible for species recognition and that is connected to all (or many of) the social behaviour systems in the animal" (Hogan 2001, p. 263). In line with this idea of a single perceptual mechanism serving several behavior systems, Morris (1969) claims that the imprinting of the following reaction is retained for life since the birds learn the species to which they belong and which sexual partner to choose. However, Lorenz' own observations of jackdaws showed that the objects of social and sexual behavior could be different. This suggests that they might develop independently at different times (Hogan 2001). The current opinion seems to be that the object-recognition mechanism for sexual behavior develops independently and separately of filial imprinting (ten Cate 1994; Hogan 2001).

Sexual imprinting provides an explanation for the development of mate preferences that, in contrast to prevailing evolutionary models, is not genetically determined (Vos 1995a). Imprinting is clearly a result of early experience (e.g., Hogan 2001), and can thus be seen as a learning process. Modern-day research seems to dissolve the boundaries between different kinds of learning and also between learning and other developmental processes (e.g., see ten Cate 1994; Hogan 2001). Nevertheless, imprinting has traditionally been distinguished from other learning processes on the grounds that it occurs early in development during a *sensitive period* and that it is *irreversible* (Eibl-Eibesfeldt 1975; Immelmann 1980). It has further been thought of as a special kind of learning since there are no obvious reinforcers, that is, no external rewards, such as food (ten Cate 1994). Sexual imprinting has the additional characteristic that the sensitive period occurs long before maturation of the associated sexual behavior system (Eibl-Eibesfeldt 1975; Immelmann 1980). Importantly, there is no sexual motivation involved in the initial learning phase of sexual imprinting (e.g., Hogan 2001). The following subsections elaborate on some of the key features of sexual imprinting.

2.3.1 What is learned?

Sexual imprinting is a process whereby sexual preferences are acquired through early experience with parents and siblings. By exposure to these individuals, a young animal can apparently learn to recognize its own species, discriminate between the sexes, and recognize kin (Vos 1995a).

Firstly, there is evidence that young animals are sexually imprinted to the species of those individuals that rear them (Vos 1994). They apparently learn the species-specific appearance of parents and siblings (Immelmann 1975; Vos 1995a; Shettleworth 1998) and at maturity a sexual preference for this species is displayed (e.g., Morris 1969; Immelmann 1980; Vos 1994). A popular way of demonstrating this is to have a pair of Bengalese finches adopt the young of a related species, the zebra finch (e.g., Bischof 1994). Klaus Immelmann was a pioneer in this line of research. He was the first to show that cross-fostered zebra finch males later developed a strong preference for Bengalese finch females, rather than conspecific females (Immelmann 1969). Kruijt et al. (1983) later showed that not only parents, but also siblings can have an impact on later sexual preferences.

Secondly, later research implies that sexual imprinting also functions to discriminate between the sexes. Wild-type males of the sexually dimorphic zebra finch, reared by one wild-type colored parent and one white parent, later developed a preference for the color morph of the mother (Vos *et al.* 1993; Vos 1994). They also showed an aversion towards females of the same color morph as the father (Vos *et al.* 1993). It has even been demonstrated that zebra finch males prefer a male of their mother's morph to a female of their father's morph (Vos 1994)!

Thirdly, inherent in the process of sexual imprinting is the learning of the appearance of kin and familiar individuals. These individuals are generally avoided as sexual partners (Vos 1995a). For instance, Friedrich Schutz (1965) raised male ducks together with chickens and various species of ducks. They would subsequently court members of the *species* with which they had been raised, but normally not those *individuals* with whom they were raised. A decade later, Bateson (1978) observed that Japanese quail, successfully imprinted to brown wild-type females, preferred unfamiliar brown females to familiar ones. However, only a slight amount of deviation from the imprinting stimulus is preferred. For instance, Bateson found that Japanese quail court first cousins rather than familiar and unfamiliar siblings, but they also preferred first cousins to unrelated individuals. Hence, the learning process of sexual imprinting, whereby future sexual preferences are acquired, seems to have built-in mechanisms for avoiding cross-breeding with other species or inbreeding with close relatives (Eibl-Eibesfeldt 1975; Vos 1995a).

2.3.2 Sensitive period

As mentioned, one of the characteristics of imprinting is that it occurs during a sensitive period. A sensitive period, or sensitive phase, refers to a limited time period during development when a certain learning process can take place (Immelmann 1980). The learning capacity ceases after the sensitive period has passed (Eibl-Eibesfeldt 1975; Immelmann 1980). Sensitive periods are not restricted to sexual and filial imprinting, but are common in many aspects of development (Hogan 2001). Examples range from the imprinting-like process of song learning in birds to the development of visual capacities in primates (Knudsen 1999).

Until recently, sexual imprinting was thought to occur during one sensitive period early in development. It was later proposed, however, that sexual imprinting is better described as a two-stage process (e.g., Bischof 1994). According to this model, the first stage is the acquisition phase (Bischof 2003). It corresponds to the traditional sensitive period and occurs when the animal is very young. This is when information about the appearance of the parents and, supposedly, siblings is stored (Bischof 1994). According to Bischof (2003), the bird acquires a *social preference* for members of his social environment and resembling stimuli. As a result of this social preference, a cross-fostered male will later prefer to court a female of the foster species rather than a conspecific female (Bischof 2003). That there is a second sensitive period is suggested by experiments (e.g., Kruijt and Meeuwissen 1991) showing that the first female an adult male is exposed to influences his final preference strongly (Clayton 1994). The second phase is thus believed to occur when the animal becomes sexually mature and has its first courting experience. This is referred to as the consolidation or stabilization phase (Bischof 1994). The validity of the stored memory of the imprinting stimulus is now thought to be tested and consolidated or slightly modified (Bischof 1994). This two-stage process resembles the process of song learning in songbirds. The song is learnt during an early memorization phase. Later on, during the so-called motor phase, the male matches his own song to the previously acquired memory or template (Clayton 1994; see also ten Cate 1994).

2.3.3 Stability of imprinting

The notion of *stability* of imprinting originally referred to the assumed fact that an animal would always remember and prefer the very first object that it was imprinted to, regardless of later experiences (Eibl-Eibesfeldt 1975). Once the early sensitive period had passed, the learnt preference was thought to remain stable (Eibl-Eibesfeldt 1975; Bischof 1994). However, we have already seen that sexual preferences can be altered in adult life. The first sexual experience consolidates or modifies the preference to some extent. Even after the consolidation phase has passed, there seems to be some room for the establishment of new preferences (see Bischof 1994; Immelmann 1975). A cross-fostered zebra finch male with a consolidated preference for the Bengalese finch might still court a zebra finch in the absence of the preferred species. Given a choice between the two species, however, he will choose the Bengalese finch. The initial preference is said to be masked by the new preference (Bischof 1994). According to Immelmann, the difference between new preferences and imprinted preferences is that the latter are more stable and are maintained indefinitely even in the absence of the preferred object, whereas the preference acquired as an adult is lost within weeks or days (Immelmann 1975). Immelmann claims that it is in fact not the capacity to acquire information that defines imprinting. Adults can acquire new information equally efficiently as the young. The truly distinguishing characteristic of imprinting is rather that the information acquired during the sensitive period is retained for life. Nevertheless, the newer data suggest that the acquired preference is not stabilized until the second sensitive period has passed. Results indicate that, at least for males, the preference remains stable after the consolidation phase (Bischof 1994).

2.3.4 Sex differences

In the early days of research into sexual imprinting it was believed that only males learned their mate preferences through imprinting, while females innately knew the appearance of their species (Vos 1995a). This conclusion was drawn from experiments with the mallard (Schutz 1965) and the zebra finch (Immelmann 1972). However, improving the experimental methods, researchers proved that females of these species were in fact sexually imprinted too (Sonnemann and Sjölander 1977; Vos 1995a).

However, the extent to which the process of imprinting is similar in males and females is still a matter of debate. For instance, it has been suggested that only males learn to discriminate between the sexes through sexual imprinting, and that imprinting is more stable in males. Part of the relevant evidence for this comes from zebra finches raised by parents of the white morph who had the color of their bills manipulated. Males were shown to prefer birds with the same bill color as their mother, irrespective of the sex of these birds. Females, on the other hand, preferred males irrespective of bill color (Vos 1995c; but see also Weisman *et al.* 1994). Vos suggests that while males learn to distinguish between the sexes on learnt morphological cues (Vos 1995c), females seem to rely more on behavioral cues, such as singing, in order to recognize males and choose a mate (Vos 1995a,c). There are indications that the song preferences of zebra finch females are acquired through a process resembling sexual imprinting (Riebel 2003). Evidence for a sex difference regarding the stability of imprinting comes from an experiment by Kendrick *et al.* (1998, 2001). They had goat females adopt newborn sheep and sheep females adopt newborn goats. As the young sheep and goats became sexually mature, both males and females first developed a preference for the maternal (as opposed to the genetic) species. In males, the imprinted preference remained stable. In females, however, the maternal influence was weaker and reversible.

2.3.5 Development

To understand how the partner recognition mechanism develops, it might be helpful to consider Hogan's exploration of behavior systems (Hogan 2001). He describes behavior systems as consisting of perceptual, motor, and central mechanisms. The perceptual mechanism of the sex system is the partner recognition mechanism. The motor mechanism is responsible for observable sexual behavior, such as various courtship displays and copulation. The central mechanism integrates perceptual input, is sensitive to internal motivational factors such as testosterone, and activates and coordinates motor output (Hogan 2001).

According to sexual imprinting theory, the perceptual mechanism of the partner recognition system develops independently of connections with central and motor mechanisms (Hogan 2001). This means that, although what is learnt during the acquisition phase has consequences for subsequent sexual preferences, the young animal has no sexual motivation at this point in development. What is learnt is not yet associated with sexual behavior of any kind or with future sexual partners (Bischof 1994). Instead, Bischof suggests that a young zebra finch, for instance, learns who is feeding it and who is competing with it for food (Bischof 1994).

It is reasonable to assume that sexual imprinting only takes place in certain situations and that there is some specific event that triggers the imprinting. Otherwise any item could become the object of sexual behavior. Without genetic guidance it would be impossible to know what to pay attention to and what to learn. A completely naïve individual surrounded by massive amounts of more or less biologically relevant stimuli would not know that it should learn about such things as sex and age. There must be something that guides the learning in the direction of conspecific individuals and probably also in the direction of opposite sex individuals. There have been suggestions that there might be some sort of predisposition for preferring conspecifics. For example, a zebra finch male brought up by a mixed species pair, consisting of one Bengalese finch and one zebra finch, was shown to imprint on the zebra finch, even when this was the male (ten Cate 1994). However, later research has shown that this is an effect of social interactions. Both zebra finch males and females interact more than a Bengalese finch foster parent with the young zebra finch, in terms of feeding behavior as well as aggression. If those interactions are prevented experimentally, so that the Bengalese finch is responsible for most of the interactions with the young bird, a preference for the Bengalese finch will be the result (ten Cate 1994). Moreover, in a clutch of cross-fostered zebra finches, the brother that begs more, and so is fed more, develops a stronger preference for females of the rearing species (Bischof 1994). It is important to note that imprinting does not seem to be only a matter of the young passively learning the features of the parents. Instead, the fact that parents respond to behaviors directed at them by the young is suggested to enhance the imprinting process (ten Cate 1994). Furthermore, some features of living things, such as movement and the fact that they provide a combination of visual, auditory, and tactile stimulation, seem to be inherently attractive and enhance learning. The explanation for this may be fairly simple, such that these features increase arousal and facilitate the focusing of attention on the object (ten Cate 1994). In a natural setting, the bird will most likely be close to parents and siblings. Since they show the kind of behavior that the bird is sensitive to, these are the stimuli the bird will imprint on (ten Cate 1994). The existence of certain sensitivities to particular behaviors does not necessarily reflect a genetically fixed, preprogrammed preference. Dispositions can often be traced to earlier external stimulation during development (ten Cate 1994). The causal factor that leads a male to use the mother's phenotype as a model for appropriate mates, and the father's phenotype as a model for inappropriate ones is not yet clear (Vos *et al.* 1993). However, Vos (1995b) speculates that the fact that both males and females seem to prefer mates resembling their mother could have something to do with the mother tending to exhibit more parental care than the father.[4]

At sexual maturity, the perceptual system becomes connected to central and motor mechanisms (Hogan 2001). That which was learnt during the acquisition phase now becomes associated with the emerging sexual behavior. This is what happens during the so-called consolidation phase. In the zebra finch male, consolidation is triggered by the appearance of a female of the rearing species, the sight of which arouses the bird. This is most likely because the female bears an obvious resemblance to the stored memory of the individuals with which the male was raised. Bischof suggests that the response of the female may also contribute to the arousal. Arousal seems to be a necessary prerequisite for consolidation. The sexually motivated and aroused male directs courting behavior toward the female. In this process, the stored memory of the

---

[4] The remainder of this paragraph is based on Bischof (1994), unless otherwise noted.

rearing species is believed to become associated with sexual behavior. If the first female that a cross-fostered and sexually mature zebra finch male is exposed to is conspecific, we have seen that he can change his preferences towards zebra finches. The reasons for this could include strong courtship motivation and the fact that the female shares some of the characteristics of the rearing species. The most popular explanation, however, seems to be that the male could after all have been imprinted to zebra finches to some extent. It is possible, for instance, that he was imprinted to the acoustic cues of his biological parents before transfer to the foster parents. There are also strong indications that cross-fostered zebra finches are imprinted to conspecific siblings. On the other hand, Kendrick *et al.* (1998) did not find any evidence of the impact of siblings on preferences in sheep and goats.

2.4 Sexual imprinting in humans

Research on imprinting has traditionally been dominated by observations of various bird species. A review by ten Cate *et al.* (1993) reveals a widespread taxonomic distribution of sexual imprinting in birds. There are also studies on a few fish species, including the Amazon Molly[5] (e.g., Immelmann 1980; Körner *et al.* 1999), indicating that sexual imprinting also exists in this phylogenetic group. Recently, imprinting in mammals has attracted some attention. To date, there is some evidence of sexual imprinting in rodents (D'Udine and Alleva 1983), sheep, goats, and primates (e.g., Kendrick *et al.* 1998). Anecdotal evidence of human-imprinted animals adds even more mammalian species to this crowd. Zookeepers often become the sexual target of many animals in their care (Wilson 1987). Pet cats and dogs are likewise known to try to mate with humans (Wilson 1987). This behavior has even been observed in chimpanzees reared by humans (Morris 1969).

The widespread phylogenetic distribution of imprinting among vertebrates is one reason for suspecting that some variant of sexual imprinting exists in humans as well. Most of what is known about developmental processes in chickens in fact applies to monkeys as well as to human beings (Hogan 2001). Furthermore, according to many theories of human behavior and personality, early experiences are fundamental to the ontogeny of perceptual mechanisms (e.g., Bowlby 1969; Bandura 1977; Money 1986). In short, it would not come as a big surprise if early experiences influence human sexual preferences through an imprinting-like process characterized by a sensitive period.

---

[5] This species consists of females exclusively. The eggs of the Amazon Molly are fertilized by males of other species and a preference for males of a certain species is learnt—or imprinted—early in life (see Körner *et al.* 1999).

Unfortunately, the exact developmental mechanisms and the impact of early experiences on human sexual preferences are poorly known and evidence for sexual imprinting in humans is sparse.

However, some behavioral observations are consistent with sexual imprinting in humans. For instance, human sex partners tend to resemble each other in many traits. The ubiquitousness of human homogamy requires an evolutionary explanation, and sexual imprinting has been suggested for this (Bereczkei *et al.* 2004). Furthermore, we have seen that as a result of sexual imprinting, ducks do not mate with individuals they grew up with and the same phenomenon has been observed in humans in different places and cultures (Wolf 2004a). It is referred to as the Westermarck effect. Westermarck argued that the deleterious consequences of inbreeding have selected for an innate tendency to develop an aversion to sexual relations with childhood associates (Wolf 2004a). For instance, it is known that children raised together in Israeli kibbutzim avoid having sexual relations with one another and instead prefer mates from outside the community (Wolf 2004a). Another demonstration of the Westermarck effect is the so-called minor marriages in Taiwan. Here, young girls are adopted into the families of their future husbands. These marriages, however, have been shown to result in relatively poor fertility and low marital stability (Wolf 2004b). Interestingly "the few Israeli kibbutzniks who chose to marry within their peer group were usually those who had entered the kibbutz after the age of six and therefore had not grown up with their future spouses. In Taiwan, girls who were adopted into families before the age of three and then married their adopted 'brother' had a lower fertility than girls adopted later" (Bateson 2004, p. 31). This suggests the existence of a sensitive period for learning that family members should be avoided as sexual partners, and possibly also a sensitive period for sexual imprinting if these phenomena are different aspects of the same learning process.

Sexual imprinting has also been invoked to explain rare "fetishistic" preferences for artifacts (e.g., Morris 1969; Eibl-Eibesfeldt 1975; Wilson 1987; Enquist *et al.* 2002), the existence and diversity of which was discussed above (see Table 2.1). Adding an artifact to the phenotype, for example glasses, is much the same thing as manipulating the phenotype of birds. Plenge and colleagues, for instance, attached a red feather to the forehead of adult Javanese manikins and showed that young birds were imprinted to this novel trait (Plenge *et al.* 2000). The artifact is learned as any other feature of the parent's appearance and becomes a preference for the next generation (Plenge *et al.* 2000; Enquist 2005). A study estimating the relative frequency of atypical sexual preferences discussed in an internet community revealed that preferences related to the human body are much more common than preferences for objects and events not usually associated with people (Scorolli *et al.* 2007). This was interpreted as showing that sexual preferences are acquired mostly through social interactions, consistent with the sexual imprinting hypothesis.

Even though many observations are consistent with the sexual imprinting hypothesis, it is problematic to prove that sexual imprinting exists in humans. The problem is that it is practically impossible to design controlled experiments to this end. It would be morally dubious, to say the least, to have gorillas adopt human babies to see which species they would later prefer to mate with. Instead, other ways of investigating the matter have to be found. One possibility is to look for retrospective correlations between people's childhood experiences and adult preferences by employing demographic records or verbal reports. Some of these studies are discussed in the remainder of this section.

A number of studies have been conducted to look for an association between parental features and sexual or partner preferences in humans. Most of these studies found a relationship between the opposite-sex parent and partner preferences in both men (Jedlicka 1980; Bereczkei *et al.* 2002; Perrett *et al.* 2002; Little *et al.* 2003) and women (Jedlicka 1980; Zei *et al.* 1981; Wilson and Barrett 1987; Perrett *et al.* 2002; Little *et al.* 2003; Bereczkei *et al.* 2004; Wiszewska *et al.* 2007). Perrett *et al.* (2002) found a predominantly maternal influence on male preferences but an influence of both the father and the mother on female preferences. Little *et al.* (2003) also found a paternal influence on the partner preferences of heterosexual men. Although sexual imprinting has been suggested as a mechanism behind the observed parental influences on sexual preferences (Zei *et al.* 1981; Wilson and Barrett 1987; Bereczkei *et al.* 2002, 2004; Little *et al.* 2003; Wiszewska *et al.* 2007), other explanations are possible. For instance, a preference for individuals resembling the opposite-sex parent could potentially be the effect of a genetically determined preference for self-similar individuals (phenotype matching) (e.g., Bereczkei *et al.* 2002; Wiszewska *et al.* 2007). Bereczekei *et al.* (2004) ruled out an explanation in terms of genetically determined preferences by looking at the preferences of adopted daughters. They showed that adopted daughters chose husbands that resembled their adoptive fathers.

Another way of eliminating an explanation in terms of genetically determined preferences is to look at rare sexual preferences for artifacts, or "fetishes". As discussed at the beginning of this chapter, such preferences are unlikely to be genetically determined. They also provide us with a "natural experiment". In a standard experiment, an experimental group is exposed to some particular treatment, such as being brought up by foster parents of another species, and the effect is observed. A natural experiment works backwards. We start with an observation, in this case a rare sexual preference, and try to infer its cause, for instance by collecting reports of childhood experiences. Drawing on the idea that sexual fetishes can be studied as a model for sexual imprinting in humans, a couple of studies of the connection between rare preferences and childhood experiences have been conducted (Enquist *et al.* in press; Aronsson *et al.* in press).

In one study, subjects who had a sexual preference for pregnant and/or lactating/breast-feeding women were surveyed (Enquist *et al.* in press). This preference turned out to be more common among men and women with younger siblings, implying that the preference is an effect of having seen one's pregnant and lactating mother. In addition, the effect was limited to birth intervals between 1.5 and 5 years. This could be interpreted as evidence for a sensitive period occurring between 1.5 and 5 years of age. The birth order effect is an indication of early learning, such as sexual imprinting, and the birth interval effect is what one would expect specifically from sexual imprinting, since a sensitive period is a unique feature of imprinting.

In another study, the relationship between parental smoking habits during a person's childhood and a sexual attraction to smoking in self-reported and gay men was investigated. Sexual attraction to smoking was found to be associated with having smoking parents irrespective of self-reported sexual orientation. When looking at cases where only the mother or the father smoked, attraction to smoking was associated with having a smoking mother, but not a smoking father, in straight men whereas having a smoking mother, but also a smoking father, was associated with attraction to smoking in gay men (Aronsson *et al.* in press). This result indicates that childhood experiences have an impact on future sexual preferences. Moreover, the parental influence seems to be sex-specific. For instance, mothers, but not fathers, seem to have an effect on individuals growing up to prefer female partners. This is reminiscent of results from sexual imprinting studies in animals, where males imprint positively to the mother, and sometimes even negatively to the father. It could be argued that the maternal influence is simply an effect of the primary caregiver, but such an explanation is contradicted by the paternal influence on gay men.

As discussed earlier, another learning mechanism has been suggested in the context of rare sexual preferences, namely sexual conditioning (Gosselin and Wilson 1980; Wilson 1987; Akins, 2004). Like imprinting, it has been hypothesized to sometimes occur in childhood (Wilson 1987). Sexual conditioning theory assumes that sexual preferences are acquired when the preferred object is experienced in conjunction with genital stimulation (Gosselin and Wilson 1980). Sexual imprinting seems like a better explanation of the above observations for two reasons. If the birth interval effect for pregnant/lactating women, for instance, is an effect of conditioning, it would require sexual motivation to be already present in 2-year-old children, and to disappear at around age 5 (see Wilson 1987 for additional arguments against early sexual conditioning). If it is an effect of imprinting, on the other hand, no sexual motivation or sexual reinforcers are required. Furthermore, the sex-specificity observed in the smoking study is not predicted by current sexual conditioning theory, while it is known from studies of sexual imprinting in animals.

2.5 Adaptationism and sexual imprinting

We may ask why evolution should favor sexual imprinting over genetically determined preferences. It is important to note that sexual imprinting in most cases generates biologically functional, adaptive preferences. That the preferences are adaptive, that is, result in successful reproduction, is not the same thing as the preferences being adaptations,[6] that is, due to natural selection for those specific preferences (see Laland and Brown 2002 for a discussion of adaptive behavior vs adaptation). A learned preference can be adaptive, but it cannot be an adaptation *per se* because there is no corresponding gene(s) that can be transferred to future generations. In contrast, the sexual imprinting *mechanism*, that is, the *ability* to be imprinted, most probably has a genetical basis and it is logical to ask if it is an adaptation. We should remember that evolution cannot be expected to generate perfection. Natural selection can only act on the variation that happens to be present at any one time (Futuyma 1998). Neither can we expect human behavioral mechanisms to be adapted to the exclusive problems faced by human beings (Laland and Brown 2002). Our genome has a much longer history than the history of mankind, and we share basic behavioral mechanisms with other vertebrates. This evolutionary history confers constraints on the available possibilities for evolution. It is also possible that there are limitations on what kind and how much information can be effectively genetically encoded. Laland and Brown (2002) question selection on properties of mind, such as partner preferences, on the basis that we do not yet have a neurobiological theory of how and whether genes influence the relevant psychological states. Indeed, it seems that normal development of all perceptual systems in birds and mammals requires environmental input (Hogan 2001). Hogan points out that developmental mechanisms do not need to be optimal but merely good enough to bring the individual to adulthood.

Nevertheless, sexual imprinting has been suggested to have certain evolutionary advantages. It has been suggested that an adaptive function of sexual

---

[6] See section 2.2 for a definition of adaptation and discussion of adaptation in relation to preferences.

imprinting is to avoid inbreeding with close relatives (exactly what your mother or brother looks like cannot be stored in your genes) as well as prevent cross-breeding with other species (Eibl-Eibesfeldt 1975; Vos 1995a). Sexual imprinting has also been proposed to be adaptive in that it guides individuals to mate with not too distantly related conspecifics, which could potentially be beneficial, for instance by preventing the loss of genes required for adaptation to a particular environment (Bateson 1983). Imprinting also ensures flexibility. The preferences of the individual become adapted to the present phenotypes of the local population. A genetically determined, fixed preference for an "ideal" partner, on the other hand, might result in the individual never finding a partner that matches the ideal (Grammer *et al.* 2003). It is hard, however, to establish that the mechanism of sexual imprinting is an adaptation. Determining what constitutes a character that is subject to natural selection is recognized as a difficult problem in evolutionary biology, as is the task of identifying adaptations (see Laland and Brown 2002). Nevertheless, we can be fairly sure that sexual imprinting is a mechanism that has arisen and been maintained in evolution and that clearly generates functional and adaptive preferences through learning. Sexual imprinting seems to be a viable alternative to the adaptationist assumption that preferences themselves are fine-tuned, genetically determined, adaptations. However, genetic determination of preferences cannot be ruled out *a priori* and whether preferences are imprinted or genetically determined is an empirical question. In Table 2.2 the two theories are compared in relation to different predictions about human sexual preferences. The table shows whether or not these predictions agree with empirical data.

Both sexual imprinting theory and adaptationist mate-quality theory agree that there should be preferences for things like sex and species identity. However, mate-quality hypotheses also predict that preferences are fine tuned to detailed information about mate quality and therefore are rational in all details (rational choice) (Enquist *et al.* 2002). Rational choice theories predict that there should be a correlation between mate quality and attractiveness. Studies have generally not found any relationship between a person's attractiveness and his/her genetic or phenotypic quality (Kalick *et al.* 1998 Shackelford and Larsen 1999) and they therefore fail to support the mate-quality hypothesis (see Enquist *et al.* 2002). Symmetry as a signal of mate quality has been questioned for a number of reasons, not least because the heritability of symmetry has been estimated to be close to zero (see Laland and Brown 2002).

Universality of preferences is another prediction made by the mate-quality hypothesis. As already noted, there is some evidence that individuals agree on sexual preferences, both within and between populations (e.g., Buss 1989; Cunningham *et al.* 1995; Rhodes *et al.* 2002). However, sexual imprinting

Table 2.2 Comparison of hypotheses for sexual preferences

Prediction	Hypothesis	
	Sexual imprinting	Rational choice (genetic determination)
Existence of preferences	Yes*	Yes*
Correlation between mate quality and attractiveness	–	Yes†
Universality (not necessarily complete)	Yes*	Yes*
Preferences for cultural innovations	Yes*	No†
Individual variation in preferences	Yes*	–
Unusual preferences (e.g., fetishism)	Yes*	No†
Geographical differences in preferences	Yes*	–

Facial Attractiveness, Gillian Rhodes. Copyright © 2001 by Ablex Publishing.
Modified from Enquist *et al.* (2002).
* Prediction agrees with empirical data.
† Prediction disagrees with empirical data.
– Prediction unknown or may go in either direction.

could also explain such universal preferences. Saxton *et al.* (2009) recently found that exposure to a particular population of faces increases the ratings of attractiveness of similar faces. To the extent that people look alike in different populations, universal preferences may derive from learning based on similar experiences rather than having the same genes (Enquist *et al.* 2002).

It is clear that there are also individual and geographical variations in preferences. If preferences are genetically determined, this must be due to genetic variation. Genetic variation could possibly explain variation in preferences for "natural" features of the human body such as hair color, skin color, and body type, but we have seen that there is a diversity of preferences for recent human inventions as well as rare fetishistic preferences. Such preferences are hard to explain from a mate-quality perspective on sexual preferences, but are predicted by the sexual imprinting hypothesis.

Based on this comparison, sexual imprinting has higher explanatory power than rival evolutionary hypotheses. It can provide an explanation in terms of the same mechanism for both rare fetishistic and typical preferences, as well as individual and cultural differences in sexual preferences.

2.6 Conclusion

This chapter has aimed to show that, as an explanation of human sexual preferences, sexual imprinting is a viable alternative to prevailing evolutionary theories, which assume that preferences are genetically determined. These theories generally assume that preferences are themselves adaptations that are rational in all aspects. It is clear that evolution must have favored a mechanism for partner recognition. However, this is likely to be a learning mechanism, meaning that sexual preferences are experience-dependent. Although sexual imprinting usually results in biologically functional preferences, it is possible for unusual, nonadaptive preferences to be learnt through the same mechanism. In order to understand the nature of human sexual preferences, we cannot only look at what problems our Pleistocene human ancestors had to solve; we must also look at the phylogenetic history we have in common with other species, as well as trying to understand the underlying developmental and neurobiological processes. This is important because it can reveal constraints on what information can be stored in our genes.

There are a number of reasons for believing that sexual imprinting may exist in humans. One is the widespread taxonomic distribution of the phenomenon. Sexual imprinting theory also agrees with the general importance of early experience for the development of perceptual mechanisms. An especially strong reason in favor of the sexual imprinting hypothesis is its high explanatory power. This is demonstrated by the agreement between theoretical predictions of the sexual imprinting hypothesis and existing patterns of human sexual preferences. Sexual imprinting provides an evolutionary explanation for both common and uncommon sexual preferences.

A few words should be said on the contribution of sexual imprinting theory to the debate over whether or not sexual preferences are innate (see Mameli 2008). The easy answer is that according to sexual imprinting theory, the learning mechanism is innate whereas the exact preferences are not since they are dependent on experience. This is in contrast to adaptationist accounts of sexual preferences where the preferences themselves are innate. However, Bateson (1991) distinguishes six separate meanings of the word "innate": (1) present at birth, (2) behavioral difference caused by a genetic difference, (3) adapted over the course of evolution, (4) unchanging throughout development, (5) shared by all members of a species, and (6) not learned. It is generally agreed that being present at birth is not a good criterion for innateness, since learning can take place before birth. For instance, hearing conspecific vocalizations in the egg or uterus could later on guide (sexual) behavior towards conspecifics. Concerning the other meanings of innateness, one could argue, following Bateson's conceptual framework, that imprinted sexual preferences are not innate because the differences in sexual preferences are not caused by genetic differences, and because they are learned.

However, some aspects of imprinted preferences can be said to be innate, in the sense that they are shared by all members of a species. For instance, the absolute majority of us are attracted to members of our own species. In the environments we are brought up in, no matter how different they are from one another, we are still surrounded by other human beings, and all of us therefore end up being imprinted to human beings. With respect to species preference, sexual imprinting makes sure our preferences are buffered against environmental variation (Hogan 2001; Mameli 2008). This might be the same process underlying the similarities in preferences between cultures.

In this context it is interesting that Laland and Brown (2002) warn researchers not to take evidence for one of the meanings of innateness as justifying the use of another, for instance taking universality to mean genetically determined adaptations. They also write that such terms as "innate" and "instinctive" are unfortunate because they are slippery and vague. Hogan (2001) claims that it is possible to discuss the development of behavioral mechanisms, including partner recognition mechanisms, without using the word "innate". Hogan prefers the term "prefunctional", which implies that functional experience (of a particular kind) is not necessary for its development. Hogan also notes that this is how Lorenz suggested "innate" be used: "It is logically consistent to talk about behaviour development that is prefunctional (or innate) versus behaviour development that is learned when the criterion is the absence or presence of functional experience." (Hogan 2001, p. 259) Motor mechanisms typically develop prefunctionally, an example is sexual behaviors such as courtship displays and copulation, while almost all perceptual mechanisms, including partner recognition, are influenced by functional experience.

Despite much evidence in favor of sexual imprinting as an explanation for human sexual preferences, it is hard to prove unequivocally that it exists in humans. Unequivocal evidence would require that an early sensitive period for acquisition of sexual preferences can be demonstrated in humans. It is a challenging task of future research to determine whether or not this is the case.

Acknowledgments

Thanks to Magnus Enquist, Stefano Ghirlanda, Johan Lind, Eva Lindström, and the editors of this volume for invaluable help with the manuscript, and to Oscar and Edvin for reminding me that there is more to life than work.

References

Akins, C.K. (2004) The role of classical conditioning in sexual behavior: A comparative, analysis of humans and nonhuman animals. *International Journal of Comparative Psychology*, 17, 241–62.

American Psychiatric Association (1994) *Diagnostic and Statistical Manual of Mental Disorders*, 4th edn. American Psychiatric Association, Washington, DC.

Andersson, M. (1994) *Sexual Selection*. Princeton University Press, Princeton, NJ.

Avital, E. and Jablonka, E. (2000) *Animal Traditions: Behavioural Inheritance in Evolution.* Cambridge University Press, Cambridge.

Bandura, A. (1977) *Social Learning Theory.* Prentice-Hall Inc., Englewood Cliffs, NJ.

Bateson, P. (1978) Sexual imprinting and optimal outbreeding. *Nature,* 273, 659–60.

Bateson, P. (1983) Optimal outbreeding. In P. Bateson (ed.), *Mate Choice.* Cambridge University Press, Cambridge, pp. 257–78.

Bateson, P. (1991) Are there principles of behavioral development? In: P. Bateson (ed.), *The Development and Integration of Behaviour: Essays in Honour of Robert Hinde.* Cambridge University Press, Cambridge.

Bateson, P. (2004) inbreeding avoidance and incest taboos. In: A.P. Wolf (ed.), *Inbreeding, Incest, and the Incest Taboo: The State of Knowledge at the Turn of the Century.* Stanford University Press, Palo Alto, CA, pp. 24–37.

Bereczkei, T., Gyuris, P., Koves, P., and Bernath, L. (2002) Homogamy, genetic similarity, and imprinting: parental influence on mate choice preferences. *Personality and Individual Differences,* 33, 677–90.

Bereczkei, T., Gyuris, P., and Weisfeld, G.E. (2004) Sexual imprinting in human mate choice. *Proceedings of the Royal Society of London, Series B,* 271, 1129–34.

Bischof, H.-J. (1994) Sexual imprinting as a two-stage process. In: J.A. Hogan and J.J. Bolhuis (eds), *Causal Mechanisms of Behavioural Development.* Cambridge University Press, Cambridge, pp. 82–97.

Bischof, H.-J. (2003) Neural mechanisms of sexual imprinting. *Animal Biology,* 53, 89–112.

Bolhuis, J.J. (1991) Mechanisms of avian imprinting: a review. *Biological Review,* 66, 303–45.

Bowlby, J. (1969) *Attachment and Loss.* Random House, Pimlico.

Bradbury, J.W., and Vehrencamp, S.L. (1998) *Principles of Animal Communication.* Sinauer Associates Inc., Sunderland, MA.

Bullough, V.L. (1988) Historical perspective. In D.M. Dailey (ed.), *The Sexually Unusual: Guide to Understanding and Helping.* Harrington Park Press, New York.

Buss, D.M. (1989) Sex differences in human mate preferences: evolutionary hypotheses tested in 37 cultures. *Behavioral and Brain Sciences,* 12, 1–14.

Cartwright, J. (2000) *Evolution and Human Behavior.* MIT Press, Cambridge, MA.

Clayton, N.S. (1994) The influence of social interactions on the development of song and sexual preferences in birds. In: J.A Hogan and J.J. Bolhuis (eds), *Causal Mechanisms of Behavioural Development,* Cambridge University Press, Cambridge, pp. 98–115.

Cunningham, M.R., Roberts, A.R., Barbee, A.P., Druen, P.B., and Wu, C.-H. (1995) "Their ideas of beauty are, on the whole, the same as ours": consistency and variability in the cross-cultural perception of female physical attractiveness. *Journal of Personality and Social Psychology,* 68 (2), 261–79.

D'Udine, B. and Alleva, E. (1983) early experience and sexual preferences in rodents. In: P. Bateson (ed.), *Mate Choice.* Cambridge University Press, Cambridge.

Eibl-Eibesfeldt, I. (1975) Imprinting and imprinting-like learning processes. In *Ethology, the Biology of Behavior.* Holt, Rinehart and Winston Inc., Austin, TX, pp. 258–71.

Enquist, M. (2005) Utseende och utseendeideal. In: C. Bunte, B.E. Berglund, and L. Larsson (eds), *Arkeologi och naturvetenskap.* Gyllenstiernska Krapperupstiftelsen, pp. 264–73.

Enquist, M., Ghirlanda, S., Lundqvist, D., and Wachtmeister, C.-A. (2002) An ethological theory of attractiveness. In: G. Rhodes and L. Zebrowitz (eds),

Facial Attractiveness: Evolutionary, Cognitive, and Social Perspectives. *Advances in Visual Cognition*, vol 1. Ablex, Westport, CT, pp. 127–51.

Enquist, M., Aronsson, H., Ghirlanda, S., Jansson, L., and Jannini, E.A. (in press) An empirical study on sexual imprinting in humans using a rare preference. *Journal of Sexual Medicine*.

Epstein, A.W. (1960) Fetishism: a study of its psychopathology with particular reference to a proposed disorder in brain mechanisms as an etiological factor. *Journal of Nervous and Mental Disease*, **130** (2), 107–19.

Epstein, A.W. (1961) Relationship of fetishism and transvestism to brain and particularly to temporal lobe dysfunction. *Journal of Nervous and Mental Disease*, **133** (3), 247–53.

Futuyma, D.J. (1998) *Evolutionary Biology*, 3rd edn. Sinauer Associates, Sunderland, MA.

Gosselin, C. and Wilson, G. (1980) *Sexual Variations*. Faber & Faber, London.

Grammer, K., Fink, B., Moller, A.P., and Thornhill, R. (2003) Darwinian aesthetics: sexual selection and the biology of beauty. *Biological Reviews of the Cambridge Philosophical Society (London)*, **78**, 385–407.

Gray A.P. (1958) *Bird Hybrids, A Check-List with Bibliography*. Commonwealth Agricultural Bureaux, Alva, Scotland.

Hogan, J.A. (2001) Development of behavior systems. In: E.M.Blass (ed.), *Handbook of Behavioral Neurobiology*, vol. 13. Kluwer Academic Publishers, New York, pp. 229–79.

Immelmann, K. (1969) Über den Einfluss frühkindlicher Erfahrungen auf die geschlechtliche Objektfixierung bei Estrildiden. *Zeitschrift für Tierpsychologie*, **26**, 677–91.

Immelmann, K. (1972) Sexual and other long-term aspects of imprinting in birds and other species. *Advances in the Study of Behavior*, **4**, 147–74.

Immelmann, K. (1975) Ecological significance of imprinting and early learning. *Annual Review of Ecology & Systematics*, **6**, 15–37.

Immelmann, K. (1980) *Introduction to Ethology*. Plenum Press, New York, pp. 104–7.

Jedlicka, D. (1980) A test of psychoanalytic theory of mate selection. *Journal of Social Psychology*, **112**, 295–99.

Kalick, S.M., Zebrowitz, L.A., Langlois, J.H., and Johnson, R.M. (1998) Does human facial attractiveness honestly advertise health? *Psychological Science*, **9** (1), 8–13.

Kendrick, K.M., Hinton, M.R., Atkins, K., Haupt, M.A., and Skinner, J.D. (1998) Mothers determine sexual preferences. *Nature*, **395**, 229–30.

Körner, K.E., Lütjens, O., Parzefall, J., and Schlupp, I. (1999) The role of experience in mating preferences of the unisexual Amazon molly. *Behaviour*, **136**, 257–68.

Knudsen, E.I. (1999) Early experience and critical periods. In: M.J. Zigmond, F.E. Bloom, S.C. Landis, J.L. Roberts, and L.R. Squire (eds), *Fundamental Neuroscience*, Academic Press, San Diego, pp. 637–54.

Kruijt, J.P. (1985) On the development of social attachments in birds. *Netherlands Journal of Zoology*, **35**, 45–62.

Kruijt, J.P. and Meeuwissen, G.B. (1991) Sexual preferences of male zebra finches: effects of early and adult experience. *Animal Behaviour*, **42** (1), 91–102.

Kruijt, J.P., Cate, C.J.T., and Meeuwissen, G.B. (1983) The influence of siblings on the development of sexual preferences of male zebra finches. *Developmental Psychobiology*, **16** (3), 233–9.

Laland, K.N. and Brown, G.R. (2002) *Sense and Nonsense: Evolutionary Perspectives on Human Behaviour*. Oxford University Press, Oxford.

Little, A.C., Penton-Voak, I.S., Burt, D.M., and Perrett, D.I. (2003) Investigating an imprinting-like phenomenon in humans. Partners and opposite-sex parents have similar hair and eye colour. *Evolution and Human Behavior*, **24**, 43–51.

Lorenz, K. (1931) Beiträge zur Ethologie sozialer Corviden. *Journal of Ornithology*, **79**, 67–120.

Lorenz, K. (1935) Der Kumpan in der Umwelt des Vogels. *Journal of Ornithology* **83**, 137–213, 289–413.

Mameli, M. (2008) On innateness: the Clutter Hypothesis and the Cluster Hypothesis. *Journal of Philosophy CV*, **12**, 719–37.

Mason, F.L. (1997) Fetishism: psychopathology and theory. In D.R. Laws and W. O'Donohue (eds), *Sexual Deviance: Theory, Assessment, and Treatment*. The Guilford Press, New York, London.

Money, J. (1984) Paraphilias: Phenomenology and classification. *American Journal of Psychotherapy*, **38** (2), 164–79.

Money, J. (1986) *Love Maps. Clinical Concepts of Sexual/Erotic Health and Pathology, Paraphilia, and Gender Transposition in Childhood, Adolescence, and Maturity*. Irvington Publishers, Inc., New York.

Morris, D. (1969) *The Human Zoo*. Jonathan Cape, London.

O'Donohue, W. and Plaud, J.J. (1994) The conditioning of human sexual arousal. *Archives of Sexual Behavior*, **23**, 321–44.

Penix, T. and Pickett, L. (2006) Other paraphilias. In: J.E. Fisher and W.T. O'Donohue (eds), *Practitioner's Guide to Evidence-Based Psychotherapy*. Springer Science+Business Media, LLC, New York, pp. 478–93.

Perrett, D.I., Penton-Voak, I.S., Little, A.C., Tiddeman, B.P., Burt, D.M., Schmidt, N., Oxley, R., and Barrett, L. (2002) Facial attractiveness judgements reflect learning of parental age characteristics. *Proceedings of the Royal Society B*, **269**, 873–80.

Pfaus, J.G., Kippin, T.E., and Centeno, S. (2001) Conditioning and sexual behavior: a review. *Hormones and Behavior*, **40**, 291–321.

Plenge, M., Curio, E., and Whitte, K. (2000) Sexual imprinting supports the evolution of novel male traits by transference of a preference for the colour red. *Behaviour*, **137**, 741–58.

Rhodes, G., Harwood, K., Yoshikawa, S., Nishitani, M., and McLean, I. (2002) The attractiveness of average faces: Cross-cultural evidence and possible biological basis. In: G. Rhodes and L. Zebrowitz (eds), *Facial Attractiveness: Evolutionary, cognitive, and social perspectives. Advances in Visual Cognition*, vol. 1. Ablex, Westport, CT, pp. 35–58.

Riebel, K. (2003) Developmental influences on auditory perception in female zebra finches – is there a sensitive phase for song preference learning? *Animal Biology*, **53**, 73–87.

Saxton, T.K., Little, A.C., DeBruine, L.M., Jones, B.C., and Roberts, S.C. (2009) Adolescents' preferences for sexual dimorphism are influenced by relative exposure to male and female faces. *Personality and Individual Differences*, **47** (8), 864–8.

Schutz, F. (1965) Sexuelle Prägung bei Anatiden./Sexual imprinting in mallard duck. *Zeitschrift für Tierpsychologie*, **22** (1), 50–103.

Scorolli, C., Ghirlanda, S., Enquist, M., Zattoni, S., and Jannini, E.A. (2007) Relative prevalence of different fetishes. *International Journal of Impotence Research*, **19**, 432–37.

Shackelford, T.K. and Larsen, R.J. (1999) Facial attractiveness and physical health. *Evolution and Human Behavior*, **20**, 71–6.

Shettleworth, S.J. (1998) Imprinting. In: *Cognition, Evolution, and Behavior*. Oxford University Press, New York, pp. 155–73.

Sonnemann, P. and Sjölander, S. (1977) Effects of cross-fostering on the sexual imprinting of the female Zebra Finch Taeniopygia guttata. *Zeitschrift für Tierpsychologie*, **45** (4), 337–48.

Steele, V. (1996) *Fetish: Fashion, Sex, and Power*. Oxford University Press, New York.

ten Cate, C. (1994) Perceptual mechanisms in imprinting and song learning. In: J.A. Hogan and J.J. Bolhuis (eds), *Causal Mechanisms of Behavioural Development*. Cambridge University Press, Cambridge, pp.117–46.

ten Cate, C., Vos, D.R., and Mann, N.(1993) Sexual imprinting and song learning: two of one kind? *Netherlands Journal of Zoology*, **43** (1–2), 34–45.

Trivers, R.L. (1972) Parental investment and sexual selection. In: B. Campbell (ed.), *Sexual Selection and the Descent of Man, 1871–1971*. Heinemann, London, pp. 136–79.

Vos, D.R. (1994) Sex recognition in zebra finch males results from early experience. *Behaviour*, **128**, 1–14.

Vos, D.R. (1995a) The development of sex recognition in the zebra finch. PhD thesis, Rijksuniversiteit Groningen, The Netherlands.

Vos, D.R. (1995b) Sexual imprinting in zebra finch females: do females develop a preference for males that look like their father? *Ethology*, **99**, 252–62.

Vos, D.R. (1995c) The role of sexual imprinting for sex recognition in zebra finches: a difference between males and females. *Animal Behaviour*, **50**, 645–53.

Vos, D.R., Prijs, J., and ten Cate, C. (1993) Sexual imprinting in zebra finch males: a differential effect of successive and simultaneous experience with two colour morphs. *Behaviour*, **126**, 137–54.

Weisman, R., Shackleton, S., Ratcliffe, L., Weary, D., and Boag, P. (1994) Sexual preferences of female zebra finches: imprinting on beak colour. *Behaviour*, **128** (1/2), 15–24.

Wilson, E.O. (1975) *Sociobiology: The New Synthesis*. Harvard University Press, Cambridge, MA.

Wilson, G.D. (1987) An ethological approach to sexual deviation. In: G.D. Wilson (ed.), *Variant Sexuality: Research and Theory*. Johns Hopkins University Press, Baltimore, MD, pp. 84–115.

Wilson, G.D. and Barrett, P.T. (1987) Parental characteristics and partner choice: some evidence for Oedipal imprinting. *Journal of Biosocial Science*, **19**, 157–61.

Wiszewska, A., Pawłowski, B., and Boothroyd, L. (2007) Father-daughter relationship as a moderator of sexual imprinting: a facialmetric study. *Evolution & Human Behavior*, **28** (4), 248–52.

Wolf, A.P. (2004a) *Inbreeding, Incest, and the Incest Taboo: The State of Knowledge at the Turn of the Century*. Stanford University Press, Palo Alto, CA.

Wolf, A.P. (2004b) Explaining the Westermarck effect, or, what did natural selection select for? In: A.P. Wolf (ed.), *Inbreeding, Incest, and the Incest Taboo: The State of Knowledge at the Turn of the Century*. Stanford University Press, Palo Alto, CA, pp. 76–92.

Zebrowitz, L.A. (1997) *Reading faces: window to the soul?* Westview Press, Boulder, CO.

Zei, G., Astolfi, P., and Jayakar, S.D. (1981) Correlations between father's age and husband's age: a case of imprinting? *Journal of Biosocial Science*, **13**, 409–18.

Chapter 3

Developmental disorders and cognitive architecture

Edouard Machery

Ever since Broca and Wernicke, studies about the nature and structure of the normal human mind have been inspired by its pathological variants. Contemporary cognitive science and evolutionary psychology subscribe to this tradition, too. Thus it is that they often refer to cognitive impairments in patients suffering from developmental disorders to support one of today's leading hypotheses about the architecture of normal cognition, i.e. the so-called 'massive modularity hypothesis'. In this chapter, I investigate whether this validation strategy can be safeguarded from Annette Karmiloff-Smith's well-known criticism. According to Karmiloff-Smith, developmental disorders are of no use in studying the architecture of normal cognition, because the abnormal minds develops abnormally, and therefore consists of cognitive systems that differ from the systems making up normal cognition. Put simply: studying mental disorders cannot provide us with any evidence about the architecture of the human mind. I argue against this conclusion by scrutinizing and debunking Karmiloff-Smith's arguments.

For the last 30 years, cognitive scientists have attempted to describe the cognitive architecture of typical human beings using, among other sources of evidence, the dissociations that result from developmental psychopathologies, such as autism spectrum disorders, Williams syndrome, and Down syndrome. Thus, in his recent defense of the massive modularity hypothesis, Steven Pinker insists on the importance of such dissociations to identify the components of the typical cognitive architecture:

> This kind of faculty psychology has numerous advantages (. . .). It is supported by the existence of neurological and genetic disorders that target these faculties unevenly,

such as a difficulty in recognizing faces (and facelike shapes) but not other objects, or a difficulty in reasoning about minds but not about objects or pictures.

(Pinker 2005, p. 4)

Similarly, Simon Baron-Cohen writes:

I suggest that the study of mental retardation would profit from the application of the framework of cognitive neuropsychology (...). In cognitive neuropsychology, one key question running through the investigator's mind is "Is this process or mechanism intact or impaired in this person?" When cognitive neuropsychology is done well, a patient's cognitive system is examined with specific reference to a model of the normal cognitive system. And, not infrequently, evidence from the patient's cognitive deficits leads to a revision of the model of the normal system.

(Baron-Cohen 1998, p. 335; see also Temple 1997)

However, in recent years, the use of developmental psychopathologies to identify the components of the typical cognitive architecture has come under heavy fire. In a series of influential articles, neuropsychologist Annette Karmiloff-Smith has argued that findings about the pattern of impairments and preserved capacities in people with developmental psychopathologies say nothing about the cognitive architecture of typical adults.[1] Thomas and Karmiloff-Smith write:

It is often assumed that similar domain-specific behavioural impairments found in cases of adult brain damage and *developmental disorders* correspond to similar underlying causes, and *can serve as convergent evidence for the modular structure of the normal adult cognitive system*. We argue that this correspondence is contingent on an *unsupported* assumption that atypical development can produce selective deficits while the rest of the system develops normally (Residual Normality).

(Thomas and Karmiloff-Smith 2002, p. 727; my emphasis)

If correct, Karmiloff-Smith's argument would have significant implications. Most significantly perhaps, it would partly undermine one of the leading hypotheses about the nature of the typical cognitive architecture—evolutionary psychologists' massive modularity hypothesis (see section 3.1)—since the evidence for this hypothesis comes in part from findings about developmental psychopathologies. More generally, researchers working on the typical cognitive architecture would have to stop relying on an important source of evidence, namely dissociations resulting from developmental psychopathologies.

---

[1] Karmiloff-Smith 1998, 2001; Paterson *et al.* 1999; Karmiloff-Smith *et al.* 2003a,b; Karmiloff-Smith and Thomas 2005; Elsabbagh and Karmiloff-Smith 2006. For further discussion of Karmiloff-Smith's work, see Gerrans 2003; Faucher 2006.

In this chapter, Karmiloff-Smith's argument is examined in detail and it is argued that it is inconclusive. Section 3.1 examines how developmental psychopathologies have been used to support hypotheses about the typical cognitive architecture, in particular the massive modularity hypothesis. Section 3.2 presents Karmiloff-Smith's argument in detail ("Karmiloff-Smith's Original Argument"). Section 3.3 shows that Karmiloff-Smith's Original Argument is deficient, and an improved argument is proposed ("Karmiloff-Smith's Improved Argument"). Section 3.4 shows that even this improved argument is unsound. It is concluded that dissociations resulting from developmental pathologies can be used to identify the components of the typical cognitive architecture and to support the massive modularity hypothesis.

Before going any further, it should be emphasized that the criticisms developed here should not obfuscate the fact that there is much to admire in Karmiloff-Smith's research. Although this point is not elaborated here, the integration of psychological, neuropsychological, developmental, and genetic perspectives on cognition that is found in her work is arguably a model for psychology.

3.1 Psychopathologies and cognitive architecture

3.1.1 Cognitive architecture

Describing the typical cognitive architecture of human beings consists in identifying the systems that make up the mind of typical individuals as well as the relations between these systems. The systems that make up the typical cognitive architecture are characterized functionally—that is, they are characterized by their outcome and by the series of operations involved in bringing about this outcome. It is thus not assumed that the components of the typical cognitive architecture are located in distinct brain areas. Two distinct systems could (but, of course, need not) be located in the same brain area. Similarly, all-in-one printers can print, scan, and fax documents. From a functional point of view, they are made up of three distinct systems, which happen to be located in the same physical object.

Controversies about the nature of the typical cognitive architecture abound. Hypotheses vary along several dimensions, three of which are important for present purposes:

– *Sparseness*: How many systems constitute the typical cognitive architecture?
– *Encapsulation*: To what extent is the functioning of each system influenced by other systems?

– *Evolution*: Are the systems that constitute the typical cognitive architecture adaptations?

A hypothesized cognitive architecture is *sparser* than another (which is more *florid*) when it is made up of fewer systems.[2] Hypotheses that postulate sparse cognitive architectures suppose that some hypothesized component systems can underwrite several competences that are characteristic of human cognition. Such systems are often said to be *domain-general*, and they stand in contrast to *domain-specific* systems. One system is more *encapsulated* than another when its functioning is influenced by a smaller number of other systems. Finally, a system is an *adaptation* if it has been selected for at some point in the past. Adaptations need not be adaptive in modern environments, and, in modern environments, they need not bring about what they evolved to do (e.g., the ethnic cognitive system discussed in Gil-White 2001 and Machery and Faucher 2005).

3.1.2 The massive modularity hypothesis

Evolutionary psychologists' massive modularity hypothesis is one of the most influential hypotheses about human cognitive architecture. Because the notion of modularity has been understood in various ways, it is useful to clarify the massive modularity hypothesis.[3] This hypothesis proposes that the typical cognitive architecture consists of numerous systems, most of which are adaptations selected for specific purposes. Thus, evolutionary psychologists' massive modularity hypothesis hypothesizes a florid cognitive architecture. In addition, the systems that make up the typical cognitive architecture are adaptations. Noteworthily, the massive modularity hypothesis, as understood here, does not propose that systems are encapsulated: some might be encapsulated, but others might not, depending on whether encapsulation allowed them to bring about the function for which they were selected.

The characterization of the massive modularity hypothesis proposed here sharply contrasts with Karmiloff-Smith's curious characterization (Karmiloff-Smith 1998, 2001, 2006; Karmiloff-Smith and Thomas 2005). According to Karmiloff-Smith, evolutionary psychologists contend that infants are born

---

[2] One might worry that the sparseness of a cognitive architecture depends on the level of description of this cognitive architecture. This worry is circumvented when one notes that for a given set of functions it is a matter of fact whether the architecture is sparse or florid.

[3] For other interpretations, see Samuels 2005; Carruthers 2006. See Barrett and Kurzban 2006; Machery and Barrett 2006 in support of the claim that the account proposed here is the relevant way to cash out evolutionary psychologists' massive modularity hypothesis.

with the set of systems that constitute the typical cognitive architecture—an extreme form of preformationism! However, evolutionary psychologists' massive modularity hypothesis in fact says nothing about the developmental schedule of the adaptations that constitute the evolved typical cognitive architecture. Some psychological adaptations might be present at birth, while others might develop later, depending (among other things) on whether it was adaptive to have these adaptations at birth. In addition, while some evolutionary psychologists, such as Steven Pinker (e.g., 1997), have claimed that the development of psychological adaptations is genetically determined (whatever that means), the massive modularity hypothesis is consistent with Karmiloff-Smith's insistence that psychological development involves a complex interaction between the environment in which the child develops and his or her genome. The reason is simply that there are numerous ways for an adaptation to develop and that adaptations often develop by relying on the regularities present in the environment (for a systematic development of this perspective, see Sterelny 2003). So, Karmiloff-Smith's construal of modularity is erroneous. Note importantly that this does not invalidate her argument against the use of developmental psychopathologies to support the massive modularity hypothesis, since, as we shall see in section 3.2, this argument does not depend at all on her peculiar characterization of this hypothesis.

3.1.3 The role of dissociations in the decomposition of the mind

Dissociations are the main source of evidence for decomposing the mind in psychology and neuropsychology. In neuropsychology, a *pure single dissociation* is found when and only when a brain lesion or an atypical developmental pattern (due, e.g., to a genetic disorder) affects the performance of patients in a first task by comparison with a control group of unlesioned or typical participants while leaving their performance intact in a second task. For instance, the study of H.M., a well-known amnesiac patient, has shown that an injury to the hippocampus leads to the loss of the anterograde, long-term, explicit memory, but not to the loss of working memory or implicit memory (Milner *et al.* 1968). An *impure single dissociation* is found when and only when, by comparison with a control group of unlesioned or typical participants, a brain lesion or an atypical developmental pattern affects the performance of patients significantly more in a first task than in a second task. Thus, it affects patients' performance more in one task than in the other. A *pure double dissociation* is found when and only when a first kind of brain lesion or atypical developmental pattern affects the performance of a first group of patients in a first task by comparison to a control group of unlesioned or typical participants while leaving their performance intact in a second task, and when a second kind of brain lesion or atypical developmental pattern affects the performance of a second group of patients in the second task by comparison to a control group of unlesioned or typical participants while leaving their performance intact in the first task. Naturally, a double dissociation in neuropsychology can also be *impure*.

Psychologists and neuropsychologists have used neuropsychological dissociations to isolate different processes involved in different tasks, appealing to the following principle: if a lesion or an atypical development affects participants' performance differently in two tasks, people are likely to solve these tasks by means of two different processes.[4] For the last 30 years, there has been a fair amount of controversy about the validity of this principle.[5] This is not the appropriate place to examine this controversy. Rather, as Karmiloff-Smith seems to do (e.g., Thomas and Karmiloff-Smith 2002, p. 729), this chapter will take for granted that at least some dissociations can be used to identify the components of the typical cognitive architecture and to support evolutionary psychologists' massive modularity hypothesis, and focus on whether the dissociations that result from abnormal neural and cognitive development can be used for this purpose.

3.1.4 The role of developmental psychopathologies

Developmental psychopathologies can support the massive modularity hypothesis in two different respects (e.g., Duchaine *et al.* 2001). Firstly, developmental psychopathologies result in dissociations, and, as we just saw, dissociations are often assumed to be evidence for distinguishing different components of the typical cognitive architecture. If dissociations resulting from developmental psychopathologies can really be used to distinguish different systems, and if developmental psychopathologies result in numerous dissociations, then developmental psychopathologies would show that the typical cognitive architecture is florid, exactly as the massive modularity hypothesis would have it.

Now, the massive modularity hypothesis does not merely propose that the typical cognitive architecture is made of many components; it also hypothesizes that these are adaptations. Can developmental psychopathologies

---

[4] Neuropsychological dissociations are also used to localize cognitive processes in the brain. Since I am interested in what type of evidence can support the decomposition of the mind, I put aside this use of dissociations.

[5] Teuber 1955; Caramazza 1986; Dunn and Kirsner 1988, 2003; Shallice 1988; Glymour 1994; Plaut 1995; Young *et al.* 2000; Van Orden *et al.* 2001; Ashby and Ell 2002; Machery 2009, chapter 5; see also the special issue of *Cortex* 2003, 39.

also provide evidence that the systems that make up the typical cognitive architecture are adaptations? Yes, indirectly. If dissociations really provide evidence for distinct systems, developmental psychopathologies can show that some systems fulfill some particular functions, *and only those*. If one would expect the human mind to include systems fulfilling these functions on the basis of evolutionary considerations, one could then argue that the fact that a system fulfills *exclusively* one of these functions is evidence that it is designed to fulfill it, and this would be evidence that it is an adaptation (Machery, in press).

Now, of course, this is not tantamount to saying that the existence of systems fulfilling evolutionary relevant functions would be *strong* evidence in support of the massive modularity hypothesis. Evidence about whether the system fulfills its function in an optimal or at least in an efficient manner would be stronger evidence. In addition, this type of argument is controversial (Richardson 2007), and its evidentiary strength is in any case weaker than other kinds of evidence that can support adaptationist hypotheses, such as cross-species comparisons. Still, evidence they do provide.

Let's consider an example of the use of a developmental psychopathology in support of the massive modularity hypothesis. Clahsen and Almazan (1998) have argued that the pattern of preserved and impaired linguistic capacities found in people with Williams syndrome provides evidence about the typical cognitive architecture of the linguistic capacity (see also Bellugi *et al.* 1994). When completing a task involves applying some syntactic rules (according to the generative-grammar framework endorsed by Clahsen and Almazan), the four teenagers with Williams syndrome they examined did equally well, if not better, than children matched for mental age and participants with specific language impairment. For instance, participants with Williams syndrome appeared to comply with principles A, B, and C of government and binding theory, and they formed the past tense of verbs and of novel verbs appropriately when those novel verbs did not sound like known irregular verbs. By contrast, when completing a task involved retrieving some specific components of the lexical entry associated with a word, participants with Williams syndrome did less well than children matched for mental age and than participants with specific language impairment. For instance, they overgeneralized the rule for forming a past tense to novel verbs that sounded like known irregular verbs, while control children and participants with specific language impairment formed the past tense of these verbs by analogy with the past tense of the known irregular verbs. Clahsen and Almazan (1998, p. 192) conclude that "[f]rom the perspective of modular linguistic theory, selective impairments such as those found in WS receive a straightforward interpretation." Adding:

> The common property shared by the unimpaired linguistic phenomena is that they involve computational knowledge of language, whereas the impaired phenomena involve (specific kinds of) lexical knowledge, i.e. the retrieval of subnode information from lexical entries. Thus, it seems that WS [Williams Syndrome] children's computational system for language is selectively spared yielding excellent performance on syntactic tasks and on regular inflection, whereas the lexical system and/or its access mechanisms required for irregular inflection are impaired.[6]
>
> (Clahsen and Almazan 1998, p. 193)

3.2 Why developmental psychopathologies provide no evidence for modularity

3.2.1 Karmiloff-Smith's Original Argument

Karmiloff-Smith and colleagues have challenged the use of developmental disorders to support the massive modularity hypothesis and, more broadly, to identify the components of the typical cognitive architecture. Karmiloff-Smith and Thomas write:

> In this chapter, we discuss why it is essential to take a neuroconstructivist approach to interpreting the data from developmental disorders and why these latter cannot be used to bolster evolutionary nativist claims. From our studies of older children and adults with the neurodevelopmental disorder, Williams syndrome, we show how processes that some claim to be "intact" actually display subtle impairments and cannot serve to divide the cognitive system into parts that develop normally and independently of parts that develop atypically.
>
> (Karmiloff-Smith and Thomas 2005, p. 307)

So, what is exactly Karmiloff-Smith's argument? It hangs on two premises. The first premise is methodological: it specifies the conditions under which one can use the dissociations resulting from developmental psychopathologies as evidence for distinct systems. The second premise is empirical: it asserts that these conditions are in fact not met. From this they conclude that developmental psychopathologies cannot be used as evidence to determine the components of the typical cognitive architecture and, *a fortiori*, to support the massive modularity hypothesis. So, her argument has the following form:

Karmiloff-Smith's Original Argument

1. One can use developmental psychopathologies to identify the components of the typical cognitive architecture only if the abnormal architecture is unimpaired *but for one element* (*residual normality*).

---

[6] For further discussion, see Thomas and Karmiloff-Smith 1999; Thomas *et al.* 2001, and Clahsen and Temple 2003.

2. The residual normality assumption is false: because the abnormal mind develops abnormally, it is made of cognitive systems that differ from the systems making up the typical cognitive architecture.
3. Hence, one cannot use developmental psychopathologies to identify the components of the typical cognitive architecture.

Let me first say a few words about the conclusion of this argument. In contrast to other arguments against the massive modularity hypothesis, Karmiloff-Smith's Original Argument does not attempt to show that this hypothesis is false or confused. It does not provide evidence that the typical cognitive architecture is sparse or that its components are not adaptations (as, e.g., Quartz 2002 and Buller 2005 do).[7] Nor does it attempt to show that this hypothesis is unclear (as, e.g., Woodward and Cowie 2004 do). Rather, Karmiloff-Smith's Original Argument supports a *methodological* conclusion. The point of her argument is not that the massive modularity hypothesis is false, but that, common wisdom notwithstanding, developmental psychopathologies provide no support for it. Karmiloff-Smith's Original Argument undermines one of the sources of evidence evolutionary psychologists and, more generally, cognitive scientists have relied on to determine the components of the typical cognitive architecture.

I now turn to Premise 1. Karmiloff-Smith and colleagues propose a necessary condition for the use of developmental psychopathologies to identify the components of the typical cognitive architecture: developmental psychopathologies can be used for this purpose only if they impair a particular cognitive system (e.g., the system underlying face recognition, numerical cognition, the formation of irregular past tenses in English, etc.) while leaving the other systems *intact*. Karmiloff-Smith and colleagues call the hypothesis that developmental psychopathologies result in an impaired system while leaving the other systems intact "the subtractivity or residual normality assumption" (Elsabbagh and Karmiloff-Smith 2006).[8] Because Karmiloff-Smith proposes that developmental psychopathologies can be used to distinguish cognitive systems only if the residual normality assumption is true, developmental psychopathologies provide evidence for the existence of distinct cognitive systems only if they result in *pure* dissociations (Elsabbagh and Karmiloff-Smith 2006).

Let's now turn to Premise 2. Premise 2 asserts that the necessary condition for the use of developmental psychopathologies to identify the components of the typical cognitive architecture is not met. It is never the case that developmental psychopathologies result in an impaired cognitive system while the

---

[7] For discussion of these arguments, see Machery and Barrett 2006 and Machery 2007.
[8] This assumption was called "subtractivity" in Saffran 1982 and "transparency" in Caramazza 1984.

other systems are intact. It would not be sufficient for Karmiloff-Smith's Original Argument to merely assert that, typically, developmental psychopathologies do not impair a single cognitive system while leaving the other systems intact. Finding a few cases where a developmental psychopathology results in such a selective impairment would be sufficient to allow the use of developmental psychopathologies to identify the components of the typical cognitive architecture, even if one were to grant Karmiloff-Smith's Premise 1. Certainly, developmental psychopathologies would then rarely be used to identify the components of the typical cognitive architecture and to support the massive modularity hypothesis, but, in principle, they could be used for this purpose.

Why do Karmiloff-Smith and colleagues believe that the necessary condition expressed by Premise 1 (residual normality) is never met? They support this premise by appealing to their theoretical views about brain development and by drawing inductively on their work on a range of developmental psychopathologies, such as Williams syndrome and language specific impairment. Let's consider Karmiloff-Smith's views about brain development here, before looking at their empirical work in section 3.2.2.

Karmiloff-Smith argues that impairments of infants' brains (and, presumably, of fetuses' as well) have cascading consequences for the development of the whole brain and, consequently, for the development of all cognitive systems. Impairments directly influence the development of the cognitive systems that immediately depend on the impaired brain parts, and the abnormal development of these cognitive systems will in turn influence the development of other systems. As Karmiloff-Smith notes (e.g., Karmiloff-Smith 2007), infants' brains are massively interconnected, and an impaired part of the brain is likely to result in abnormal inputs being sent to numerous other brain parts, influencing their developmental trajectory. In addition, impairments are likely to result in abnormal experiences that will influence the developmental trajectories of every cognitive system. Finally, during development, brains can reorganize themselves to bring about important cognitive functions. As a result, it is extremely unlikely that any cognitive system can remain intact in developmental psychopathologies, although the extent to which a cognitive system will be impaired is likely to vary. Different kinds of early impairments (perceptual impairments of different types, impaired attention, etc.) are likely to result in different profiles of severely and subtly impaired cognitive systems, giving rise to the different syndromes identified by psychiatrists and psychologists (Williams syndrome, Down syndrome, etc.). Karmiloff-Smith and Thomas express this holistic view of brain development as follows:

> Because the brain develops as a whole system from embryogenesis onwards, we believe it to be highly unlikely that children with genetic disorders will end up with a patchwork of neatly segregated, preserved and impaired cognitive modules.
>
> (Karmiloff-Smith and Thomas 2005, p. 308)

Adding:

> People with genetic disorders do not, in our view, have normal brains with parts preserved and parts impaired. Rather, they have developed an atypical brain throughout embryogenesis and subsequent postnatal growth, so we should expect fairly widespread impairments across the brain rather than a very localised one.
>
> (Karmiloff-Smith and Thomas 2005, p. 312)

3.2.2 Face recognition in Williams syndrome

In addition to appealing to their holistic views about brain development, Karmiloff-Smith and colleagues argue that the empirical work on a large number of developmental psychopathologies inductively supports Premise 2 of Karmiloff-Smith's Original Argument. For the sake of space, I will focus on Williams syndrome. Williams syndrome is a rare neurological disorder with a well-studied genetic etiology (Donnai and Karmiloff-Smith 2000; Karmiloff-Smith 2007; Martens *et al.* 2008). This syndrome is characterized by a distinctive pattern of severely impaired capacities that coexist with apparently preserved capacities. Specifically, while spatial cognition is severely impaired among people with Williams syndrome, other psychological competences, such as face recognition and syntactic processing, are *apparently* preserved (Bellugi *et al.* 1988, 1994). People with Williams syndrome perform within the normal range on many tasks meant to evaluate people's syntactic competence and people's face recognition capacity (e.g., the Benton Facial Recognition Task and the Rivermead Face Memory Task).

Because Williams syndrome appears to be characterized by a pattern of intact and impaired capacities, it has often been adduced as evidence for the existence of distinct cognitive systems. Thus, Bellugi and colleagues (1988) argued that Williams syndrome and prosopagnosia (i.e. the incapacity to recognize faces, including the faces of relatives and acquaintances) constitute a double dissociation, providing evidence that face recognition is underwritten by a dedicated cognitive system, while Pinker (1999) used Williams syndrome to argue for the separability of language and intelligence. As we saw earlier, Clahsen and Almazan (1998) have proposed that Williams syndrome shows that irregular and regular past-tense formations are underwritten by two distinct systems.

By contrast, in keeping with their views about neurobiological and cognitive development, Karmiloff-Smith and colleagues have argued that Williams syndrome is a *pervasive* disorder that affects all aspects of the cognitive life of people with Williams syndrome. To provide evidence for such a claim, Karmiloff-Smith and colleagues have mostly focused on the three capacities that have been said to be preserved in individuals with Williams syndrome: face recognition, language processing, and social cognition. They argue that, appearances to the contrary notwithstanding, these three capacities are impaired in people with Williams syndrome. Because they are impaired, it is not the case that the systems underlying them are intact. Thus, people with Williams syndrome do not illustrate the residual normality assumption, which provides the basis for the inductive argument in support of Premise 2 of Karmiloff-Smith's Original Argument. For the sake of space, I focus on face recognition in this chapter.

A large body of evidence shows that face recognition in typical individuals involves the processing of configural cues (e.g., Farah *et al.* 1998; Maurer *et al.* 2002; more on the nature of configural cues in section 3.4.2).[9] In addition to the facial features themselves (e.g., a nose of a particular shape), the *spatial* configuration of these features is used to recognize faces. The importance of this spatial configuration for face recognition is illustrated by the well-known face-inversion effect: it is difficult to decide whether a picture of a face is identical to or different from an inverted picture because the configural information is not available in inverted faces (e.g., Yin 1969; Valentine 1988).

Karmiloff-Smith and colleagues have argued that, in contrast to typical individuals' face processing, the system that underlies face recognition in individuals with Williams syndrome does not process configural cues; rather, it relies exclusively on the processing of facial features. In Karmiloff-Smith (1997), adolescent and adult participants were asked to decide whether or not two pictures represented the same face. Participants with Williams syndrome had less difficulty than control participants in identifying inverted faces. Furthermore, they pointed to specific facial features when asked to explain their decision. This suggests that, in contrast to typical individuals, they do not rely on configural cues to recognize faces. Similarly, Deruelle *et al.* (1999, Experiment 2) have provided some evidence that participants with Williams syndrome are less sensitive to the inversion effect, suggesting that they might not encode and recognize faces configurally (but see the inversion effect reported in Mills *et al.* 2000). They asked participants to complete a same–different task with inverted and upright faces and houses. Participants were simultaneously presented with two pictures and asked to determine whether or not they were the same. In half of the cases, the pictures were upright; in half of the cases, they were inverted.

---

[9] Face perception consists in recognizing an object as a face, while face recognition consists in recognizing the identity of a perceived face.

As expected, Deruelle and colleagues found an interaction effect for the control participants' performance: the deterioration of their performance for the inverted pictures of faces by comparison to the upright pictures of faces was significantly larger than the deterioration of their performance for the inverted pictures of houses by comparison to the upright pictures of houses. By contrast, while the performance of participants with Williams syndrome deteriorated to a greater extent for inverted faces by comparison to upright faces than for inverted houses by comparison to upright houses, the interaction did not reach significance. Deruelle and colleagues concluded this study as follows:

> We are then inclined to suggest that the WS subjects were less disturbed than the control subjects by the change of face orientation because they are incapable of encoding faces in terms of configural information and encode both upright and inverted faces through local characteristics.[10]
>
> (Deruelle *et al.* 1999, p. 288)

In addition to this processing difference, electrophysiological differences in brain activation during face processing by people with Williams syndrome and by comparison participants have also been reported. In Mills *et al.* (2000), participants were asked to decide whether the second picture of a pair of sequentially presented face pictures was the same as the first picture. It was found that the event-related potentials (ERP) waveform elicited by face recognition differs in controls and participants with Williams syndrome. In controls, there is a clear difference in the ERP waveform between upright and inverted faces. In addition, the waveform is asymmetric for upright faces, with the right hemisphere being more strongly activated. By contrast, participants with Williams syndrome's ERP waveform did not show any difference between upright and inverted faces nor did they display a hemispherical asymmetry.

Although this study is often cited uncritically, it should be noted that it does not clearly support Karmiloff-Smith's claim about processing differences in people with Williams syndrome and typical participants. Mills and colleagues found that participants with Williams syndrome displayed an inversion effect, a hallmark for the use of configural cues to identify faces. This seems to suggest that the electrophysiological differences in brain activation might not be related to the hypothesized difference between face processing in typical individuals and in people with Williams syndrome. Similarly, Mills and colleagues noted that children do not show the ERP waveform typically elicited by face recognition in typical adults. The problem is that children are known to use most types of

---

[10] For critical discussion of Karmiloff-Smith 1997 and Deruelle *et al.* 1999, see Tager-Flusberg *et al.* 2003.

configural cues in face recognition (Maurer *et al.* 2002). Again, this suggests that these electrophysiological differences might be unrelated to the processing or absence of processing of configural cues.

Be that as it may, Karmiloff-Smith and Thomas summarize their work on face recognition in people with Williams syndrome as follows:

> In sum, people with WS do not present with a normally developed "intact" face processing module and an impaired space processing module, as nativists would claim. Rather, from the outset they have followed an atypical developmental trajectory such that both facial and spatial processing reveal a similar underlying impairment in configural processing. It is simply because the problem space of face processing lends itself more readily to featural analysis than spatial analysis does, so that it merely seems normal in the older child and adult.[11]
>
> (Karmiloff-Smith and Thomas 2005, p. 314)

In what follows, I will take for granted the claim that people with Williams syndrome have difficulty using configural cues, even though, as noted earlier, some articles report the existence of an inversion effect in people with Williams syndrome (e.g., Mills *et al.* 2000).

Finally, Karmiloff-Smith and colleagues have argued that, far from being a peculiar case, face recognition in people with Williams syndrome is typical of the cognition of people with developmental disorders. They contend that, for each of the syndromes they have looked at (including specific language impairment and dyslexia), patients' performance is different from controls' in *every* domain examined, although divergences from controls' performance are sometimes subtle and can thus be missed when insufficiently sensitive tasks are used.

This body of research is meant to support Premise 2 of Karmiloff-Smith's Original Argument. It is supposed to show that, in keeping with Karmiloff-Smith's holistic view of cognitive development, developmental psychopathologies never result in pure dissociations: they do not affect particular cognitive capacities while leaving others unimpaired. Thus, it is not the case that the residual normality assumption is true: developmental psychopathologies do not leave some systems intact while impairing others. If it is also true that developmental psychopathologies can be used to identify the components of the typical cognitive architecture and to support the massive modularity hypothesis only if the residual normality assumption is true, then Karmiloff-Smith is right to conclude that developmental psychopathologies provide no support whatsoever for the massive modularity hypothesis.

---

[11] It is noteworthy that this idea was discussed in the early 1990s (Bellugi *et al.* 1994). Thus, it isn't the case that researchers have blindly assumed that similar processes underwrite similar performances, as Karmiloff-Smith and colleagues are prone to suggest.

3.3 The epistemology of developmental dissociations

In reply to Karmiloff-Smith, some psychologists have argued that atypical developments due to psychopathologies do result in pure dissociations—thus granting Premise 1, but rejecting Premise 2 of Karmiloff-Smith's Original Argument. Clahsen and colleagues provide a good example of this kind of reply:

> Evidence against subtractivity in linguistic domains would come from cases in which the underlying functional architecture of the language system itself had altered with the development of language modules that do not exist in the normal brain. However, there is no such empirical evidence in relation to language development in either adults or children.
>
> (Clahsen *et al.* 2004, p. 222)

In this section, I develop a different line of argument. I examine the two premises of Karmiloff-Smith's Original Argument, and I argue that Premise 1 is erroneous, while Premise 2 is ambiguous. As a result, her argument cannot be accepted as it stands.

3.3.1 No need for pure dissociations

As we saw in section 3.2, Karmiloff-Smith and colleagues contend that developmental psychopathologies can be used to identify the components of the typical cognitive architecture and, *a fortiori*, to support the massive modularity hypothesis only if developmental psychopathologies give rise to *pure* dissociations. I now argue that this claim is mistaken: *impure* dissociations can also be used to support hypotheses about the components of the typical cognitive architecture.

The use of neuropsychological dissociations for distinguishing processes rests on the following assumption: that the performance of patients and controls is similar in one task (e.g., reading regular words) but different in another task (e.g., reading irregular words) is (non-conclusive) evidence that these two tasks are solved by two distinct systems (section 3.1.3). Although this assumption is controversial, it has been taken for granted in this chapter because Karmiloff-Smith's Original Argument does not hang on challenging it. This assumption entails that impure dissociations between two functions provide defeasible evidence that these functions are fulfilled by two distinct systems. If these two tasks were solved by a single system, then, plausibly (but not necessarily), patients' impairment (i.e. the difference between their performance and that of the controls) would be the same in both tasks.

In the case of a neuropsychological impure dissociation, the inference that two tasks (1 and 2) are solved by two distinct systems is stronger the more similar the performances of controls (e.g., typical individuals) and patients (e.g., amnesiac patients) are in task 1 and the more different their performances are in task 2. One has only weak evidence that two tasks are solved by two distinct processes when the difference between the performances of control and patients in task 1 is similar (although not identical) to the difference between the performances of controls and patients in task 2—that is, if the difference between the two differences is small. In this respect, a pure dissociation is merely a limiting case of an impure dissociation: in a neuropsychological pure dissociation, patients and controls perform identically in task 1, while patients' performance is strongly impaired in task 2.

The upshot is clear: *pace* Karmiloff-Smith, it is not the case that dissociations resulting from developmental psychopathologies can be used only if the residual normality assumption is true, and Premise 1 of Karmiloff-Smith's Original Argument should be rejected.

3.3.2 The ambiguity of Premise 2

I now argue that Premise 2 can be understood in two different ways. Premise 2 could first say that the cognitive architecture of people with, for example, Williams syndrome and the cognitive architecture of typical individuals are made of the *same cognitive systems*, but that *all* the systems making up the former architecture are *impaired* to a smaller or greater extent. I will call this reading of Premise 2 the "weak reading." According to this reading, people with Williams syndrome have the same face recognition system as typical individuals, but this system is impaired, resulting in distinct performances in face recognition tasks. The second reading, which I will call "the strong reading," asserts that the cognitive architecture of people with, for example, Williams syndrome and the cognitive architecture of typical individuals are made of *different cognitive systems*. According to this reading, people with Williams syndrome do have a system for face recognition, but it is a different system from the face recognition system of typical individuals—not just an impaired version of their system.[12]

To grasp the distinction between these two readings, it might be useful to consider the following analogy. Suppose that the typical mind is like an all-in-one printer—a single physical object that fulfills several functions. According to the weak reading of Premise 2, the mind of an individual with Williams syndrome would be like an all-in-one printer that consists of the same processes (for faxing, printing, and copying) as the typical all-in-one printer, but these processes are damaged to a smaller or greater extent: for instance,

---

[12] It is likely that Karmiloff-Smith and colleagues intended Premise 2 to be understood in this way. In addition, they have often been read in this way (e.g. see Clahsen *et al.* 2004).

faxing might not work anymore, while printing and copying might work to some extent. By contrast, according to the strong reading of Premise 2, the mind of an individual with Williams syndrome would be like an all-in-one printer in which each function (faxing, printing, and copying) is fulfilled by a process of a distinct kind—a process that differs from the corresponding process in typical all-in-one printers. This all-in-one printer would fulfill some of the functions that typical all-in-one printers fulfill, but, as a consequence of its atypical developmental trajectory (so to speak!), it would fulfill them in its own way.

As we saw above, it is not the case that dissociations resulting from developmental psychopathologies can be used only if the residual normality assumption is true. Now, suppose that the weak reading of Premise 2 is true: developmental psychopathologies such as Williams syndrome result in more or less impaired versions of the same systems that compose the typical cognitive architecture. Developmental psychopathologies would then result in impure dissociations, providing relevant evidence for identifying the components of the typical cognitive architecture and for evaluating the massive modularity hypothesis. Thus, the weak reading of Premise 2 fails to support Karmiloff-Smith's rejection of the evidential role of developmental psychopathologies. By contrast, if the strong reading of Premise 2 is true (that is, if developmental psychopathologies result in systems that differ in kind from the systems making up the typical cognitive architecture), the dissociations resulting from developmental psychopathologies would be of no use in identifying the components of the typical cognitive architecture. The reason is simple: the impure dissociations caused by these psychopathologies would not result from greater or smaller impairments of the processes that make up the typical cognitive architecture, but from the greater or smaller efficiency of a distinct kind of processes. As a result, Karmiloff-Smith would be right to reject the evidential role of developmental psychopathologies.

Consider again the analogy between the mind and an all-in-one printer. If the strong reading of Premise 2 is correct, the mind of an individual with Williams syndrome is like an all-in-one printer that consists of a distinct kind of processes. Now, because the processes that make up this all-in-one printer are a distinct kind of process, it is impossible to draw any conclusion about the processes that make up a typical all-in-one printer from the structure of this atypical printer. For instance, the fact that printing and copying are fulfilled by two distinct processes in the atypical all-in-one printer provides no evidence about whether these two functions are also fulfilled by two distinct processes in typical printers.

3.3.3 Reformulating Karmiloff-Smith's Original Argument

Because only the strong reading of Premise 2 supports Karmiloff-Smith's intended conclusion, it is necessary to reformulate her argument. This can be done as follows:

Karmiloff-Smith's Improved Argument

1. Pure and impure dissociations can be used to identify the components of the typical cognitive architecture only if they result from (perhaps damaged to a smaller or a greater extent) the very systems that make up the typical cognitive architecture.

2. Developmental psychopathologies result in systems that differ in kind from the systems making up the typical cognitive architecture (strong reading of Premise 2).

3. Hence, it is not possible to use developmental psychopathologies to identify the components of the typical cognitive architecture.

3.3.4 A difficulty

Before examining the empirical support for the strong reading of Premise 2, I should acknowledge that the distinction between the two readings of Premise 2 supposes that one can set apart the situation where two token systems (in two individuals) are instances of two different types of system from the situation where two token systems (in two persons) are instances of the same type of systems, but one of them is impaired. Earlier, I contrasted the case where an all-in-one printer includes a different system for printing from the typical printer and the case where an all-in-one printer includes a damaged version of the typical system for printing. I thereby assumed that the distinction at hand could be drawn for printing systems. When it comes to cognitive systems, this distinction turns out to difficult to draw, since there are no agreed-upon individuation criteria for cognitive systems (Machery 2009, chapter 5). The lack of such criteria might look like an insuperable difficulty for the argumentative strategy followed here, but it should be kept in mind that Karmiloff-Smith's argument against the use of dissociations resulting from developmental psychopathologies also relies on such a distinction (see Premise 2 of Karmiloff-Smith's Improved Argument).

How can the distinction at hand be drawn? Suppose that, in a typical cognitive architecture, doing A (e.g., identifying someone's gender) can be done on the basis of various probabilistic cues, x, y, or z (shape of the face, pilosity, etc.). Suppose now that someone is able to do A, but only imperfectly: she cannot use some of the cues used by controls (say, she cannot use x). As a result, she tends to be less reliable than controls, and she might be unable to do A when using x is crucial for doing A. Now, I propose that the more cues she can use to do A (among the cues used by a control), the more likely it is that her system for doing A is a damaged version of the typical system for doing A rather than an altogether different system. The rationale for this proposal is that if the system for doing A were an instance of a different type of system (rather than a damaged version of a typical system), the fact that it would use many of the cues used by a typical system would be puzzling. I also propose that it would be strong evidence that her system for doing A would be of a different type if she used cues that controls do not use.

Note that these proposals are not meant to *define* what having two token systems of the same type consists in. It is conceivable that two different token systems (one of which is a typical system) might be of the same type (the atypical system being a damaged version of the typical system), although the atypical system uses very few of the cues used by the typical system because it is a very damaged system. Rather, these proposals are meant to characterize *the evidence* one can defeasibly use to decide whether or not two token systems are of the same type.

3.4 Evaluation of the strong reading of Premise 2

3.4.1 Karmiloff-Smith's theoretical argument in support of Premise 2

Karmiloff-Smith and colleagues support Premise 2 of her original argument by two types of consideration: (1) plausible theoretical considerations about brain development and (2) empirical evidence about the pervasiveness of the impairments resulting from developmental psychopathologies. I now argue that the theoretical considerations adduced by Karmiloff-Smith provide no support for either the weak or the strong reading of Premise 2 and thus cannot be used to support the improved argument presented at the end of section 3.3.

Karmiloff-Smith's point about brain development is simple. Brain development in infancy and early childhood is holistic: brain areas are massively interconnected, so the developmental changes in one of these areas have consequences in numerous other areas. As a result, even minor developmental brain problems have cascading effects that affect the development of the whole brain and of all cognitive capacities.

This argument might seem plausible, but in fact it carries little weight. The reason is that Karmiloff-Smith's description of brain development is at odds with an important feature of biological development—its modularity (e.g., Schlosser and Wagner 2004). In the context of developmental biology, "modularity" refers to the property that the developmental pathway leading to the development of a trait is shielded from the changes that might happen to other developmental pathways. Karmiloff-Smith's argument in support of Premise 2 amounts to denying that this typical feature of biological development also applies to brain development. But because biological development is typically modular, it would be surprising if brain development was not modular to some degree too. We should thus refrain from putting too much weight on Karmiloff-Smith and colleagues' theoretical considerations about brain development. Thus, neither the strong nor the weak reading of Premise 2 is clearly supported by these considerations. Consequently, the plausibility of the strong reading of Premise 2 (the reading Karmiloff-Smith needs, as argued in section 3.3) hangs on the empirical evidence drawn from her research on Williams syndrome and other developmental psychopathologies. I discuss the significance of this research in the remainder of this section.

3.4.2 What is the nature of Williams syndrome's face-processing impairment?

I now argue that the evidence about face recognition suggests that the face recognition system in people with Williams syndrome is a damaged version of the typical system for face recognition rather than a system of a different type. If this is correct, Karmiloff-Smith's work on face recognition in Williams syndrome provides no support for the strong reading of Premise 2.[13] Given that this strong reading is also not supported by her theoretical considerations about brain development, we have little reason to believe that developmental psychopathologies result in a cognitive architecture made up of altogether different systems rather than in a cognitive architecture made up of more or less damaged versions of the cognitive systems that compose the typical cognitive architecture. As a result, the impure dissociations that result from developmental psychopathologies can be used as evidence about the typical cognitive architecture.

The face recognition system in people with Williams syndrome is a damaged version of the typical system for face recognition because people with Williams syndrome use many of the cues used by typical individuals to identify faces and because they do not seem to use cues that are not used by typical individuals. Three kinds of cues are commonly distinguished in the literature: individual facial features (e.g., a nose of a particular shape), the holistic relation between cues, and the configuration of cues (but see Maurer *et al.* 2002 for further distinctions). While the distinction between the first kind of cues and the two other kinds of cues is clear, the distinction between holistic and configural cues is unclear. Karmiloff-Smith and colleagues (Karmiloff-Smith *et al.* 2004, p. 1259) contend that "the term 'holistic' is deemed to cover the gluing together

---

[13] Research on syntactic processing would support a similar conclusion (Clahsen and Temple 2003).

of facial features (and hairline) into a gestalt, without necessarily conserving the spatial distances between features", while a cue is configural if it is constituted by the spatial distances between features. Thus, Karmiloff-Smith and colleagues seem to have in mind the following distinction. An isosceles triangle constituted by a pair of specific eyes and a nose would be a holistic cue since it is not constituted by the specific distances between the center of the eyes and between the eyes and the tip of the nose. By contrast, an isosceles triangle that is constituted by a pair of specific eyes and a nose, whose base is 71 millimeters and whose altitude is 41 millimeters would be a configural cue.

Research shows that people with Williams syndrome are perfectly able to use featural cues *and* holistic cues, even though they seem to have difficulty using configural cues. Karmiloff-Smith's own work suggests that people with Williams syndrome are able to use featural cues to identify faces, just like typical people in at least some contexts. In addition, Tager-Flusberg and colleagues (2003) have shown that people with Williams syndrome can use holistic cues.[14] They asked participants with Williams syndrome and controls to complete a part-whole task. In the whole-face condition, participants are presented with pictures of faces before being asked to decide which of two pictures they were presented with (the distractor picture differs only by one feature). In the isolated-part condition, participants are presented with pictures of faces before being asked which of two features (e.g., which of two noses) they were presented with. Half of the stimuli are presented upright, while half of the stimuli are inverted. It was found that, although participants with Williams syndrome do less well overall than control participants, the two groups answer similarly. Particularly, "both groups were more accurate in the whole-face than in the isolated-part test condition for upright faces, but not for inverted faces" (Tager-Flusberg *et al.* 2003, p. 18). This finding suggests that people with Williams syndrome do encode the holistic pattern formed by facial features.

Karmiloff-Smith and colleagues (2004) have criticized Tager-Flusberg's work. They looked at the *development* of the use of configural cues on the following grounds:

> [O]thers maintain that people with the syndrome display normal face processing (e.g., [. . .] Tager-Flusberg, Plesa-Skwerer, Faja, & Joseph, 2003). This is of course tantamount to claiming that face processing develops normally in WS.
>
> (Karmiloff-Smith *et al.* 2004, p. 1259)

---

[14] Their work is not formulated in terms of holistic versus configural cues, since this distinction was developed by Karmiloff-Smith and colleagues (2004) in response to Tager-Flusberg and colleagues (2003).

Arguing that the performance of people with Williams syndrome in tasks requiring the use of configural cues develops differently from that of controls, they conclude that face recognition in people with Williams syndrome is abnormal.

There are two problems with this reply. First, Karmiloff-Smith and colleagues are wrong to state that the crucial dispute is about the development of face recognition in Williams syndrome.[15] Rather, the crucial question is whether people with Williams syndrome acquire an impaired or a preserved system for face recognition and, if it is impaired, whether it is a damaged version of the typical face recognition system or a system of a distinct kind. And the abnormality of the development of a cognitive system is weak evidence that the developing system is abnormal, since biological development is often *robust*: there are often several developmental pathways to the same endpoint. Thus, even if Karmiloff-Smith and colleagues were right that face processing develops differently in people with Williams syndrome and typical people, it could still be that their developmental trajectories lead to the same endpoint.

Second, Tager-Flusberg *et al.* (2003) and Karmiloff-Smith *et al.* (2004) seem to concur that people with Williams syndrome can use two of the three types of cues that typical individuals use. Given our discussion of system individuation above, this suggests that face recognition in Williams syndrome is underwritten by a moderately damaged version of the typical system for face recognition rather than by a system of a distinct kind. The strong reading of Premise 2 is thus not supported, and Karmiloff-Smith's Improved Argument fails.

3.5 **Conclusion**

Dissociations resulting from developmental psychopathologies have played an important role for supporting or undermining the main hypotheses about the typical cognitive architecture, such as evolutionary psychologists' massive modularity hypothesis. If Karmiloff-Smith were right, our confidence in these hypotheses would have to be re-evaluated. Fortunately, her argument against the use of developmental psychopathologies to study the typical cognitive architecture fails. Because biological development is often modular, it is unclear whether we should expect developmental psychopathologies to affect every aspect of our mental life. Furthermore, supposing they do, if developmental psychopathologies result in more or less impaired versions of

---

[15] This is not to say that the developmental trajectory of face processing in people with Williams syndrome is unimportant or uninteresting.

the systems that make up the typical cognitive architecture, the resulting impure dissociations would be an appropriate source of evidence for identifying the components of the typical cognitive architecture. Impure dissociations would not be an appropriate source of evidence only if developmental psychopathologies resulted in systems that differ in kind from the systems making up the typical cognitive architecture. However, the empirical evidence most often discussed by Karmiloff-Smith and colleagues—the alleged incapacity of individuals with Williams syndrome to use configural cues in face recognition—fails to support this strong claim. Abnormal development might affect many (perhaps all) cognitive systems, but it does not produce different kinds of systems.

References

Ashby, F.G. and Ell, S.W. (2002) Single versus multiple systems of learning and memory. In J. Wixted and H. Pashler (eds), *Stevens' Handbook of Experimental Psychology: Volume 4 Methodology in Experimental Psychology*. Wiley, New York, pp. 655–92.

Baron-Cohen, S. (1998) Modularity in developmental cognitive neuropsychology. In J.A. Burack, R.M. Hodapp, and E. Zigler (eds), *Handbook of Mental Retardation and Development*. Cambridge University Press, Cambridge, pp. 334–48.

Barrett, H.C. and Kurzban, R. (2006) Modularity in cognition: Framing the debate. *Psychological Review*, 113, 628–47.

Bellugi, U., Sabo, H., and Vaid, J. (1988) Spatial deficits in children with Williams syndrome. In J. Stiles-Davis, M. Kritchevsky, and U. Bellugi (eds), *Spatial Cognition: Brain Bases and Development*. Lawrence Erlbaum Associates, Hillsdale, NJ, pp. 273–98.

Bellugi, U., Wang, P.P., and Jernigan, T.L. (1994) Williams syndrome: An unusual neuropsychological profile. In S.H. Broman and J. Grafman (eds), *Atypical Cognitive Deficits in Developmental Disorders: Implications for Brain Function*. Lawrence Erlbaum Associates, Hillsdale, NJ.

Buller, D.J. (2005) *Adapting Minds: Evolutionary Psychology and the Persistent Quest for Human Nature*. MIT Press, Cambridge, MA.

Caramazza, A. (1984) The logic of neuropsychological research and the problem of patient classification in aphasia. *Brain and Language*, 21, 9–20.

Caramazza, A. (1986) On drawing inferences about the structure of normal cognitive systems from the analysis of patterns of impaired performance: The case for single-patient studies. *Brain & Cognition*, 5, 41–66.

Carruthers, P. (2006) *The Architecture of the Mind*. Oxford University Press, New York.

Clahsen, H. and Almazan, M. (1998) Syntax and morphology in Williams syndrome. *Cognition*, 68, 167–98.

Clahsen, H. and Temple, C. (2003) Words and rules in children with Williams syndrome. In Y. Levy and J. Schaeffer (eds), *Language Competence Across Populations*. Lawrence Erlbaum Associates, Mahwah, NJ, pp. 323–52.

Clahsen, H., Ring, M., and Temple, C. (2004) Lexical and morphological skills in English-speaking children with Williams Syndrome. In S. Bartke and J. Siegmueller (eds), *Williams Syndrome Across Languages*. John Benjamins, Amsterdam, pp. 221–44.

Deruelle, C., Mancini, J., Livet, M.O., Cassé-Perrot, C., and de Schonen, S. (1999) Configural and local processing of faces in children with Williams syndrome. *Brain and Cognition*, 41, 276–98.

Donnai, D. and Karmiloff-Smith, A. (2000) Williams syndrome: From genotype through to the cognitive phenotype. *American Journal of Medical Genetics: Seminars in Medical Genetics*, 97, 164–71.

Duchaine, B.C., Cosmides, L., and Tooby, J. (2001) Evolutionary psychology and the brain. *Current Opinion in Neurobiology*, 11, 225–30.

Dunn, J.C. and Kirsner, K. (1988) Discovering functionally independent mental processes: The principle of reversed association. *Psychological Review*, 95, 91–101.

Dunn, J.C. and Kirsner, K. (2003) What can we infer from double dissociations? *Cortex*, 39, 1–7.

Elsabbagh, M. and Karmiloff-Smith, A. (2006) Modularity of mind and language. In K. Brown (ed.), *The Encyclopedia of Language and Linguistics*, Vol. 8. Elsevier, Oxford, pp. 218–24.

Farah, M., Wilson, K., Drain, M., and Tanaka, J. (1998) What is "special" about face perception? *Psychological Review*, 105, 482–98.

Faucher, L. (2006) What's behind a smile? The return of mechanism: Reply to Schaffner. *Synthese*, 151, 403–9.

Gerrans, P. (2003) Nativism and neuroconstructivism in the explanation of Williams syndrome. *Biology & Philosophy*, 18, 41–52.

Gil-White, F.J. (2001) Are ethnic groups biological "species" to the human brain? *Current Anthropology*, 42, 515–54.

Glymour, C. (1994) On the methods of cognitive neuropsychology. *British Journal for the Philosophy of Science*, 45, 815–35.

Karmiloff-Smith, A. (1997) Crucial differences between developmental cognitive neuroscience and adult neuropsychology. *Developmental Neuropsychology*, 13, 513–24.

Karmiloff-Smith, A. (1998) Development itself is the key to understanding developmental disorders. *Trends in Cognitive Sciences*, 2 (10), 33–8.

Karmiloff-Smith, A. (2001) Elementary, my dear Watson, the clue is in the genes... or is it? *Proceedings of the British Academy*, 117, 525–43.

Karmiloff-Smith, A. (2007) Williams syndrome. *Current Biology*, 17 (24), R1035–36.

Karmiloff-Smith, A. and Thomas, M. (2005) Can developmental disorders be used to bolster claims from Evolutionary Psychology? A neuroconstructivist approach. In J. Langer, S. Taylor Parker, and C. Milbrath (eds), *Biology and Knowledge Revisited: From Neurogenesis to Psychogenesis*. Lawrence Erlbaum Associates, Mahwah, NJ, pp. 307–22.

Karmiloff-Smith, A., Brown, J.H., Grice, S., and Paterson, S. (2003a) Dethroning the myth: Cognitive dissociations and innate modularity in Williams syndrome. *Developmental Neuropsychology*, 23, 227–42.

Karmiloff-Smith, A., Scerif, G., and Ansari, D. (2003b) Double dissociations in developmental disorders? Theoretically misconceived, empirically dubious. *Cortex*, 39, 161–3.

Karmiloff-Smith, A., Thomas, M., Annaz, D., Humphreys, K., Ewing, S., Brace, N., Van Duuren, M., Graham, P., Grice, S. and Campbell, R. (2004) Exploring the Williams syndrome face-processing debate: The importance of building developmental trajectories. *Journal of Child Psychology and Psychiatry*, 45 (7), 1258–74.

Machery, E. (2007) Massive modularity and brain evolution. *Philosophy of Science*, 74, 825–38.

Machery, E. (2009) *Doing Without Concepts*. Oxford University Press, New York.

Machery, E. (in press) Discovery and confirmation in evolutionary psychology. In J.J. Prinz (ed.), *Oxford Handbook of Philosophy of Psychology*. Oxford University Press, Oxford.

Machery, E. and Barrett, H.C. (2006) Debunking adapting minds. *Philosophy of Science*, 73, 232–46.

Machery, E. and Faucher, L. (2005) Social construction and the concept of race. *Philosophy of Science*, 72, 1208–19.

Martens, M.A., Wilson, S.J., and Reutens, D.C. (2008) Williams syndrome: A critical review of the cognitive, behavioral, and neuroanatomical phenotype. *Journal of Child Psychology and Psychiatry*, 49 (6), 576–608.

Maurer, D., Le Grand, R., and Mondloch, C.J. (2002) The many faces of configural processing. *Trends in Cognitive Sciences*, 6 (6), 255–60.

Mills, D.L., Alvarez, T.D., George, M.S., Bellugi, U., and Neville, H. (2000) Electrophysiological study of face processing in Williams syndrome. *Journal of Cognitive Neuroscience*, 12 (Supplement), 47–64.

Milner, B., Corkin, S., and Teuber, H.L. (1968) Further analysis of the hippocampal amnesia syndrome: 14 year follow-up study of H.M. *Neuropsychologia*, 6, 215–34.

Mobbs, D., Garrett, A.S., Menon, V., Rose, F.E., Bellugi, U., and Reiss, A.L. (2004) Anomalous brain activation during face and gaze processing in Williams syndrome. *Neurology*, 62, 2070–6.

Paterson, S.J., Brown, J.H., Gsdöl, M.K., Johnson, M.H., and Karmiloff-Smith, A. (1999) Cognitive modularity and genetic disorders. *Science*, 286, 2355–8.

Pinker, S. (1997) *How the Mind Works*. W.W. Norton & Company, New York.

Pinker, S. (2005) So how *does* the mind work? *Mind & Language*, 20 (1), 1–24.

Plaut, D.C. (1995) Double dissociation without modularity: Evidence from connectionist neuropsychology. *Journal of Clinical and Experimental Neuropsychology*, 17, 291–321.

Quartz, S.R. (2002) Toward a developmental evolutionary psychology: Genes, development, and the evolution of the human cognitive architecture. In S.J. Scher and F. Rauscher (eds), *Evolutionary Psychology: Alternative Approaches*. Kluwer, Dordrecht.

Richardson, R.C. (2007) *Evolutionary Psychology as Maladapted Psychology*. MIT Press, Cambridge, MA.

Saffran, E.M. (1982) Neuropsychological approaches to the study of language. *British Journal of Psychology*, 73, 317–37.

Samuels, R. (2005) The complexity of cognition: Tractability arguments for massive modularity. In P. Carruthers, S. Laurence, and S. Stich (eds), *The Innate Mind: Structure and Content*. Oxford University Press, Oxford, pp. 107–21.

Schlosser, G. and Wagner, G.P. (2004) *Modularity in Development and Evolution*. Chicago University Press, Chicago.

Shallice, T. (1988) *From Neuropsychology to Mental Structure*. Cambridge University Press, Cambridge.

Sterelny, K. (2003) *Thought in a Hostile World*. Blackwell, New York.

Tager-Flusberg, H., Plesa-Skwerer, D., Faja, S., and Joseph, R.M. (2003) People with Williams syndrome process faces holistically. *Cognition*, **89**, 11–24.

Temple, C.M. (1997) Cognitive neuropsychology and its applications to children. *Journal of Child Psychology and Psychiatry*, **38**, 27–52.

Teuber, H.-L. (1955) Physiological psychology. *Annual Review of Psychology*, **6**, 267–96.

Thomas, M. and Karmiloff-Smith, A. (1999) Quo vadis modularity in the 1990s? *Learning and Individual Differences*, **10**, 245–50.

Thomas, M. and Karmiloff-Smith, A. (2002) Are developmental disorders like cases of adult brain damage? Implications from connectionist modelling. *Behavioral and Brain Sciences*, **25**, 727–88.

Thomas, M. and Karmiloff-Smith, A. (2003) Modeling language acquisition in atypical phenotypes. *Psychological Review*, **110**, 647–82.

Thomas, M., Grant, J, Barham, Z., Gsödl, M., Laing, E., Lakusta, L, Tyler, L. K., Grice, S., Paterson, S., and Karmiloff-Smith, A. (2001) Past tense formation in Williams syndrome. *Language and Cognitive Processes*, **2**, 143–76.

Valentine, T. (1988) Upside-down faces: a review of the effects of inversion upon face recognition. *British Journal of Psychology*, **79**, 471–91.

Van orden, G.C., Pennington, B.F., and Stone, G.O. (2001) What do double dissociations prove? *Cognitive Science*, **25**, 111–72.

Woodward, J. and Cowie, F. (2004) The mind is not (just) a system of modules (just) shaped by natural selection. In C. Hitchcock (ed.), *Contemporary Debates in Philosophy of Science*. Blackwell, Oxford.

Yin, R.K. (1969) Looking at upside-down faces. *Journal of Experimental Psychology*, **81**, 141–5.

Young, M.P., Hilgetag, C.-C., and Scannell, J.W. (2000) On imputing function to structure from the behavioural effects of brain lesions. *Philosophical Transactions of the Royal Society of London*, B**355**, 147–61.

Chapter 4

On the role of ethology in clinical psychiatry: what do ontogenetic and causal factors tell us about ultimate explanations of depression?

Erwin Geerts[†] and Martin Brüne

In this chapter, we argue that there are good reasons to reintroduce traditional ethological research in evolutionary approaches to mental disorders. By focussing on the relationship between early attachment relationships and non-verbal behaviour in depressive patients, we show that answering proximate questions about mental disorders has important implications for ultimate explanations in psychiatry. For example, ethological observations demonstrate that high levels of support-seeking behavior in depressive patients are associated with an unfavourable diagnosis, thus disproving certain adaptationist theories of depression. Contrary to earlier dramatic statements about ethology being swallowed by sociobiology or behavioural ecology, we show that reports of ethology's death have been greatly exaggerated, and that it rightly deserves its place in psychiatry.

4.1 Introduction

Ethology, as conceptualized by Niko Tinbergen (1952, 1963), deals with the causation, ontogeny, survival value, and evolutionary trajectories of behavior.

[†] Erwin Geerts (born June 2, 1966) passed away on August 30, 2010.

The first two issues concern the so-called proximate causes of behavior, including the processing of internal and external stimuli, and the ontogenetic development of behavior over an individual's life-span. For example, questions related to ontogeny inquire how behavior changes with aging, and which early experiences are necessary for a given behavior to develop properly. The latter two aspects, referred to as the ultimate causes, address the historical processes that have shaped behavior over many generations. The question about the selective advantages or biological function of behavior includes the analysis of how behavior impacts on an individual's inclusive fitness (i.e., survival and reproductive success). The second ultimate aspect other traces back the phylogeny of behavior. This aspect involves cross-species comparison of behavior and concerns how the behavior might have arisen and been modified in the course of evolution. Both proximate and ultimate causes are regarded as complementary in the explanatory power of behavior, that is, a full understanding of the behavior under study requires an answer to each one of these questions that does not contradict any of the answers to the other questions (e.g., Curio 1994). By combining proximate and ultimate factors, ethology bridges the gap between traditional comparative psychology and psychiatry on one hand, and sociobiology, behavioral ecology, and evolutionary psychology on the other.

Several authors have stressed the relevance and usefulness of an ethological approach to psychiatric disorders. Accordingly, a full appreciation of all four questions raised by Tinbergen not only has the potential to advance insights into psychopathology, but also to constitute an empirically testable framework for the understanding of abnormal behaviors (e.g., Lorenz 1937; Tinbergen 1974; Dixon 1998; Troisi 1999; Bouhuys 2003; Brüne 2008). At first sight, this claim seems to be counterintuitive for several reasons. First, psychopathological conditions can hardly be attributed adaptive values; rather it is their maladaptive features that are intrinsically part of the definitional criteria of pathology. Second, some scientists, including E.O. Wilson (1975), predicted that ethology, together with comparative psychology, would be at risk of being cannibalized by neurophysiology on the one hand, and behavioral ecology and sociobiology on the other. Accordingly, ethological explanations of abnormal behaviors would become superfluous. In fact, one can argue that Wilson's prediction, at least with respect to an ethological approach of psychopathology, has come true. Many of the original ideas and theoretical frameworks on behavior as developed by ethology (including Tinbergen and Tinbergen's theory on autism; Tinbergen and Tinbergen 1972), have been refuted (e.g., Bateson and Klopfer 1987). Furthermore, if evolutionary scenarios are considered useful at all to explain psychopathology (which is rarely the case in the mainstream literature on psychological problems), then sociobiological and evolutionary psychological explanations of psychopathology are much more popular than ethological observation studies on the causation and ontogeny of psychopathology. However, as we will argue in this chapter, this does not imply that the core characteristic of ethology—the approach of behavioral phenomena from the perspective of Tinbergen's four questions—is no longer valid, neither in general, nor specifically with respect to psychopathology. Moreover, as Bolhuis and Wynne (2009) have pointed out, an evolutionary interpretation of cognitive processes can provide clues to their underlying mechanisms, yet they do not explain how they work. From the perspective of treatment of malfunctioning cognition and emotions, effective psychiatric interventions can only be brought about if one has an understanding of the "how". This requires the experimental approach as designed by comparative psychology and ethology.

Affective disorders, foremost depression, may serve as an example of how ethological research can contribute to the clinical understanding and treatment of such maladaptive traits. From a sociobiological perspective, several adaptive functions of depression have been proposed that explain how depression may have prevailed in the human species over the course of evolution. For instance, depression has been interpreted as an (involuntary) subordinate strategy in agonistic interactions (e.g., Price *et al.* 1994), as unbalanced altruistic behavior (McGuire *et al.* 1994), as a conservation withdrawal strategy to reallocate and save necessary resources in unpropitious situations (Nesse 2000), and as a strategy to seek support and to decrease social threat (Allen and Badcock 2006). The core idea of these hypotheses is that depression is an adaptive response to actual or impending defeat in social contest, and that in our evolutionary history such situations were relevant enough in terms of consequences for an individual's inclusive fitness to create selective pressures on behavior. However, an important argument against an adaptionist explanation of depression is that there is substantial empirical evidence that indicates that depression is associated with reduced inclusive fitness. It has been shown that when compared to normal controls depressed people have a shortened life expectancy (e.g., Koopmans *et al.* 1997) and reduced fertility (Keller and Miller 2006). Furthermore, depression is associated with a decreased functioning of the immune system (Herbert and Cohen 1993). Finally, clinical depression and (even mild) depressive symptoms induce rejection in others (Segrin and Dillard 1992). Taking the direct and indirect negative effects of depression on inclusive fitness into account, Keller and Miller (2006) argue that on the basis of evolutionary genetic models it is unlikely that depression is a trait that has been selected for in the course of evolution. One of the main problems of adaptationist views on psychopathology lies in the fact that many theoreticians in the field overlook that psychopathological conditions are anything but "disease entities"—a mistake that already led Emil Kraepelin's evolutionary perspective into a scientific dead end. Rather, psychopathological conditions—unlike neurological disorders—are almost exclusively distinct from a statistical norm by degree (quantitatively), not by kind (qualitatively). In other words, if one accepts that psychopathological conditions such as depression are maladaptive extremes of variation, distinct from a statistical norm by degree, not by kind, it can make sense to analyse symptoms and syndromes within an adaptationist framework (Brüne 2008). For instance, Nettle (2004) argues that depression may be at the end of a normal distribution of affective reactivity that has been selected for. Likewise, Gilbert (2001) interprets depression as the consequence of a maladaptive expression of innate natural defenses (e.g., the fight–flight system) that have been selected for in the course of evolution. According to these approaches depression itself does not serve a biological function but the natural defenses do. In this chapter we will argue that ethological research in psychiatry adds to the mounting criticism of adaptationist approaches to depression.

Conversely, a point of critique that one can make against ethological studies is that they fail to link their findings on the causal factors of depression to ontogenetic and ultimate approaches of depression. In other words, these studies make use of the ethological techniques of behavioral observation (i.e., the analysis of all behavior as displayed by the subject from the perspective of the context in which the behavior is displayed and of the consequences of the behavior), but fail to integrate their findings within the framework of Tinbergen's four questions on behavior. In this chapter we seek to address this critique. We choose Bowlby's attachment theory as a guide to behavioral studies into depression as a model disorder (Bowlby 1969, 1988). First, we will explore in how far early attachment and early parental rearing styles are associated with depression. We then describe the results of ethological observations in depressed patient populations and explore the extent to which these findings fit with findings on developmental processes. Finally, we will integrate the findings from this review of studies with proximate factors in depression in an ultimate approach to depression.[2]

---

[2] For the sake of clarity and conciseness, we will deliberately focus on behavioral studies in psychiatric populations based on classic ethological methodology, thus excluding the extensive literature on facial expressions of emotion in depression, about which excellent reviews exist (e.g., Gaebel and Wölwer 2004).

4.2 Ontogenetic processes: early attachment relationships, parental rearing styles, and their relationship with depression

Early interactions between parents and their offspring are considered to play an important role in the social development of the offspring and the child's social functioning in later life. Within the framework of an ethological approach of social behavior Lorenz (1937) was the first to recognize this in his work on "imprinting" in greylag geese. In primates, including humans, early mother–child interactions are involved in the formation of attachment, i.e., the emotional bond, between the mother and the child.

Attachment is considered to be a basic biological aspect of human nature (Bowlby 1969, 1988; Gilbert 2005). It is hypothesized that the biological function of attachment is protection of the child against dangerous situations (e.g., Bowlby 1969). The innate tendency of the child to stay in proximity to the attachment figure increases the child's chances of survival and, as a consequence, his or her reproductive fitness. Parental rearing styles such as care and protection may serve a similar function.

Based on the observable behavioral responses of children to their mothers after a brief period of separation from their attachment figure, termed the "strange situation" experiment (e.g., Ainsworth *et al.* 1978), four types of attachment styles have been identified: secure attachment, anxious avoidant attachment, anxious resistant attachment, and disorganized attachment. Securely attached children are confident that the caregiver will be available, responsive, and helpful in aversive or frightening situations. This assurance allows them to explore the environment and to feel competent in dealing with it. Based on his or her early experiences with the primary caregiver the child builds a mental representation (or "inner working model") of the self and others. In securely attached children this representation of the self is to be lovable and competent, and the representation of others is to be reliable, consistent, and warm (e.g., Bowlby 1988; Bartholomew and Horowitz 1991).

Attachment relationships between mothers and newborn children, once established, are considered to be stable and "transferable" to other relationships, including the intimate partner. Moreover, according to attachment theory, the early attachment experiences serve as a prototype for the child's later expectations about social interactions. Indeed, early attachment styles have been shown to be associated with the child's behavior in later situations with other people (see Bowlby 1988). The attachment styles of adults have also been found to be associated with the adult's recollections of their parents' rearing styles during their childhood (Gittleman *et al.* 1998). In adolescence and adulthood insecurely attached people may lack the social relationships necessary to buffer them against stressful events and to provide support in aversive situations. In addition, insecurely attached adult relationships themselves may be a stressor that is involved in the onset of mental illness (e.g., Gittleman et al. 1998).

Insecure attachment styles (ambivalent, avoidant, and disorganized), which are partly based on unfavorable parental rearing styles (e.g., lack of care or overprotection), are assumed to increase the risk for psychopathology later in life. Indeed, several studies have demonstrated that independent of current mood depressed people recall their mother as less caring then healthy control people (for a meta-analysis see Gerlsma et al. 1990). To a lesser extent, there is also evidence that depressed patients recall higher levels of overprotection by their parents (e.g., Gerlsma et al. 1990). Carnelley et al. (1994) have also shown that depressed women more often show insecure attachment styles compared to healthy controls. Prospective studies have demonstrated that recalled parental rearing styles, in particular poor maternal care, predict an unfavorable subsequent course of depression (Parker 1998; Geerts et al. 2009).

It is assumed that the association between insecure attachment and unfavorable parental rearing styles and the risk of psychopathology in later life is effected via the social interactions of the child in adult life (e.g., Bowlby 1988; Parker et al. 1992). Various studies support the assumption that disturbed social interactions and interpersonal functioning are involved in the onset and course of depression. For instance, interpersonal stress, negative social interactions, and lack of social support are considered to be important risk factors in depression (e.g., Bos et al. 2007), particularly if these interpersonal situations are entrapping or humiliating (Brown et al. 1995; Gilbert and Allen 1998).

A large part of human communication is effected via nonverbal behavior. For instance, Cahn and Frey (1992) demonstrated that people's feelings of being understood or misunderstood appear to be related more to their perception of their partners' nonverbal behavior than to the perception of their verbal behavior. Burgoon (1985) cites a meta-analytic study by Philpott (1983) that demonstrates that verbal behavior accounted for approximately 31% of the variance in social meaning in human interactions. According to Burgoon about 60–65% of the meaning in social interactions is communicated nonverbally. One can thus argue that an ethological analysis of psychiatric patients' nonverbal interactions with others may contribute to a better understanding of the role of interpersonal processes in the onset and course of the disorder.

4.3 The ethological analysis of deviant behavior

Behavioral observation is used as a research tool in psychiatry and clinical psychology, comparative psychology, and ethology. In psychiatry the level of these observations can be described as molar (e.g., Troisi 1999). In clinical psychological interactions the observations are described in terms of the physician's impression, and quantification of observed behavior mostly takes place on interval scales (e.g., the five-point interval scale to quantify agitation or retardation in the Hamilton Rating Scale for Depression (Hamilton 1969)). In comparative psychology and ethology behavioral observations take place on a more molecular level and behavior is quantified in terms of the frequency and duration of different behavioral elements. An important distinction between comparative psychology and ethology is that the first often investigates single elements of behavior (or a limited set of single elements of behavior) in an experimental setting whereas the focus of ethology is on the whole set of behaviors as displayed by a subject in its "naturalistic" environment.

A basic assumption that underlies the ethological study of behavior is that the behavior of the species under study is hierarchically organized. Several elements of behavior may share a common causal factor and a common (biological) function. Whether or not elements of behavior share a causal factor and/or a common (biological) function is based on contextual and consequential evidence (e.g., Eibl-Eibesfeldt 1989; Troisi 1999), that is, elements of behavior that occur in one context but not in others are assumed to share a similar cause that is related to this context. In addition, behavioral elements that produce a similar effect are assumed to be functionally homologous. Based on contextual evidence and consequential evidence, these elements are pooled into a higher-order factor. In turn, several higher-order factors together can constitute the behavioral repertoire subserving biological motivation (e.g., reproduction, territorial behaviour, social behavior, etc.). In the observation and registration of behavior ethologists make use of so-called ethograms. These are catalogues of discrete elements of behavior that make up part of the behavioral repertoire of the species under study. The behavioral elements that constitute an ethogram need to be sufficiently distinguishable from other elements of behavior (i.e., sufficiently uniform and recurrent) in order to allow the registration of objective parameters (e.g. frequency and duration).

Most ethological studies into psychopathological conditions have used the ethogram developed by Grant (1968), or variants of it. This ethogram was developed for the coding of human behavior during interviews and comprises more that 100 different codable simple or complex movements such as eyeblink, gaze direction, and facial movements, as well as body posture. These behavioral elements are allocated to one of four categories, namely flight, assertion, contact, and relaxation. In other words, every movement is ascribed a meaning in the process of interpersonal communication. For example, drawing back the corners of one's mouth represents an involuntary and unconscious expression of submission and is often motivated by a tendency to escape a situation. In contrast, an eyebrow flash invites social interaction and is accordingly grouped with other affiliation signals as "contact" behavior (Grant 1968). Based on Grant's ethogram and its variants as employed by Polsky and McGuire (1980), and Schelde and co-workers (e.g., Schelde *et al.* 1988), Troisi has refined and simplified this ethogram in the Ethological Coding System for Interviews (ECSI; Troisi 1999). Table 4.1a presents this ethogram. It consists of 37 behavioral elements that are grouped according to their meaning.

Bouhuys and co-workers have set out a series of studies on the role of nonverbal interpersonal processes in the onset and course of depression. Because the ECSI assesses only patient behavior, they developed an ethogram to describe the behavior of both depressed patients and their interviewers during a clinical interview. This ethogram is based on findings that single elements of the observable behavior of depressed patients are associated with the patients' clinical state (see Bouhuys *et al.* 1991). In addition, they based their analyses on contextual and consequential evidence that behavior as displayed during speaking may have a different cause and a different impact on the conversation partner when compared to the same behavior as displayed while listening (e.g., gaze and gestures). The ethogram by Bouhuys and co-workers consists of 12 elements of patient and interviewer behavior that are described in terms of frequency and duration during speaking, and frequency and duration during listening. Based on factor analyses the resulting behavioral elements are pooled into six factors that describe the patient's behavior and seven factors that describe the interviewer's behavior (Bouhuys *et al.* 1991; Bouhuys and van den Hoofdakker 1991; Geerts 1997). Table 4.1b presents the patient and interviewer ethograms as developed by Bouhuys and co-workers.

4.4 Association between observable behavior and depression

The first ethological observation studies on depression focused on the question of whether the observable behavior of depressed patients differed from that of healthy controls and from that of patients with other psychiatric disorders. For instance, Polsky and McGuire investigated whether or not ethological methods can be used to identify different types of psychopathology (Polsky and McGuire 1980). They observed the behavior of patients with depression, schizophreniform disorder, and personality disorder in settings on the ward. Patients with depression displayed nonsocial behaviors the most and invited others to participate in social interaction less than patients with schizophrenia

Table 4.1a Definitions of behavioral patterns included in the ECSI and instructions for calculating the composite scores of behavioral categories (Troisi 1999)

Behaviour patterns

1. Look at. Looking at the interviewer.
2. Head to side. The head is tilted to one side.
3. Bob. A sharp upwards movement of the head, rather like an inverted nod.
4. Flash. A quick raising and lowering of the eyebrows.
5. Raise. The eyebrows are raised and kept up for some time.
6. Smile. The lip corners are drawn back and up.
7. Nod. The normal affirmative gesture.
8. Lips in. The lips are drawn slightly in and pressed together.
9. Mouth corners back. The corners of the mouth are drawn back but not raised as in smile.
10. Look away. Looking away from the interviewer.
11. Look down. Looking down at feet, lap or floor.
12. Shut. The eyes are closed.
13. Chin. The chin is drawn in towards the chest.
14. Crouch. The body is bent right forward till the head is near the knees.
15. Still. A sudden cessation of movement, a freezing.
16. Shake. The normal negative gesture.
17. Thrust. A sharp forward movement of the head towards the interviewer.
18. Lean forward. Leaning forward from the hips towards the interviewer.
19. Frown. The eyebrows are drawn together and lowered at the centre.
20. Shrug. The shoulders are raised and dropped again.
21. Small mouth. The lip corners are brought towards each other so that the mouth looks small.
22. Wrinkle. A wrinkling of the skin on the bridge of the nose.
23. Gesture. Variable hand and arm movements used during speech.
24. Groom. The fingers are passed through the hair in a combing movement.
25. Hand-face. Hand(s) in contact with the face.
26. Hand-mouth. Hand(s) in contact with the mouth.
27. Scratch. The fingernails are used to scratch part of the body, frequently the head.
28. Yawn. The mouth opens widely, roundly and fairly slowly, closing more swiftly. Mouth movement is accompanied by a deep breath and often closing of the eyes and lowering of the brows.
29. Fumble. Twisting and fiddling finger movements, with wedding ring, handkerchief, other hand, etc.
30. Twist mouth. The lips are closed, pushed forward and twisted to one side.
31. Lick lips. The tongue is passed over the lips.
32. Bite lips. One lip, usually the lower, is drawn into the mouth and held between the teeth.
33. Relax. An obvious loosening of muscle tension so that the whole body relaxes in the chair.
34. Settle. Adjusting movement into a more comfortable posture in the chair.
35. Fold arms. The arms are folded across the chest.
36. Laugh. The mouth corners are drawn up and out, remaining pointed, the lips parting to reveal some of the upper and lower teeth.
37. Neutral face. A face without expression and without particular muscular tension. It is the basic awake face.

Scoring instructions

Add items 2–6 to get AFFILIATION; add items 7–9 to get SUBMISSION; add items 10–15 to get FLIGHT; add items 16–22 to get ASSERTION; add items 24–32 to get DISPLACEMENT; add items 33–37 to get RELAXATION

Table 4.1b Description of the registered behavior of depressed patients and of interviewers during a clinical interview and overview of how the behavioral elements constitute the patients' and interviewers' behavioral factors (Bouhuys et al. 1991, Bouhuys and van den Hoofdakker 1991; Geerts et al. 1995)

Summary of the registered behavioral elements and abbreviations

1. Vocalizations:	sp = speech
	bch = verbal backchannel: "yes yes", "hmm hmm", emitted to show one is listening
2. Head movements:	yes = yes nodding
	no = no shaking
	head = head movements, other than yes nodding and no shaking
3. Looking	look = looking in the direction of the other's face
4. Leg movements:	leg = leg movements
5. Hand movements:	gest = gesticulating
	botol = light body touching; when only fingers or hands are manipulating the body
	botoi = intensive body touching; when also wrist or forearm are moving
	obto = object touching: hands or fingers are manipulating an object
	otha = hand movements, other then object- or body touching or gesturing

d = duration; f = frequency; /sp = during speaking; /li = during listening; 1, ..., 1/4 = weight by which the elements are summed into the behavioral factors.

Constituent behavior of the patients' factors 2

Restlessness-1	Restlessness-2	Speech	Active Listening	Eagerness	Speaking Effort
1/2 * dleg/sp	1 * dobto/sp	1 * dps	1/2 * dotha/li	1/2 * dyes/sp	1/3 * dlook/sp
1/2 * dbotol/sp	1/2 * fotha/sp	1 * fsp	1/2 * dbotoi/li	1/2 * dno/sp	1/3 * dhead/sp
1/2 * fleg/sp	1/2 * fbotoi/sp	1/3 * fyes/i	1 * fobto/sp	1/2 * fyes/sp	1/3 * dgest/sp
1/2 * fbotol/sp	1 * dobto/li	1/3 * fno/li	1 * dbotoi/li	1/2 * fno/sp	1/2 * flook/sp
1/2 * dleg/li	1 * fobto/li	1/3 * fsp/li	1/2 * dhead/li	1/2 * dyes/li	1/2 * fhead/sp
1/2 * dbotol/li		1 * dspli	1/2 * dotha/li	1/2 * dno/i	1 * fgest/sp
1/2 * fbotol/li			1/4 * flook/li		1 * dlook/li
			1/4 * fhead/li		
			1/4 * fbotoi/li		
			1/4 * fotha/li		

(continued)

Table 4.1b (continued) Description of the registered behavior of depressed patients and of interviewers during a clinical interview and overview of how the behavioral elements constitute the patients' and interviewers' behavioral factors (Bouhuys et al. 1991, Bouhuys and van den Hoofdakker 1991; Geerts et al. 1995)

Constituent behavior of the patients' factors 2

Restlessness-1	Restlessness-2	Speech	Active Listening	Turn Taking	Change Looking
1/2 * dbotol/sp	1/3 * dhead/sp	1/4 * dsp	1/2 * dbotoi/li	1/2 * dlook/sp	1 * fhead/sp
1/2 * dbotoi/sp	1/3 * dotha/sp	1/4 * dyes/sp	1/2 * dotha/li	1/2 * dleg/sp	1 * flook/sp
1/2 * dhead/li	1/3 * dobto/sp	1/4 * dno/sp	1/2 * fbotoi/li	1/2 * dleg/li	1/2 * flook/li
1/2 * dbotol/li	1/3 * dlook/li	1/4 * dgest/sp	1/2 * fotha/li	1/2 * dgest/li	1/2 * fhead/li
1/3 * fbotoi/sp	1/3 * dno/li	1 * dsp/li		1/2 * fleg/li	
1/3 * fotha/sp	1/3 * dobto/li	1/4 * fsp		1/2 * fgest/li	
1/3 * fbotol/sp	1/2 * fleg/sp	1/4 * fyes/sp			**Encouragement**
1 * fbotol/li	1/2 * fobto/sp	1/4 * fno/sp			1/2 * dvbc/li
	1 * fobto/li	1/4 * fgest/sp			1/2 * dyes/li
		1/2 * fsp/li			1/2 * fvbc/li
		1/2 * fno/i			1/2 * fyes/li

Reprinted from Psychiatry Research, 57, Geerts, Bouhuys, Meesters, and Jansen, Observed behavior of patients with seasonal affective disorder and an interviewer predicts response to light treatment, Copyright (1995), with permission from Elsevier

or personality disorders. They also found that during hospitalization the interpersonal space between the patient and others changed in patients with schizophrenia and in patients with personality disorder, but not in depressed patients. Overall, the amount of social behavior in patients with schizophrenia increased with clinical improvement (Polsky and McGuire 1980). Differences in observed social behavior between patients who would improve and those who would not emerged in the second week of hospitalization. However, behavior as observed during the first week did not predict subsequent clinical improvement. The findings by Polsky and McGuire have been replicated by others (e.g., Fossi *et al.* 1984; Schelde 1998).

Jones and Pansa (1979) obtained comparable results from behavioral observations of depressed patients, schizophrenic patients, and controls during a clinical interview at hospital admission. Compared to healthy controls depressed patients demonstrated both shorter durations and lower frequencies of smiling and looking at the interviewer. In the early phase of the interview gaze duration was also shorter than in patients with schizophrenia. Furthermore, when compared to patients with schizophrenia as well as to healthy controls, depressed patients demonstrated a shorter duration and lower frequency of body-focused hand movements (e.g., self-touching). With clinical improvement, the frequency and duration of smiles during the first 2 minutes of the interview increased. Hence, the findings of these early ethological observational studies of depressed patients showed that depressed patients can be distinguished from patients with other types of psychopathology and from controls. It turns out that, compared to schizophrenia and personality disorders, depression particularly affects the social behavior of patients. These findings supported the suggestion that an ethological approach may contribute to a better understanding of clinical depression.

4.5 Disturbed interpersonal behavior as a possible causal factor in depression

The findings cited above indicate that depression is associated with low levels of social interaction. However, the findings from ward observations failed to demonstrate an association between lowered levels of social behavior and the subsequent course of depression. Focusing on nonverbal behavior during a clinical interview, Ranelli and Miller (1981) found that depressed patients who did not improve during a follow-up displayed a high frequency of body-focused adaptors (e.g., body touching), posture shifts, and speech pauses. Patients who did improve displayed long speech pauses and head aversions. Using the ESCI, Troisi *et al.* (1989) demonstrated that nonresponse to a 5-week treatment with amitriptyline was predicted by high levels of Submissive Behaviour, Affiliation, and Assertive Behaviour (see Table 4.1a). Bouhuys and co-workers investigated the predictive quality of their ethogram with respect to the course of depression over a 10-week follow-up in a population of hospitalized major depressed patients. They found that high levels of the patients' Restlessness-1 and Speaking Effort and low levels of Active Listening (see Table 4.1) were associated with a poor outcome of depression (Bouhuys and van den Hoofdakker 1993). Moreover, they found that the observable behavior of the interviewer was also associated with the subsequent course of depression: when compared to patients with a favorable course of depression, interviewers displayed relatively lower levels of Restlessness-1 and Active Listening, and relatively higher levels of Encouragement (see Table 4.1b) toward patients who showed a poor outcome (Bouhuys and van den Hoofdakker 1993). They furthermore demonstrated that the response predicting behaviors of patients and interviewers were interrelated. These findings were replicated in a population of outpatients with seasonal affective disorder (SAD; see Geerts 1997). The findings in patients with SAD resembled those in patients with major depression mostly with respect to high levels of patients' Speaking Effort and of interviewers' Encouragement as predictors of poor outcome of depression. Based on the elements of behavior that constitute these factors, Bouhuys and co-workers interpreted them as displays of the patients' involvement in the interview and their seeking for support and of the interviewer's involvement and giving support.

Hale *et al.* (1997) investigated the interaction between depressed patients and their partner and between these patients and an unfamiliar person. They showed that the partners of patients who did not improve during the follow-up period spoke more during the interactions than those of patients who did improve. Neither the behavior of the patients in these interactions, nor that of unfamiliar people was associated with the subsequent course of the depression.

Taken together, the findings from clinical interviews demonstrate that nonverbal behavior of depressed patients, as well as the behavior of their interviewer, is associated with the subsequent course of depression. When one combines the findings from the different studies, it appears that high levels of social involvement and support seeking of patients predict an unfavorable course of depression. In addition, high levels of involvement and support as displayed by interviewers are also associated with an unfavorable prognosis.

Geerts and co-workers investigated the nature of the interrelationship between the patients' Speaking Effort and their interviewers' Encouragement. They experimentally demonstrated that depressed patients attune the levels of their nonverbal support-seeking behavior to the levels of nonverbal support that they receive (Geerts 1997). These findings indicate that, while in interaction, patients converge the amount of their displayed nonverbal seeking for support to the amount of nonverbal support as displayed by the interviewer. In a later study, Geerts *et al.* (2000) used a pseudo-interaction paradigm (e.g., Bernieri and Rosenthal 1991) to statistically confirm the causal interrelatedness in clinical interviews. In two independent patient populations they demonstrated that the levels of the patients' nonverbal seeking for support and those of the interviewers' giving it were significantly more similar than may be expected from chance levels.

In healthy subjects and in remitted depressed patients nonverbal convergence between conversation partners during an interaction underlies satisfaction with the interaction (e.g., Cappella and Palmer 1990, 1992; Geerts *et al.* 2006). Furthermore, it has been demonstrated that nonverbal convergence underlies rapport (e.g., feelings of mutual understanding, positivity, and mutual willingness to cooperate; Bernieri 2005). Geerts (1997) hypothesized that lack of nonverbal convergence may be a mechanism that is involved in the onset and course of depression. This hypothesis has been tested in five independent populations of depressed patients so far (see Geerts 1997 for an overview and Geerts *et al.* 2009). In all but one study it was found that lack of nonverbal convergence between the patients' Speaking Effort and the interviewers' Encouragement predicted an unfavorable short-term outcome of depression. Moreover, in remitted depressed patients lack of nonverbal convergence between patients and an interviewer has been shown to predict depression relapse within a 2-year follow-up (Geerts *et al.* 2006). In the same patient sample Bos and co-workers (2005) found that patients who reached low levels of nonverbal convergence with their interviewer were at higher risk of getting involved in negative interpersonal events during the follow-up. They furthermore demonstrated that these negative events mediated nonverbal convergence to recurrence of depression. Geerts and Bouhuys (2002) demonstrated that the findings on the association between lack of nonverbal convergence and poor outcome of depression can also be generalized to the interactions of patients with their partner and with unfamiliar people. Taken together, the findings so far indicate that the causal interrelationship between depressed patients' nonverbal support-seeking behavior and the nonverbal support-giving behavior by the people with whom they interact plays an important role in the subsequent course of depression. These findings favor the hypothesis that lack of nonverbal convergence between patients and conversation partners is a mechanism that underlies negative interpersonal events, which in turn provoke (an unfavorable course of) depression. It has been demonstrated that lack of nonverbal convergence cannot be explained by the severity of depression or by patients' personality traits such as Neuroticism and Extraversion, or their cognitive interpretation of facially expressed emotions (Geerts 1997; Bos 2005). Hence, the nonverbal communications between depression-prone people and people from their social environment on the one hand and personality traits and social cognition on the other appear to play independent roles in the onset and course of depression.

To summarize, behavioral observations of depressed patients during a clinical interview confirm findings from ward observation that impaired social behavior is involved in depression. Moreover, in contrast to ward observations, the nonverbal behavior registered from clinical interviews does predict the subsequent course of depression. One can easily see why these ethological findings are in contrast with an adaptationist interpretation of depression. For instance, they show that high levels of support-seeking behavior and high levels of received nonverbal support are associated with an unfavorable prognosis. Moreover, they show that the communication style of depression-prone patients may generate the negative interpersonal events that in turn provoke depression. These findings are in line with the findings by Segrin and Dillard (1992) that (even mild symptoms of) depression induces rejection in other people. Moreover, they provide a mechanistic explanation of why depressed people induce rejection in others. Hence, the findings from ethological observations indicate that depression is a consequence of malfunctioning interpersonal processes, rather than a strategy to ameliorate stressful interpersonal situations.

4.6 Are causal factors of depression linked to adverse early experiences?

Evidence exists that adult interpersonal interactions may mediate or modify the association between early attachment and parental rearing styles on the one hand, and the course of depression on the other (Parker *et al.* 1992; Gittleman *et al.* 1998), that is, insecure early attachment and negative experiences with parental rearing may predispose to psychopathology through the effects of adult social functioning, especially when adult social interactions go wrong. However, adult social interactions that deviate from the experiences based on early attachment and parental rearing may also modify this association (Bowlby 1988; Parker *et al.* 1992). For instance, supportive social interactions in later life may buffer against the negative effects of early insecure attachment and anomalous parenting (Gittleman *et al.* 1998). It has been shown that behavioral similarity between mothers and their children underlies secure attachment (e.g., Isabella and Belsky 1991; Jaffe *et al.* 2001). Based on these findings, Geerts *et al.* (2009) hypothesized that nonverbal convergence as observed in the dyadic interactions of depressed patients is associated with how the patients recalled parental rearing styles. This hypothesis was confirmed: the higher the levels of maternal care the patients recalled, the more the patients and interviewers converged on similar levels for their Speaking Effort and Encouragement. Moreover, they found that nonverbal convergence modified the association between recalled maternal care and the subsequent course of depression: the observed correlation between high level of recalled maternal care and a favorable subsequent course of depression was confined to patients who reached high levels of nonverbal convergence with the interviewer. Although these findings clearly need to be replicated, they are in line with the suggestion that early parent–child interaction is involved in the interpersonal functioning of depressed patients, and with the suggestion that adult interpersonal functioning links these early parent–child experiences to the course of depression. To the best of our knowledge, the hypothesized link between early attachment and recalled parental rearing on the one hand and nonverbal behavior in adult depressed patients has not been investigated in other studies.

Apart from the question of whether or not adult nonverbal interpersonal functioning can be linked statistically to early attachment and parental experiences, an important question that needs to be answered is how early attachment and parental rearing styles exert their effects on adult interpersonal functioning. Animal experiments have shown that the care-giving behavior of mothers affects hypothalamic gene expressions in pups. Variation in maternal care is associated with variation in (the development of) the pups' functioning of the hypothalamic–pituitary–andrenocortical axis (HPA axis; for a review see Meaney 2001). This axis represents the biobehavioral stress–response system (e.g., fight–flight, see Korte et al. 2005). The effect of early maternal care on the expression of genes is one example of the epigenetic effects of environmental contingencies. A crucial aspect in the light of natural selection theory is that via epigenesis the same gene can lead to different phenotypical outcomes depending on the way it is activated or deactivated by maternal care. One can hypothesize that the HPA axis is also involved in human support-seeking strategies in stressful circumstances. Recent studies indicate that this effect of early maternal care on the expression of genes in the hypothalamus also occurs in humans (McGowan et al. 2009).

4.7 Possible evolutionary explanations of depression

If one accepts that (1) depression is the consequence of malfunctioning interpersonal processes (and other biobehavioral responses to stress) that find an ontogenetic source in early parent–child interactions and (2) attachment and bonding are innate capacities as part of human nature (e.g., Gilbert 2005), the question arises why natural selection did not work against early attachment and parental rearing styles that predispose to depression. Indeed, in the light of evolution and natural selection the high prevalence of psychiatric disorders, including depression, seems counterintuitive, given the strong negative effects that these disorders appear to have on an individual's inclusive fitness. Depression is common and affects about 120 million people worldwide (WHO 2008). It is the number one cause of disability, and according to the World Health Organization by 2020 it will be the second most important disorder in terms of burden of disease.

Bowlby (1969) originally believed that secure attachment and a trustful internal working model was the only one designed by evolution, and that other forms of attachment were deviations from the norm. Contemporary (evolutionary) approaches of attachment, however, hypothesize that, depending on the socioeconomical and environmental contingencies, insecure types of attachment may also have an adaptive value (e.g., Belsky 2002). Indeed, from an evolutionary perspective it is plausible to assume that fluctuations in environmental conditions may have selected for a set of flexible adaptive behavioral responses. From the perspective of parent–offspring conflict theory (Trivers 1974) one can argue that the biological interests of parents and their offspring differ. From the perspective of the parents' inclusive fitness it may not always be in their best interest to provide sensitive care (Belsky 2002). Several behavioral ecological studies in avian and mammalian species have shown that parents adjust their parental investment to environmental circumstances. Unfavorable rearing conditions (from the perspective of the parent) may force parents to reduce their parental investment, resulting in less favorable attachment and rearing styles. In the child these unfavorable conditions may affect the biobehavioral and psychological stress-response system such that behavior later in life is more adapted to the unfavorable environmental circumstances. Within the contemporary framework of attachment theory one can thus speculate that depression (and other types of psychopathology such as, for instance, antisocial personality disorder and borderline personality disorder) are the costs of insecure attachment styles. Natural selection may have favored insecure attachment as long as the benefits in terms of inclusive fitness outweigh the costs. However, cross-fostering experiments in animals have shown that attachment styles can be "inherited" via early life experience (see Meaney 2001). These experiments indicate that a nongenetic mechanism of inheritance may also underlie the transmission of attachment styles to next generations. Such a nongenetic route of transmission can have a serious impact on natural selective forces that act upon genes.

4.8 Discussion: why psychiatry needs ethology

In this chapter we have reviewed ethological (semi-)naturalistic observational studies of depression in light of Bowlby's attachment theory. In summary, the studies cited above are in line with attachment theory by demonstrating that early attachment and parental rearing styles, and nonverbal interpersonal processes may play a causal role in depression. Moreover, the behavioral processes that underlie poor outcome of depression are similar to the processes in early mother–child interactions that underlie insecure attachment. However, we only found one study in which observed interpersonal behavior modified the association between early parental rearing style and adult depression. Hence, the hypothesis that adult social interactions link early attachment to psychopathology (or modifies this relationship) needs to be tested in future studies. From an ultimate (or evolutionary) perspective, the findings indicate that depression itself does not serve a biological function, but may be the consequence of a malfunctioning of natural defense mechanisms that serve a biological function in coping with stress.

We have argued that Tinbergen's (1963) four questions that address the causes of behavior at the proximate and the ultimate level can be a useful and informative tool for behavioral observations in psychiatric populations and are a necessary tool for a full and comprehensive understanding of psychopathology in an evolutionary perspective. We are convinced that this is a fruitful approach to understand abnormal behavior for several reasons.

First, ethology is the only scientifically valid framework that systematically categorizes behavioral elements in terms of their species-specific communicative meaning. Accordingly, drawing on cross-species as well as cross-cultural issues, behaviors observed in psychiatric patients can be understood within an empirically testable concept of behavior. For example, behaviors seen in depression, such as crouching postures, averted gaze or displacement activities, often reflect defensive strategies or motivational ambivalence (see, e.g., Troisi 2002; Gilbert 2005). Accordingly, changing patterns of behavior can be linked to questions about clinical improvement or deterioration, even before the patient is subjectively aware of it. For example, an increase in displacement activities can alert clinicians to check for clinical deterioration. Such behavior may indicate increasing motivational conflict and ambivalence, which can be a sign of impending suicidal behavior. These examples of behavioral analyses based on ethological methodology explicitly assume that behaviors found in clinical conditions are not qualitatively distinct from behaviors in healthy individuals but different by degree, that is, intensity, frequency, or contextual inappropriateness (Brüne 2008).

Second, behavioral observation is often much more reliable than subjective report because it is much less under conscious control compared to verbal communication such that an individual's "real" motives cannot so easily be concealed (Troisi *et al.* 1998). For the same reason an ethological approach is at least as valuable as (evolutionary) psychological approaches based on questionnaires (Daly and Wilson 1999). We therefore cannot conceive of any fruitful psychiatric research without referring to patients' actual behavior. Third, standard rating scales utilized in clinical assessments critically depend on the clinician's impression of patients' nonverbal behavior. Even though manuals available for clinical rating scales hardly ever recur on ethological methodology, clinicians intuitively use their species-specific endowments for deciphering nonverbal expressions in therapist–client interactions. The extent to which actual clinical judgments rely on unconsciously perceived communicative signals sent by patients compared to their subjective report is a highly under-researched topic in clinical psychiatry (Brüne *et al.* 2008).

Finally, we believe that a fruitful approach could be to combine research into cognitive deficits with studies of nonverbal behavior, as has recently been demonstrated by Brüne and co-workers (2009). All these issues are clearly within the domain of an ethological approach that cannot be addressed using more "sterile" sociobiological theorizing.

Conversely, we do not want to downplay the disadvantages of ethological observation in psychiatric populations. We believe that ethological methodology is still outside psychiatric mainstream research because it is extremely time-consuming and because it requires a lot of training that medical or psychology students usually do not receive (because ethology is not part of the curricula). Thus, ethological terminology is unfamiliar to most clinicians. Moreover, behavioral observation in psychiatric populations has not been linked with physiological measures such as neurotransmitter activity or genetic variation.

In this respect we feel that a critical note should also be addressed to human ethological approaches of psychopathology. So far, most ethological studies on nonverbal interpersonal processes in depression have focused on the question on causation. In researching this chapter, we found that little is known on ontogenetic processes, that is, what factors are involved in the development of the impairments in nonverbal interpersonal communication. In this respect, human ethology can benefit from joint research with developmental psychology, child psychology and psychiatry. The future of psychiatry, at least in our opinion, will be to find answers to all four questions proposed by Tinbergen (1952, 1963) with regard to behavior, emotion, and cognition.

Curio (1994) argued that a full ethological approach (i.e., the study of behavior within the framework of the four questions) is useful not only to investigate whether ultimate interpretations make sense in the light of a proximate approach but also because answers on proximate factors can provide new and unexpected ideas on the possible biological function of behavior and on its evolutionary history. In line with Curio's argument, the studies cited in this chapter shed a different light on the role of natural selection and evolutionary processes in psychopathology and depression, specifically when compared to adaptationist approaches. From the perspective of evolutionary theory we made an attempt to provide an explanation of depression that does anticipate on biological adaptation as a starting point. The recent findings on the epigenetic effects of early parental behavior emphasize the need for a better understanding of environment–gene interactions in the evolutionary explanations of behavior. We think that Tinbergen's questions provide a useful and proven successful framework for approaching these environment–gene interactions. Finally, human ethology may benefit from research into psychiatric populations because studying pathological variants of behavior may be more informative regarding the physiological function of the behavior than examining the physiological correlate (Lorenz 1973), an approach that may return to the roots in the line of thinking and methodology of Darwin's *The Expression of the Emotions in Man and Animals* (Darwin 1872).

References

Ainsworth, M.D., Blehar, M.C., Waters, E. and Wall, S. (1978) *Patterns of Attachment: Assessed in the Strange Situation and at Home.* Lawrence Erlbaum Associates, Hillsdale, NJ.

Allen, N.B. and Badcock, P.B.T. (2006) Darwinian models of depression: a review of evolutionary accounts of mood and mood disorders. *Progress in Neuro-Psychopharmacology and Biological Psychiatry*, 30, 815–26.

Bartholomew, K. and Horowitz, L.M. (1991) Attachment styles among young adults: a test of a four-category model. *Journal of Personality and Social Psychology*, 61, 226–44.

Bateson, P.P.G. and Klopfer, P.H. (1987) *Perspectives in Ethology: Alternatives.* Plenum Press, New York.

Belsky, J. (2002) Modern evolutionary theory and patterns of attachment. In J. Cassidy and P.R. Shaver (eds), *Handbook of Attachment.* Guilford Press, New York, pp. 141–61.

Bernieri, F.J. (2005) The expression of rapport. In V. Manusov (ed.), *The Sourcebook of Nonverbal Measures: Going Beyond Words.* Lawrence Erlbaum Associates, New York, pp. 347–59.

Bernieri, F.J. and Rosenthal, R. (1991) Interpersonal coordination: behavior matching and interactional synchrony. In R.S. Feldman and B. Rimé (eds), *Fundamentals of Behaviour.* Cambridge University Press, Cambridge, pp. 401–32.

Bol, J.J. and Wynne, C.D.L. (2009) Can evolution explain how our minds work? *Nature*, 458, 832–33.

Bos, E.H. (2005) Interpersonal mechanisms in recurrence of depression. http://dissertations.ub.rug.nl/faculties/medicine/2005/e.h.bos/. University of Groningen, Groningen.

Bouhuys, A.L. (2003) Ethology and depression. In P. Philippot, R.S. Feldman, and E.J. Coats (eds), *Nonverbal Behavior in Clinical Settings.* Oxford University Press, New York, pp. 233–62.

Bouhuys, A.L. and van den Hoofdakker, R.H. (1991) The interrelatedness of observed behavior of depressed patients and of a psychiatrist: an ethological study on mutual influence. *Journal of Affective Disorders*, 23, 63–74.

Bouhuys, A.L. and van den Hoofdakker, R.H. (1993) A longitudinal study of interaction patterns of a psychiatrist and severely depressed patients based on observed behavior: an ethological approach of interpersonal theories of depression. *Journal of Affective Disorders*, 27, 87–99.

Bouhuys, A.L., Jansen, C.J., and van den Hoofdakker, R.H. (1991) Analysis of observed behaviors displayed by depressed patients during a clinical interview: relationships between behavioral factors and clinical concepts of activation. *Journal of Affective Disorders*, 21, 79–88.

Bowlby, J. (1969) *Attachment and loss. Volume 1: Attachment.* Basic Books, New York.

Bowlby, J. (1988) Developmental psychiatry comes of age. *American Journal of Psychiatry*, 145, 1–10.

Brown, G.W., Harris, T.O., and Hepworth, C. (1995) Loss, humiliation and entrapment among women developing depression: a patient and non-patient comparison. *Psychological Medicine*, 25, 7–21.

Brüne, M. (2008) *Textbook of Evolutionary Psychiatry. The Origins of Psychopathology.* Oxford University Press, Oxford.

Brüne, M., Sonntag, C., Abdel-Hamid, M., Lehmkämper, C., Juckel, G., and Troisi, A. (2008) Non-verbal behavior during standardized interviews in patients with schizophrenia spectrum disorders. *Journal of Nervous and Mental Disease*, 196, 282–88.

Brüne, M., Abdel-Hamid, M., Sonntag, C., Lehmkämper, C., and Langdon, R. (2009) Linking social cognition with social interaction: Non-verbal expressivity, social competence and "mentalising" in patients with schizophrenia spectrum disorders. *Behavioral and Brain Functions*, 5, 6.

Burgoon, J.K. (1985) Nonverbal signals. In: M.L. Knapp and G.R. Miller (eds), *Handbook of Interpersonal Communication*. Sage Publications, Beverly Hills, pp. 344–90.

Cahn, D.D. and Frey L.R. (1992) Listeners' perceived verbal and nonverbal behaviors associated with communicators' perceived understanding and misunderstanding. *Perceptual and Motor Skills*, 74, 1059–64.

Cappella, J.N. and Palmer, M.T. (1990) Attitude similarity, relational history, and attraction: The mediating effects of kinesic and vocal behaviors. *Communication Monographs*, 57, 161–83.

Cappella, J.N. and Palmer, M.T. (1992) The effect of partners' conversation on the association between attitude similarity and attraction. *Communication Monographs*, 59, 180–89.

Carnelley, K.B., Pietromonaco, P.R., and Jaffe, K. (1994) Depression, working models of others and relationship functioning. *Journal of Personality and Social Psychology*, 66, 127–40.

Curio, E. (1994) Causal and functional questions: how are they linked. The Niko Tinbergen Lecture 1992. *Animal Behaviour*, 47, 999–1021.

Daly, M. and Wilson, M. (1999) Human evolutionary psychology and animal behaviour. *Animal Behaviour*, 57, 509–19.

Darwin, C. (1872) *The Expression of the Emotions in Man and Animals*. Murray, London.

Dixon, A.K. (1998) Ethological strategies for defence in animals and humans: their role in some psychiatric disorders. *The British Journal of Medical Psychology*, 71, 417–45.

Eibl-Eibesfeldt, I. (1989) *Human Ethology*. Aldine de Gruyter, Hawthorne, NY.

Fossi, L., Faravelli, C., and Paoli, M. (1984) The ethological approach to the assessment of depressive disorders. *Journal of Nervous and Mental Disease*, 172, 332–41.

Gaebel, W. and Wölwer, W. (2004) Facial expressivity in the course of schizophrenia and depression. *European Archives of Psychiatry and Clinical Neuroscience*, 254, 335–42.

Geerts, E. (1997) An ethological approach of interpersonal theories of depression. http://dissertations.ub.rug.nl/faculties/medicine/1997/e.a.h.m.geerts/. University of Groningen, Groningen.

Geerts, E. and Bouhuys, N. (2002) *Nonverbal communication in depressed patients' daily social interaction*. Paper presented at the 16th International conference of the International Society of Human Ethology, Montreal, Canada.

Gerlsma, C., Emmelkamp, P.M. and Arrindell, W.A. (1990) Anxiety, depression, and perception of early parenting: a meta-analysis. *Clinical Psychology Review*, 10, 251–77.

Geerts, E., Bouhuys, N., and Bos, E. (2000) *Nonverbal coordination in dyadic interactions between depressed patients and interviewers*. Paper presented at the 15th international conference of the International Society of Human Ethology. Salamanca, Spain.

Geerts, E., van Os, T., Ormel, J., and Bouhuys, N. (2006) Nonverbal behavioral similarity between patients with depression in remission and interviewers in relation to satisfaction and recurrence of depression. *Depression and Anxiety*, 23, 200–9.

Geerts, E., van Os, T., and Gerlsma, C. (2009) Nonverbal communication sets the conditions for the relationship between parental bonding and the short-term treatment response in depression. *Psychiatry Research*, 165, 120–27.

Gilbert, P. (2001) Evolutionary approaches to psychopathology: the role of natural defences. *The Australian and New Zealand Journal of Psychiatry*, 35, 17–27.

Gilbert, P. (2005) Social mentalities. A biopsychosocial and evolutionary approach to social relationships. In M.W. Baldwin (ed.), *Interpersonal Cognition*. The Guilford Press, New York, pp. 299–334.

Gilbert, P. and Allen, S. (1998) The role of defeat and entrapment (arrested flight) in depression: an exploration of an evolutionary view. *Psychological Medicine*, **28**, 585–98.

Gittleman, M.G., Klein, H.H., Smider, N.A., and Essex, M.J. (1998) Recollections of parental behaviour, adult attachment and mental health: mediating and moderating effects. *Psychological Medicine*, **28**, 1443–55.

Grant, E.C. (1968) An ethological description of non-verbal behaviour during interviews. *British Journal of Medical Psychology*, **41**, 177–83.

Hale, W.W., Jansen, J.H.C., Bouhuys, A.L., Jenner, J.A., and van den Hoofdakker, R.H. (1997) Nonverbal behavioral interactions of depressed patients with partners and strangers: the role of behavioral social support and involvement in depression persistence. *Journal of Affective Disorders*, **44**, 111–22.

Hamilton, M. (1967) Development of a rating scale for primary depressive illness. *British Journal of Social and Clinical Psychology*, **6**, 278–96.

Herbert, T. and Cohen, S. (1993) Depression and immunity: a meta-analytic review. *Psychological Bulletin*, **113**, 472–86.

Holmes, J. (1993) Attachment theory: a biological basis for psychotherapy? *British Journal of Psychiatry*, **163**, 430–8.

Isabella, R.A. and Belsky, J. (1991) Interactional synchrony and infant mother attachment: a replication study. *Child Development*, **62**, 373–84.

Jaffe, J., Beebe, B., Feldstein, S., Crown, C.L., and Jasnow, M.D. (2001) Rhythms of dialogue in infancy. In W.D. Overton (ed.), *Monographs of the Society for Research in Child Development*, **66** (265), 52–65.

Jones, I.H. and Pansa, M. (1979) Some nonverbal aspects of depression and schizophrenia occurring during the interview. *Journal of Nervous and Mental Disease*, **167**, 402–9.

Keller, M.C. and Miller, G. (2006) Resolving the paradox of common, harmful, heritable mental disorders: Which evolutionary genetic models work best? *Behavioural and Brain Sciences*, **29**, 385–452.

Koopmans, P.C., Sytema, S., and Giel, R. (1997) Affectieve stoornissen. In *Omvang En Gevolgen Van Psychische Stoornissen in Nederland*. Department of Social Psychiatry. University of Groningen, Groningen.

Korte, S.M., Koolhaas, J.M., Wingfield, J.C., and McEwen, B.S. (2005) The Darwinian concept of stress: benefits of allostasis and costs of allostatic load and the trade-offs in health and disease. *Neuroscience and Biobehavioral Reviews*, **29**, 3–38.

Lorenz, K. (1937) Über die Bildung des Instinktbegriffs. *Naturwissenschaften*, **25**, 289–300.

Lorenz, K. (1973) *Die Rückseite des Spiegels: Versuch einer Naturgeschichte menschlichen Erkennens*. Piper, München.

McGowan, P.O., Sasaki, A., D'Alessio, A.C., Dymov, S., Labonté, B., Szyf, M., Turecki, G., and Meany, M.J. (2009) Epigenetic regulation of the glucocorticoid receptor in human brain associates with childhood abuse. *Nature Neuroscience*, **12**, 342–8.

McGuire, M.T., Gawzy, I.F., Spar, J.E., and Weigel, R.M. (1994) Altruism and mental disorders. *Ethology and Sociobiology*, **15**, 299–321.

Meany, M.J. (2001) Maternal care, gene expression, and the transmission of individual differences in stress reactivity across generations. *Annual Review of Neurosciences*, **24**, 1161–92.

Nesse, R.M. (2000) Is depression an adaptation? *Archives of General Psychiatry*, **57**, 14–20.

Nettle. D. (2004) Evolutionary origins of depression: a review and reformulation. *Journal of Affective Disorders*, **81**, 91–102.

Parker, G. (1998) The Parental Bonding Instrument. Black Dog Institute. http://www.blackdoginstitute.org.au/docs/THEPARENTALBONDINGINSTRUMENT.pdf.

Parker, G., Barret, E.A., and Hickie, I.A. (1992) From nurture to network: examining links between perceptions of parenting received in childhood and social bonds in adulthood. *American Journal of Psychiatry*, **149**, 877–85.

Polsky, R.H. and McGuire, M.T. (1980) Observational assessment of behavioral changes accompanying clinical improvement in hospitalized psychiatric patients. *Journal of Behavioral Assessment*, **2**, 207–23.

Price, J., Sloman, L., Gardner, R., and Gilbert, P. (1994) The social competition hypothesis of depression. *British Journal of Psychiatry*, **164**, 309–15.

Ranelli, C.J. and Miller, R. (1981) Behavioral predictors of amitriptyline response in depression. *American Journal of Psychiatry*, **138**, 30–4.

Schelde, T. (1998) Major depression: behavioural markers of depression and recovery. *Journal of Nervous and Mental Disease*, **186**, 133–40.

Segrin, C. and Dillard, J.P. (1992) The interactional theory of depression: a meta-analysis of the research literature. *Journal of Social and Clinical Psychology*, **11**, 43–70.

Simpson, J.A. (2000) Attachment theory in modern evolutionary perspective. In J. Cassidy and P.R. Shaver (eds), *Handbook of Attachment*. Guilford Press, New York, pp. 115–40.

Tinbergen, N. (1952) "Derived" activities; their causation, biological significance, origin, and emancipation during evolution. *Quarterly Review of Biology*, **27**, 1–32.

Tinbergen, N. (1963) On aims and methods of ethology. *Zeitschrift für Tierpsychologie*, **20**, 410–33.

Tinbergen, N. (1974) Ethology and stress diseases. *Science*, **185**, 20–7.

Tinbergen, E.A. and Tinbergen, N. (1972) *Early Childhood Autism: An Ethological Approach*. Parey, Berlin

Trivers, R.L. (1974) Parent-offspring conflict. *American Zoologist*, **14**, 249–64.

Troisi, A. (1999) Ethological research in clinical psychiatry: the study of nonverbal behavior during interviews. *Neuroscience and Biobehavioral Reviews*, **23**, 905–13.

Troisi, A. (2002) Displacement activities as a behavioral measure of stress in nonhuman primates and human subjects. *Stress*, **5**, 47–54.

Troisi, A., Pasini, A., Bersani, G., Grispini, A., and Ciani, N. (1989) Ethological predictors of amitriptyline response in depressed outpatients. *Journal of Affective Disorders*, **17**, 129–36.

Wilson, E.O. (1975) *Sociobiology: The New Synthesis*. Harvard University Press, Harvard.

World Health Organisation (2008) Mental health: Depression. Available at: http://www.who.int/mental_health/management/depression/definition/en/.

Part 2
Evolutionary theory and the concept of mental disorder

Chapter 5

Darwin, functional explanation, and the philosophy of psychiatry

Jerome C. Wakefield

The evolution-based harmful dysfunction analysis (HDA) of mental disorder holds that a disorder is a harmful failure of an evolved function. The HDA has been widely acknowledged to have considerable explanatory power. Nevertheless, the specific relationship between evolved functions and disorders postulated by the HDA remains in conflict with some other evolutionary views of mental disorder. Moreover, the HDA's evolutionary approach clashes with culture-based accounts of the concept of disorder that emphasize cross-cultural variation and social construction of disorder categories. In this chapter, I first summarize the HDA and explain the rationale for its main features. Then, I review a selection of the various objections to the HDA that have appeared in the recent literature, either to the evolutionary emphasis in general or to the HDA's specific construal of the relation between evolution and disorder. I consider the extent to which the HDA can be defended from these objections.

In a volume about Darwinian theory's relationship to psychiatry, I can't help but observe that to me perhaps the oddest thing about philosophy of psychiatry today is the degree of anti-Darwinism there is in thinking about the concepts of biological function and medical disorder. Resistance is surprisingly fierce to identifying biological functions with naturally selected effects, and to identifying dysfunctions of the kind underlying medical disorders, including psychopathology, with failures of such naturally selected functions. This resistance is not generally from scholars who reject Darwin's theory, but rather from scholars–philosophers, and also psychiatrists and other mental health professionals—who think that the concepts of "function," "dysfunction", and "disorder" have meanings that diverge from the Darwinian interpretation.

Of course, the meanings of concepts like "function", "dysfunction", and "disorder" cannot involve direct, explicit reference to evolutionary theory. After all, Aristotle, Galen, and Harvey, as well as today's religious fundamentalists, share these concepts and share most disorder judgments with the rest of us, yet they either don't know about or reject the theory of natural selection. Rather, the more abstract notions of function and biological design, and the medical notion of dysfunction—failure of how things are supposed to work according to biological design—have been shared since antiquity. In our era, evolutionary theory has offered an explanation for the essence of biological design, so we now have a scientific underpinning for the concepts that have been understood for millennia, analogously to discovering that water—a concept shared since antiquity—is in fact essentially composed of the H_2O molecule. Due to the discovery of the essence of biological design, we now have a scientifically anchored way of conceptualizing health and disorder—although the scientific anchoring is only "in principle" when it comes to mental disorders because we know so little about the mechanisms, functions, and dysfunctions of mental processes. Or so I have argued (Wakefield 1992a,b; 1993, 1999a,b).

With the widespread resistance to this sort of picture in mind, in my initial remarks I start far from psychiatry with an encomium to Darwin in which I try to state from a conceptual perspective what Darwin achieved, and why I think there is a direct link of this achievement to the philosophy of medicine. I argue that Darwin in fact is best understood as offering a solution to the age-old problem of the origin and nature of biological functions and dysfunctions. After briefly stating my evolution-based view of disorder—the harmful dysfunction (HD) analysis—I then embark on the central concern of this chapter, which is to consider just a few of the objections to construing Darwin's achievement in the way I suggest. I focus particularly on Gert and Culver's (2004) "no distinct sustaining cause" account of disorder (based on earlier work by Clouser *et al.* 1981, 1997) as presented in a recent article in which they critique the HD analysis and specifically reject its evolutionary account of disorder, and Lennart Nordenfelt's (2003) extended critique of an evolutionary view of disorder. In each case, I explain why I think the Darwinian approach not only survives the respective objections but is superior in explanatory power to any offered alternative.

5.1 Functional explanation: Aristotle, Lucretius, Darwin

5.1.1 Aristotle and the mystery of biological functions

Darwin's (1964) discoveries articulated with the tradition of thinking about biology that came down from ancient philosophy. A central problem in philosophy of biology, especially since Aristotle–indeed, *the* central mystery at the heart of biology (aside from the existence of life itself, but these two puzzles ultimately virtually coincide)—was the mystery of the design-like, beneficial, complexly structured adaptive features of organisms. The Greeks already had the concept of a natural or biological function as one subcategory of a broader concept of function (*ergon*) that, like our own general term *function*, applied to social roles and artifacts as well as biological structures. What is common to these phenomena is that various processes seem shaped for their ends. When a thing's nature must be explained by reference to what it is shaped to do, that is commonly referred to as *teleology* or *teleological explanation*.

The fact that artifacts and social roles have features that are shaped for their ends is not all that mysterious because human intentions that represent the end can be influential in shaping the means in both cases, so the source of the teleology can be readily identified. Note that, contrary to a widespread misunderstanding, teleology in itself, construed as a generic explanatory notion expressed by certain uses of the term "function", is not necessarily mysterious or mystical, as the artifact and social-role instances reveal. Biological functions, however, pose an explanatory challenge of the highest magnitude, one that has frequently led in desperation to the embrace of mysticism. The eye can't just accidentally enable us to see, any more than it is just a happy accident that clothes fit our bodies or that violinists are capable of producing music that we enjoy, but how could the usefulness of the ability to see have shaped the structure of the eye, given that human intentions were plainly not involved in the eye's construction, unlike clothes and violinist's skills?

Because there have been many mystical and theistic accounts of the source of biological teleology, teleological explanation has an unscientific ring to it for many, and is often dismissed out of hand. However, it is not teleology per se but rather the specific mystical or theistic explanations of teleology that are unscientific. The basic problem—namely, that some biological features seem to require explanations that refer to their effects as part of the explanation for why they are the way they are—is a real problem that cannot be dismissed and is not inherently unscientific. Unless seeming teleology can be explained away as an illusion, the realities of the biological facts pose the challenge of formulating a scientifically adequate account of such teleology, and thus vindicating the intuition that teleological explanation is a legitimate form of explanation that is especially appropriate to the biological sciences.

Aristotle's (1971, 1989) characterization of biological functions as effects that are organized and prompted by "final causes" was an attempt to address, or at least describe, the central mystery of how it is that, say, acorns turn into oaks, the eyes enable us to see, the hands enable us to grasp, thirst pushes us to drink the water we need, fear keeps us away from danger, and so on, where the sources of the apparent teleology are anything but obvious, yet teleological explanation seems required. These mechanisms are too well orchestrated and their effects too beneficial and unlikely to have come about accidentally for non-teleological mechanical explanation to suffice. The mechanisms involved must, it would seem, have been shaped by the very beneficial effects they have, just as social roles, artifacts, and individual actions are shaped to accomplish certain ends. But how to explain this shaping without human or divine intentional action or the assumption of some mysterious kind of teleology infused throughout the universe?

Aristotle's "final causes" are explanatory of the mechanisms and processes that lead to them. When Aristotle defines the four types of cause in *Metaphysics*, final causes are modeled on examples of human intentionality:

> 'Cause' means...(d) The same as 'end'; i.e., the final cause; e.g., as the 'end' of walking is health. For why does a man walk? 'To be healthy,' we say, and by saying this we consider that we have supplied the cause.
>
> (Metaphysics 1013a)

However, although presented through the example of intentional action, the notion of a final cause as an effect that explains the mechanisms that give rise to it is a generic notion not inherently tied to intentionality that also applies, for example, to the biological development of creatures such as oak trees and human beings:

> For it is that which is yet to be—health, let us say, or a man—that, owing to its being of such and such characters, necessitates the pre-existence or previous production of this and that antecedent; and not this or that antecedent which, because it exists or has been generated, makes it necessary that health or a man is in, or shall come into, existence.
>
> (*Parts of Animals* I.1)

The great mystery in the biological realm then, is how it can be that there is apparent teleology that requires explanation by final causes—where the means needed to yield the end come into existence in a way that requires explanation in terms of the end—and yet there is no apparent intentionality involved:

> This is most obvious in the animals other than man: they make things neither by art nor after inquiry or deliberation. That is why people wonder whether it is by intelligence or by some other faculty that these creatures work—spiders, ants, and the like... It is absurd to suppose that purpose is not present because we do not observe the agent deliberating. Art does not deliberate. If the ship-building art were in the wood, it would produce the same results by nature. If, therefore, purpose is present in art, it is present also in nature.
>
> (*Physics* II.8)

Aristotle's postulation of final causes did not go very far towards solving this mystery. Indeed, it could be considered just a relabeling of the problem along with some speculative hypothesizing of an unknown solution. But it does clearly reveal the nature of the mystery: How do we explain the shaping of a mechanism for an end—that is, quasi-design—without a designer?

5.1.2 Lucretius on natural selection

Thinking about how to construe what Darwin distinctively contributed to the resolution of this traditional puzzle becomes a conceptually interesting exercise once one realizes that natural selection as an explanation for the remarkable adaptive traits of organisms is an idea that has been around since ancient times. For example, the Roman poet-philosopher Lucretius, in his *De Rerum Natura* (*On the Nature of Things*) (1992; see also Campbell 2003), was inspired by the work of the Greek philosopher Epicurus, who in turn adopted Democritus's atomistic and mechanistic approach to scientific explanation. Lucretius thus attempted to replace Aristotle's final causes with a mechanistic explanation of teleological phenomena. Selection was the mechanism that enabled him to provide a cogent explanation of adaptive traits without bringing mysterious final causes or some predestined overall adaptive essence in the universe into the story.

Lucretius, following Empedocles's doctrine that all biological features are due to chance, believed that all organisms were originally generated spontaneously from the earth, randomly coming into existence in all possible structures and shapes. (Phenomena like the growth of bacteria and mold, and even the appearance of insects seemingly out of nowhere in food were often cited in ancient times to support the notion of spontaneous generation.) Once randomly created, the species with the adaptive traits were the ones that survived, according to Lucretius—so the effects of their features on survival and reproduction explained why the species was one of the few that continued to exist.

However, Lucretius, like almost all other biological theoreticians before Darwin, assumed rigid species essences—that is, essential features that constrain species reproduction over time and keep the species in certain respects the same and unmalleable. Species seem so different and so differently adapted (the continuities between them, we now know, are hidden by the pruning of natural selection) that this view of rigid species essences must be considered the "commonsense" view, comparable to the view in astronomy that the sun must circle the earth. Aristotle appears to have held that species essences along with reproductive equipment that duplicates the essence in new organisms are somehow themselves determined by some teleological principle, although this remains obscure. Lucretius, taking rigid species essences as a given while allowing them to come into existence by chance, assumed that each spontaneously generated species breeds true, if perchance it happens to have the equipment to breed at all.

Although Lucretius (and other classical writers) already understood the Darwinian insight that the adaptiveness of organisms must be due to the selection of those with advantageous traits, he lacked two crucial Darwinian notions. First, he did not understand the importance and extraordinary subtlety and variety over time of spontaneous variation within a species. Second, he failed to grasp the consequent malleability of species over time, and the potential for the gradual transformation of a species' basic adaptive features over time as a result of natural selection working on spontaneous variations, even until a new species might emerge.

Lucretius fell short, then, in holding that the competition is between fixed types of organisms with ordained packages of traits determined by an unmalleable species essence (that is, by a species essence that could not change over time within the lineage into some other essence) rather than—as in Darwin—between varying traits themselves that formed a malleable thus potentially transitory species essence within the lineage. By conceiving of the scope of natural selection in a radically extended way that potentially encompassed all of the organism's features, Darwin was able to explain the "origin of species" and the selectionist thesis was able to penetrate to the very core of a species' nature. Darwin thus solved the ancient mystery of biological adaptation, adding it to the list of teleologically explainable processes once random mutation and natural selection replaced spontaneous generation as the basic originating mechanism. Whereas before Darwin each species' unique set of basic adaptations remained mysterious, after Darwin the origins of the full adaptive structure of each species could in principle be comprehended as shaped by selection for their effects.

5.1.3 Darwin and teleology

Darwin's insights directly addressed the problem with which Aristotle had wrestled. Darwin rejected all the mysticism associated with explanations of biological teleology in terms of divine intervention or vital principles or even final causes as active, efficient causes. However, he accepted that the challenge was to explain teleology—namely, the fact that features of the organism are clearly in some sense shaped to accomplish their adaptive ends.

Darwin himself appears to have been aware of this link of his theory to the philosophical tradition. When Asa Gray, an eminent Harvard botanist, published a brief appreciative comment on Darwinism in *Nature* in 1874, he noted "...Darwin's great service to Natural Science in bringing back to it Teleology: so that instead of Morphology versus Teleology, we shall have Morphology wedded to Teleology." Darwin responded in a letter to Gray that "What you say about Teleology pleases me especially and I do not think anyone else has ever noticed the point" (both quoted in Lennox 1993, p. 409). And indeed, Darwin (1964) occasionally lapses into explicit teleological language (e.g., final causes, ends) in *Origin of Species*. Whereas Lucretius thought that his mechanistic explanation by selection completely eliminated teleology, Darwin saw more accurately that teleology was an essential problem in biology having to do with a distinctive form of explanation, and that selection offered simultaneously a mechanistic (i.e., divorced from mystical or divine or vital causes) and a teleological explanation of adaptation and natural functions. In the conversation stretching from Aristotle to Darwin, "function" is an inherently teleological notion, and the power of the selection account is that it is teleologically explanatory of features by their effects, and yet mechanistic.

For those who steadfastly identify teleology with literal intended purposes represented in an agent's mind, another term, *teleonomy*, has been coined for teleology-like processes in biology to avoid any hint of association with theistic and other mystical, non-scientific explanations of such animal behavior. However, I am using the term "teleology" to refer to a form of explanation that need not be intentional or agentic and can encompass teleological processes that are ultimately mechanical in nature.

The essential explanatory puzzle posed by function attributions within biology has tended to get lost in recent debates over the concept of function. Analyses of "function" created for quite other reasons have been appropriated into the discussion of biological function without adequate attention to the problem of functional explanation. For example, the fact that a feature, via its function, has a characteristic beneficial contribution to a containing system does nothing to explain the feature via the beneficial contribution. Thus, Cummins's (1975) "functional analysis" account of functions, in which a function is just an effect of interest that a feature has on a larger containing system, was originally created to address the nature of functionalism in philosophy of mind, where causal relations among brain states are claimed to determine their content. Such an account of function does not attempt to elucidate how functional relations are explanatory of those causal relations, and so is irrelevant to the teleological puzzle in biology.

A second example is Christopher Boorse's (1975, 1976, 1977) influential analysis that places cybernetic accounts of goal-seeking at the heart of the account of natural biological functions. A common form of biological teleology involves an organized system of reactions and adjustments shaped to achieve or preserve a certain goal, such as temperature homeostasis in warm-blooded animals, that is, where the end (e.g., maintaining a certain body temperature) must be invoked to explain the nature of the system's behavior (e.g., sweating). However, if one is going to satisfy the explanatory component of function descriptions, then the fact that a feature contributes to a goal at a species-typical level is not enough as an analysis. The fact that a feature *contributes* to a goal in the cybernetic sense, while interesting in its own right, does not imply that its contribution to the goal *explains* why the feature exists. Nor is it enough to say that survival and reproduction are the ultimate goals to which features of the organism contribute. To elucidate the explanatory notion of function one must show how the concept somehow requires that the ends of survival and reproduction shape and explain the existence of the systems that contribute to them, a task that goes beyond the description of sheer contributions in cybernetic terms.

5.1.4 Black box essentialist account of function

One thing that is clear is that nobody, ancient or contemporary, actually means "naturally selected effect" by "biological function". After all, people can disagree about this equation, and religious fundamentalists certainly do disagree with evolutionists. Plus, many people around the world—or in the past (including Aristotle and Augustine)—have or had no notion of evolutionary theory, yet understood that the function of the eye is to see, and understood disorder concepts that emerge from function concepts. Remember, Hippocrates got it almost invariably right, judged by our standards, when he labeled a condition as a disorder, even though we share almost nothing in the way of common scientific knowledge about underlying mechanisms. So, there must be some more abstract concept of "function" and thus "dysfunction" that people share, and this shared concept allows them to disagree about the theoretical origins of functions. The concept, then, has to be neutral between, for example, Darwinian and theistic accounts of the origins of biological teleology. Just as scientists were able to argue over whether fire is phlogiston liberation or oxidation only because they shared a theory-neutral concept of "fire" along the lines of "whatever is essentially like that phenomenon in the fireplaces and forest fires", so we must understand the concept of "function" in a theory-neutral way that allows people to disagree about whether functions are due to natural selection or God's intentions or final causes. Darwin discovered the nature of biological functions, but one needs to identify a mediating concept that picks out the phenomenon of which he found the nature.

I identify such a mediating concept in my "black box essentialist" (Wakefield 1999b) account of "function". I am not going to reiterate or defend that account here, except to say: "function" was originally identified by reference to certain clear cases of remarkable, obviously non-accidental, and quite beneficial capacities and other effects imparted by bodily and psychological features, such as the hand being able to grasp, eyes enabling us to see, thirst prodding us to obtain the water our bodies need, fear keeping us away from danger, and so on. The origin of such beneficial, complexly structured, design-like effects—effects that the feature itself was clearly somehow shaped to cause—was a mystery, so the concept of "biological function" could only be framed in a quasi-essentialist manner, as follows: a function of an organismic feature is any effect of the feature that is explained by the same sort of essential process or processes as the base set of design-like, complexly structured, beneficial effects that somehow explain the feature that causes them. Darwin provided the best scientific theory we have of what that essential process is, thus (if he is correct) explained what functions are.

5.1.5 From philosophy of biology to philosophy of medicine

So much for philosophy of biology. Now, what about the link to philosophy of medicine/psychiatry? At about the time that ancient philosophers from Plato through to Lucretius were worrying about the explanation for the existence of biological functions within the life sciences, a parallel and equally momentous development was occurring within the professions, namely, the establishment, initially manifested most notably in the writings of Hippocrates and Aristotle through to Galen, of medicine—including a nascent psychiatry—as an independent profession and quasi-scientific intellectual discipline that drew on the biological sciences. The concern about teleology in biology and the birth of medicine as a formal intellectual discipline at about the same time makes sense. Medicine was interested in the ways that biological functioning went harmfully wrong in humans (and in some animals humans wanted to be healthy) and thus was firmly rooted in the biological theory of natural functioning. So, for example, if we are biologically designed to have a balance among four humours as a function of some unknown design features, then a disorder, such as depressive disorder, will be due to a dysfunction that involves an imbalance of the humours, such as an excess of black bile.

Thus, simultaneously with solving the 2500-year-old biological mystery of what explains the nature of species and their typical adaptive features, and solving thereby the 2500-year-old mystery of the nature and origins of biological functions that seem to be design-like, Darwin was also solving the 2500-year-old medical question of the ultimate nature of biological dysfunctions. If medical disorder is the failure of some form of order, what is that form of order? The Darwinian answer is that the order that fails in medical disorder is the order imposed by natural selection—so disorders are what might be called evolutionary dysfunctions.

Or so I believe. But many philosophers of psychiatry and medicine and biology don't believe one or another part of this story. So many, in fact, and with such diverse opinions as to why this story of Darwin's triumph simultaneously in both philosophy of biology and philosophy of medicine must be incorrect, that I can't possibly begin to address even a small proportion of them. But even a rebuttal of a thousand critics begins with single interlocutor, so in the remainder of this chapter, after a presentation of my own evolution-based HD analysis of the concept of disorder, I select some initial anti-evolutionary philosophers of psychiatry that happen to have come to my attention recently due to their objections to the HD approach, and show why they are wrong—or at least nowhere near as right as an evolutionary view. I think in grappling with the critics, the nuanced power of the evolutionary account of disorder emerges, so although the focus here on a few selected thinkers may seem arbitrary, the insights that emerge have a broader power, I hope.

5.1.6 The harmful dysfunction analysis

So, what does the term "mental disorder" mean? I argue that a disorder is a harmful dysfunction (Wakefield 1992a). Thus, two criteria are necessary (and jointly sufficient) for a condition to be a disorder. One is a factual biological criterion, the other a value criterion (hence this is a "hybrid" analysis).

By "dysfunction" I mean a failure of some internal mechanism to be capable of performing a function for which it was evolutionarily designed. For convenience, such failures can be referred to as evolutionary dysfunctions. Because this article is focused on Darwin and the criticisms of the evolutionary component of the HD analysis, I do not address the value component of the HD analysis here and why I think Darwin is not quite enough for philosophy of medicine, and values too are necessary. Some evolutionary psychologists attempt to account for disorders in terms of selection for the disorder itself, which has hidden adaptive aspects. The HD analysis is incompatible with such explanations, and predicts that if what is now considered a disorder is shown to be a selected feature, then our intuitions would change and we would come to consider it a non-disorder, re-conceptualizing it as a normal variation—as has happened with fever, for example.

The fact that Darwinian theory is the best theory we have of the essence of functions does not mean that our pretheoretical intuitions about disorder, dysfunction, and function will necessarily fit Darwinian theory perfectly. They should fit in clear, prototypical cases. However, once a theory accounts for an essence, the theory itself may have complexities that are unanticipated by the conceptual structure and may pose challenges to that structure.

5.2 Culver and Gert on distinct sustaining causes

In this part of the chapter I consider a prominent view of the concept of disorder that was put forward prior to the appearance of my 1992 article on harmful dysfunction. It should have been addressed in that original article, but was not. It is also a view that is not only critical of my evolutionary approach but has a constructive alternative account of the distinction between disorder and non-disorder to offer—and one that is also a hybrid fact/value account.

5.2.1 Culver and Gert's hybrid account of dysfunction as harm with no distinct external sustaining cause

Clouser *et al.* (1981, 1997; see also Culver and Gert 1982) present what might be considered one of the few hybrid fact-value accounts of disorder other than mine—as harm (a value criterion) caused by an internal condition for which there is no distinct sustaining cause (DSC; a factual criterion; see below). They define "malady", a term they use to cover all medical disorder categories such as "injury", "illness", "sickness", "disease", and so on (1997, p. 177), but I will use my term "disorder" and I will consider their analysis as an attempt to define medical disorder, which appears to be their intention.

Clouser *et al.* (1997) define malady/disorder as follows:

> Individuals have a malady if and only if they have a condition, other than their rational beliefs or desires, such that they are incurring, or are at a significantly increased risk of incurring, a harm or evil (death, pain, disability, loss of freedom, or loss of pleasure) in the absence of a distinct sustaining cause.
>
> (Clouser *et al.* 1997, p. 190)

A DSC can be any environmental situation outside the individual. These authors offer a non-evolutionary account of the concept of dysfunction as any feature internal to the individual that causes the individual harm (other than rational beliefs and desires), in the absence of a distinct (external) sustaining cause.

In a recent discussion of the DSC view by Gert and Culver, they defend at length their claim that their account is superior to and more explanatorily powerful than the HD analysis's evolutionary approach. It is primarily this article and thus these authors on which I will focus. They dismiss an evolutionary view on the grounds that some undesirable conditions might nonetheless be the result of natural selection, and assert that their own view better captures the notion of dysfunction:

> Perhaps nature designed people to deteriorate and die to allow for the species to develop. Regardless of nature's design, if the person is suffering or at a significantly increased risk of suffering death, pain, disability, or an important loss of freedom or pleasure, and there is no distinct sustaining cause, he has a dysfunction.
>
> (Gert and Culver 2004, p. 420)

5.2.2 The basic idea of the DSC view, and *prima facie* counterexamples

The basic idea of the DSC view is easy enough to understand. There is an old saw in medicine that when a patient comes in for pain in his foot, first look to see if there is a pebble in his shoe—if there is, of course that wouldn't be a medical disorder, but a different kind of problem. The DSC account of disorder essentially elevates the "pebble in the shoe" example of a cause outside the patient that is responsible for sustaining the patient's distress into a universal account of the concept of disorder.

Note that the cause must sustain the harm for the condition to be a non-disorder according to the DSC. For example, after a while the pebble in the shoe might cause a lesion or inflammation that is autonomous of whether the pebble continues to press against the foot. That would then be a pathology according to the DSC because, although the pebble was the external cause of the harm, it did not continue to sustain the harm.

This simple idea does correctly classify a considerable number of cases, for it is true that many non-disorder kinds of problems are due to problematic environments. This makes sense because many mental mechanisms are biologically designed to respond to the environment, and thus a response, even if distressful or otherwise problematic, to an ongoing environmental situation is often normal. For example, anxiety due to an ongoing real threat is not a disorder because anxiety is designed to occur under such external circumstances, and a child who acts antisocially out of self-protection in a threatening urban environment in which one must join a gang to survive need not be disordered. So, the Gert–Culver proposal may initially look quite intuitive. Bengt Brulde (personal communication 2009) tells me that he illustrates this view to his classes by noting that if someone has trouble breathing at an altitude of 10,000 feet, but then can breathe okay when brought down to 3000 feet, one assumes it was not a lung disorder but an environmentally caused reaction.

On a more careful look, there are lots of serious *prima facie* problems with this approach, if taken at face value. Just to take Brulde's example, we can ask: is it really the DSC or some more complex implicit theory of what is biologically designed that guides our judgments in this example? To test this, imagine that someone has trouble breathing at sea level, but can breath okay when in a high-pressure barometric environment (the equivalent of being, say, 1000 feet *below* sea level). Such people are commonly thought to have a disorder, yet the harm is every bit as DSC-dependent as in Brulde's example. The reason our judgment changes is obvious: we think people are biologically designed to be capable of breathing okay at around sea level (modulo some complexity introduced by respiratory adaptation to high elevations). So, the DSC view is not as simple or as adequate as it initially looks.

The fact that the DSC is in fact a superficially appealing approximation to a superior evolutionary account emerges quite clearly as Gert and Culver are forced to make one *ad hoc* amendment after another to their initial view in order to preserve its plausibility. These amendments may help keep the view from being falsified, but they are only explicable if placed within an HD-like context of biological design.

5.2.3 **Failure of the "rational beliefs or desires" clause to eliminate abundant *prima facie* counterexamples**

Why do Gert and Culver modify their formula by eliminating from dysfunction status any harms caused by the agent's rational beliefs and desires? The addition of the "rational belief or desire" clause is Gert and Culver's strategy for addressing a tidal wave of obvious counterexamples to a simple formulation of the DSC view. If any harmful (or harm-risking) internal state without a DSC is to be considered a disorder, then every instance of harmful ignorance, false belief, unsatisfied desire, desire for risky activities, or overly ambitious aspirations could be a disorder—and so on, for any decision or act or feature with a downside.

It is not entirely clear exactly what Gert and Culver mean by this clause, but if, as seems plausible, the clause means "rational" in something like the sense of "based on reasonable evidence and instrumental reasoning in light of individual preferences", then the problem is that non-disordered desires and beliefs are sometimes irrational. Indeed, this clause offers a good test of the DSC view against the HD analysis because the two accounts diverge. A proneness to possess many irrational beliefs and desires is built into us biologically, and when such irrational beliefs or desires occur and are potentially harmful, the DSC suggests they should be seen as disorders, whereas the HD view says they should not be. In fact, such beliefs and desires are not considered disorders.

For example, the exaggerated sense of the virtues of a lover or child can often lead us to make miscalculations ("love is blind"), sexual desire for those other than our partners can lead to violation of social norms and destruction of valued relationships and of reputation, human group loyalty can lead us to our deaths in wars, and the programmed human taste preference for fat and sugar can lead us in our food-rich environment to eat in a way that increases the risk of heart disease. Yet none of these risk-increasing internal conditions are considered disorders in and of themselves because they are judged to be part of how we are biologically designed.

Aside from such counterexamples based on biologically shaped irrationalities, there are many internal states other than beliefs and desires that increase the risk of harm and are independent of any external sustaining cause, yet are not considered dysfunctions or disorders. These are not touched by the "rational belief or desire" clause. Lack of talent, for example, causes loss of pleasures that would have accrued due to the realization of one's ambitions.

(Gert and Culver seem at one point to flirt with accepting the implication of their view that extreme ambition for the realization of which one has inadequate talent is a disorder, but this misclassification of the stuff of a normal life would be a *reductio ad absurdum* of their account, in my view, except in the unlikely event that it could be shown that there is some biologically designed psychological tendency to adjust one's ambitions to one's talents even when one has misjudged one's talents). Similarly, a great variety of normal-range personality and temperamental traits (e.g., greater than average shyness, lower than average IQ, unkempt appearance, temperamental nature) can cause pain and decrease pleasure yet are normal variations. Just being short in stature due to being on the extreme end of the normal curve of distribution of height (*not* due to hormone deficiency) would presumably be a disorder on the DSC account, given that we know that short statue is associated with less success in life. So, both within the belief–desire system, and in features other than beliefs and desires, there is a vast array of harmful internal features without DSCs that are not disorders, each one a counterexample to Gert and Culver's analysis. Placing the system of rational beliefs and desires in the "non-disorder" category, even when harmful, is a safe bet, according to the HD analysis. This system is a biologically designed feature that is characteristic of human beings, recognized in Aristotle's dictum that human beings are "rational animals".

5.2.4 Instability of the distinct sustaining cause intuition

To return for a moment to the prototypical pebble-in-your-shoe type example of the intuition behind the DSC, consider a slight variation: What if your foot hurts when and only when you walk on it, with no pebble in your shoe? Just as with the pebble's impingement on the foot, the pressure of the pavement now becomes a DSC of the harm—the pain—and yet, contrary to what Culver and Gert's view implies, surely this *would* be judged evidence of a likely disorder. In the mental health domain, we do have situational disorders in which a specific environmental feature triggers an emotionally disordered reaction—even to the extent of a situational psychotic disorder.

Indeed, the pebble-in the-shoe example only works as a non-disorder because we understand that pain receptors are biologically designed to yield pain sensations when something impinges on the skin in the way and with the level of pressure that a pebble impinges on a foot in a shoe when one walks on it. To translate the pebble example to another part of the body (in case the pebble is considered essential to the non-disorder judgment), what if a smooth pebble resting lightly in the palm of one's hand caused severe pain? This too would surely suggest disorder, the pebble being a DSC of the pain notwithstanding, because the pressure exerted by a pebble in the hand rather than in a shoe is not believed to be within the biologically designed functioning of human beings that should trigger a pain response. In contrast, we believe we are biologically designed so that walking should be relatively free of pain, and that it must be a biological failure of some kind if there is pain just from walking.

So, we have three cases of DSCs: pain when walking due to pressure of a pebble in a shoe (no disorder), pain when walking due to pressure of the ground (possible disorder), and pain when a pebble is held resting loosely in the hand (possible disorder). All three examples involve external sustaining causes of the harm (pain), yet two would suggest disorders. These examples, I would argue, show that it is not the existence of a sustaining cause, but the failure of biologically expectable design parameters that indicates disorder. The adequacy of the DSC view thus seems to collapse with just a little prodding of the examples that make it seem appealing. In fact, these examples suggest that there is an implicit background judgment of what is biologically natural that undergirds Culver and Gert's account and makes it appear plausible only as long as it is consistent with the evolutionary account.

The DSC account is consistent with our intuitions about disorder some of the time and not others, and both the successes and failures are explained by the harmful dysfunction analysis's evolutionary account. It is an illusion that the existence of a DSC in and of itself has anything logically to do with the presence or absence of dysfunction. A DSC suggests non-disorder when it offers an explanation of harm that is consistent with biologically designed functioning (e.g., pebble in shoe causing pain, inability to breath at 10,000 feet), but is quite consistent with disorder when such a cause is not understandable as part of biological design (pebble resting lightly in hand causing pain, pressure of sidewalk when walking causing pain, difficulty breathing at sea level).

5.2.5 Retreat to the statistical criterion and why it does not save the distinct sustaining cause view

Gert and Culver are quite aware that there are conditions with external sustaining causes that are apparent disorders, thus are apparent counterexamples to a simple DSC formulation. They themselves offer the examples of a phobic response to some feared object in which the intense anxiety occurs only when the object is present, and allergic reactions that are sustained by the allergen's presence in the air. The way they deal with these seeming counterexamples would also apply to the pebble and altitude examples (as well as the situational psychotic reaction) noted above, so might show these initial worries to be misplaced. This further amendment to their view (not presented in their usual formal statement of the DSC criterion) therefore merits careful examination.

Their answer is basically that there can be DSCs of disorders after all, as long as the response to the external cause is *not* a universal human trait. That is, *atypical* harmful responses to DSCs are exceptions to the DSC rule and are to be classified as disorders.

Thus, Culver and Gert in effect transform their DSC criterion into a statistical criterion. They presume that such atypicality is the case for phobias and allergies, and we may presume it is also the case with respect to the hypothetical examples above of painful responses to walking or to having a pebble rest lightly in one's hand—unlike the universal pain reaction to walking with a pebble in one's shoe. So, invoking statistical atypicality of the response to a DSC as indicating disorder, and typicality of response as indicating non-disorder, does help to properly classify many conditions that otherwise would be counterexamples to the DSC view.

One might construe this turn to a statistical criterion as a way to mimic an evolutionary approach, for the species-typical responses to external causes are likely to be biologically designed responses. If so, that offers a formula for two kinds of possible counterexamples to the revised statistical version of the DSC view if the HD view is correct: (1) when species-typical responses to external causes are not biologically designed, they may still be disorders, and (2) when atypical responses are part of biological design, including normal variation, then they may still be non-disorders.

Both kinds of counterexamples are readily available. First, the fact that a sustaining effect is universal in human beings does not imply that the reaction is normal. If toxic fumes of a certain kind cause almost everyone to experience irregular heartbeat, weakness, and faintness—or psychotic ideation—for as long as the fumes are present, that is still a disorder. For that matter, if there is a pollen to which everyone is allergic, that allergy is still a disorder, and if there were a situation so horrific that it typically induced a situational psychosis, that too would be a disorder despite the typicality of the response—just as, despite the fact that a certain pressure applied to a bone will typically cause it to break, that expectable outcome is still a disorder because bones were selected for structural support and integrity, not breaking. Indeed, one such example in the mental realm is the calculation by military psychiatrists that almost every soldier undergoing combat conditions will suffer a breakdown after a certain number of days of continuous combat, and the fact that pilots of combat aircraft typically experience a mental breakdown after a certain number of sorties. Yet these are sustaining causes—taken out of the theater of war, these individuals quickly recover and can be sent back into battle.

Second, even atypical responses are excluded from diagnosis when seen as part of the range of naturally selected variation in the workings of our various mechanisms. This is true even in the extreme case of what might otherwise be considered psychotic symptoms of schizophrenia. For example, hallucinations involving hypnagogic imagery when falling asleep, or hallucinating that a recently deceased and yearned-for loved one is appearing to one, are not diagnosed as disorders, even though both are not so common as to be typical responses.

Atypicality of outcome in response to a DSC, while sometimes epistemologically useful in evaluating whether or not a condition is a disorder, is simply not a viable part of the concept of disorder. In some individuals, running can be a DSC of cardiac pain (angina is sometimes defined as "chest pain that comes with physical exertion and eases with rest"), and in other individuals running can be a DSC of pain in the abdomen (because they are pregnant). Neither of these are universal effects, yet in one case there is a disorder, in the other case not.

5.2.6 Normal inability versus pathological disability

The basic problem for the DSC view is that it does not provide a fine-grained enough account of disorder. Even if it correctly eliminates from the disorder category those conditions that are harmful due to either rational beliefs and desires or DSCs with atypical responses, that still allows the enormous domain of internal harmful conditions to be categorized as pathological, even though many such conditions are normal. The DSC view has no resources for drawing the needed distinctions. Even on so basic a question of how to distinguish a normal inability from a pathological disability, the view falters because it cannot refer to what is biologically designed.

If one allows the fullest scope to the notion of harm, then we all have an unlimited number of disorders according to the DSC view because we have endless features that could be better and thus confer harm. The fact that we are unable to fly, for example, is a harmful effect of our bodily structure, thus we must all be suffering from a flying disorder. It is easy to deal with this sort of problem within an evolutionary framework: dysfunctions are only failures of abilities that we are naturally selected to possess. For the DSC view, however, this becomes a major problem.

In the face of this problem, Culver and Gert retreat once again to a statistical criterion for what is a normal ability. Using their statistical approach, they try to make the cut between those harmful inabilities that are normal and those that are disordered, as follows:

> That humans cannot fly is a clear example of an inability. No humans can fly. Further, there are some extraordinary abilities that very few humans have, but that does not mean that the rest of us, who do not have these abilities, are therefore disabled. It is at this point that one must look to what is the norm for the species. The lack of an ability to run a mile in four minutes is an inability rather than a disability. Even though there are a handful of humans who can actually run that fast, it is so far from the norm that there is no question that a person is not disabled because he or she cannot do it. Labeling is more difficult with more normal abilities. Just how far should one be able to walk or run...
>
> (Clouser, Culver, & Gert, 1997, p. 191)

It is true that statistical frequency can sometimes be useful as a guide to what is probably biologically designed, thus to normality versus disorder, but not everyone who is statistically toward the bottom of some distribution (e.g., athleticism, beauty) is disordered, even though there might be extremes that do represent failure of the basic functions along a dimension. Conceptually, the reason why inability to fly is not a disorder among humans is not because none of us can fly—that conceivably could be due to some sort of universal virus or toxic agent that, thalidomide-like, has for millennia prevented the development of our biologically designed wings. So, what is average can itself be a disorder—as would be the case, for example, if there was general lead poisoning in the human race over a period of time that lowered IQ (as some have postulated was the case among the aristocracy in the Roman Empire due to lead in the vessels used to serve wine). The reason is rather that human beings are not biologically designed to fly.

The statistical approach to disability versus normal inability also does not work because of the statistical unusualness of some normal abilities and thus the possibility of disorder that pathologically lowers the level of ability in those with average abilities. According to the HD analysis, it is not the statistical normality of a level of ability but rather the causal process that leads to the level of ability that determines whether or not there is a disorder. For example, if your child has lost 20 points off her IQ due to lead in the paint in your home, is that instance of lead poisoning any the less a disorder if she is a naturally highly gifted child and thus after the harmful effects is still somewhat above average? Culver and Gert's view implies that there is no disorder in such an instance because their statistical approach to inability does not allow that an average person could be experiencing pathological disability, yet there clearly is a disorder in such a child because there is a dysfunction (the lead acting on brain tissue) that is causing a harm consisting of the lowering of her ability. In summary, the concept of disability itself, as linked to disorder, is one that requires the notion of evolutionary dysfunction and cannot be defined statistically.

5.2.7 Concluding comments on Culver and Gert

The DSC account of disorder is appealing because it is often the case that DSCs signal a normal reaction rather than a biological dysfunction. However, the DSC as a stand-alone criterion is grossly overly inclusive regarding disorder. The counterexamples occur whenever the DSC view deviates from the evolutionary view. The way Culver and Gert handle the tidal wave of counterexamples is to add various additions to the DSC view. Obviously many of the routine beliefs and desires of human beings create greater harms or risks of certain kinds, which would be seemingly clear cases of disorder according to the DSC, so Culver and Gert add a stipulation that a condition is not a disorder if it is part of the system of rational beliefs and desires of human beings. Obviously, there are harms maintained by DSCs that are disorders, such as allergies and phobias, again offering seemingly definitive counterexamples, so Culver and Gert add a stipulation that if the DSC does not have the same effect universally in all human beings, then the condition is a disorder anyway. Aside from the fact that one can construct further counterexamples that require new stipulations, the basic problem is that these amendments are *ad hoc* in the sense that there is no explanation of why they emerge in any coherent way from the DSC. Rather, they are plausible claims on intuitive grounds quite independently of the DSC because they are consistent with the HD analysis. The net result of adding these clauses is that the DSC becomes more of an approximation to an evolutionary view, and where it still deviates, further counterexamples crop up.

5.3 The designed-defense objection

5.3.1 The objection that biologically designed defenses can constitute a disorder

Evolutionary psychiatrists commonly distinguish between biological *defects*, that is, failures of some mechanism to be capable of performing as evolutionarily selected, which I have labeled "dysfunctions", and biologically designed *defenses* that can themselves be very unpleasant and even harmful in various ways as the body or mind sacrifices other normal functions in a biologically designed way to defend itself against some threat (e.g., vomiting in response to ingested toxins, sneezing in response to dust in nasal passages, fever in response to an infection). An evolutionary account of disorder implies that biologically designed defensive reactions in and of themselves, despite the associated discomfort, are not disorders if they perform as they were biologically designed to perform and do not unduly disrupt other functions that they were not designed to override. (Of course, defenses, too, can break down in various ways and become disordered, as, for example, in an auto-immune disorder or out-of-control fever of unknown origin, and these are true disorders.) Relying on this defect/defense distinction, Tengland (2001) offers an objection to the HD analysis as follows:

> Let me show what the problem with the concept of dysfunction is. The most typical illness or disease we come across is a viral infection. In this disorder there seems to be no dysfunction. On the whole the body seems to function optimally, all the "defense mechanisms" are utilized in order to fight off the invasion of micro-organisms. The symptoms which these defense mechanisms give rise to are fever, a sore throat, a running nose, etc. These mechanisms were selected through evolution. And, since they were selected because they have a function to perform we must according to Wakefield conclude that no infection, of whatever sort it may be, is a disorder. However, even though we have no dysfunction, I would still claim that an infection is paradigmatically a disease. Thus, Wakefield's theory fails to take all diseases into account.
>
> (Tengland 2001, p. 86)

The objection is: very often, all the unpleasant symptoms of a viral infection are manifestations of the workings of biologically designed defenses against the virus; thus the entire disorder consists of defenses; thus, there is no dysfunction, yet there is a disorder; thus, a disorder can consist of biologically designed processes and need not involve evolutionary dysfunction.

In all fairness, it should be said that I have picked Tengland out arbitrarily—the same objection has been posed by many other commentators both before and after Tengland's book. For example, psychologist Scott Lilienfeld with coauthor Lori Marino (1999) object to the HDA on the grounds that the flu—a clearly consensual disorder—is not a dysfunction in itself and consists entirely of symptoms that are defensive reactions, so the flu is just a set of defenses with no dysfunction:

> Wakefield (1999[a])...argues that (a) sneezing, coughing, and fever are symptoms of flu, rather than disorders per se... and (b) flu involves a dysfunction and thus does not represent a counterexample to the HDA. But where is the dysfunction in the flu?...[T]he underlying flu cannot be the disorder according to the HDA, because the flu itself is not a failure of a system to perform its designed function and therefore is not a dysfunction.
>
> (Lilienfeld and Marino 1999, p. 407)

Philosopher Christian Perring, in a recent post on his blog, reviews Allan Horwitz's book *The Creation of Mental Disorder*, in which Horwitz explicitly notes that the book's argument is based on the HD analysis. Perring objects as follows:

> For example, a knife wound needs to be treated by a doctor, yet it need not involve any failure of the internal functions of the body. Indeed, the normal functioning of the

body is what leads to the healing of the wound. Similarly, a person with a common cold is fighting off a virus in a normal way, but is not suffering from an internal dysfunction. Applying Horwitz's approach to such cases would have the unacceptable implication that they are not medical conditions.

(I leave aside Perring's puzzling "knife wound" example—clearly, a gash in the skin and the underlying tissue that causes tissue injury and allows blood to gush from a wound and pathogens to enter the body is a clear failure of multiple selected functions, notwithstanding the fact, cited by Perring in a *non sequitur*, that the body is designed to then naturally heal the wound and correct the dysfunctions.)

For the sake of argument, I grant the central premise of these commentators that all of the symptoms of some consensual infectious disorders are biologically designed defenses. In addressing this concern, one must distinguish between what one thinks about the overall condition versus what one thinks about a specific symptom. There is no question that the flu as an overall condition will continue to be seen as a disorder. However, the way that specific symptoms are seen may change. Consider fever, a prototypical symptom of the flu, which was long considered in and of itself a clearly pathological condition. As people have come to recognize that fever is a designed response to a disease, the attitude has changed from fever being a consensual disorder to fever being understood to be generally a normal reaction to an underlying pathology. So, even if it is the case now that people tend to see specific symptoms as disorders, one possibility is that once people know the symptoms are designed defenses, as in the case of fever they will change their minds and come to believe that these symptoms too are non-disordered designed defensive responses to something else.

But what about the entire condition? Why would we say that there is a disorder if, as per the hypothesis, each and every symptom is a designed defense? Nor can such judgments be entirely due to ignorance of the nature of the symptoms because most physicians surely already know that such symptoms are defensive responses, yet consider infectious diseases like the flu and colds to be clear cases of disorder nonetheless. What is the basis for such a judgment?

Defenses in these instances are reactions to the threat posed by viral infection. Viral infections of the kind that trigger defensive symptoms are clear cases of dysfunction at the cellular level. A virus enters a cell and reproduces itself within the cell and spreads to other cells by causing dysfunction of the cellular machinery that are certainly not normal functions of the cell. When a virus invades a cell and co-opts its genetic machinery for its own purposes, that is a clear dysfunction because the cell's mechanisms are no longer working as designed. Indeed, many viruses must cause the cells they infect to rupture and die so that freshly manufactured copies of the virus can move on to infect other cells. Viruses are designed to cause such dysfunctions.

The intuitive idea that such reactions represent some underlying pathological condition is vindicated by what we know about infectious diseases. I have a hard time understanding which part of the process of a viral infection these critics see as the non-dysfunction: is it the forcible entry into the cell, often involving physical damage, or is it the takeover of cellular genetic machinery to churn out viral replicates—surely not the way our genetic material is biologically selected to function!—or is it the part characteristic of most viral infections where the virus, having replicated within the cell, needs to escape to infect other cells and thus to continue the disease process and so breaks open the cell, generally killing it, so as to flood the intercellular medium with virus and start the process anew with many further cells?

For example, here is a simple explanation of the terminology used to describe flu viruses, as in the H1N1 swine-flu virus:

> All flus are named for the shapes of hemagglutinin and neuraminidase displayed on the virus's shell. Hemagglutinin is sometimes called the "spike": the virus uses to enter a cell, while neuraminidase is the "helicopter blade" that chops off receptors, allowing newly made virus to escape.
>
> (McNeil 2009, p. A30)

Except for defenses and other happenstance barriers, even a cold virus could kill one via the killing of the infected cells (the happenstance is that cold viruses don't seem to do well in the warmer temperatures deeper in the body beyond the nasal passages). The broader point is that simultaneous and persistent triggering of an array of defenses that have a substantial cost in terms of discomfort, weakness, and other symptoms generally occurs in response to some dysfunction or potential dysfunction, and we are correctly inclined to infer a disorder, whereas when defenses are a response to a non-dysfunction threat and stay within a plausibly naturally selected range of response, such as when we sneeze in response to dust in the air, we do not infer a dysfunction.

One might try to evade this response by taking the cold/virus argument and extrapolating counterfactually from it to create a potentially legitimate counterexample, as follows. Imagine a mechanism within individuals that scanned the environment for viruses and, when it recognized a dangerous virus in the vicinity that might potentially infect the individual, started the defensive reactions as a precautionary measure. So here we would have no dysfunction and lots of biologically designed defenses. But it might appear the same as a cold, say. So wouldn't we call this a disorder, and wouldn't this be a counterexample to the HD analysis?

The fact that such a mechanism does not exist makes this less than compelling because it remains to be seen how such a condition might be judged and whether people would indeed consider this a disorder, but that need not be the end of the argument. There do exist such "pre-emptive" defensive reactions prior to any actual dysfunction in other domains, such as the startle response in reaction to a looming entity that might be a predator or other injurious object, and the nausea or even vomiting response in reaction to a terrible smell that, if the rotting food were eaten, would be likely to cause damage from toxins. Even after eating spoiled food, we have mechanisms that appear to detect toxins in the gut before the toxins have done any damage, and cause a quite serious regurgitation response to rid us of the toxins before they do cause damage. Now, here is the interesting point: we do not consider any of these unpleasant reactions to be disorders if there is no internal tissue damage. Of course, once the toxin has an impact on the body, there is an internal injury and perhaps broader illness if the bacteria get a hold.

Note also that here as elsewhere in thinking about medical disorder, as when one has a lump that is harmless at the present time but malignant and likely to cause harm later on, one may add the proviso "harmful, or a significant risk of harm, due to a dysfunction." Perhaps it might be necessary (though the examples so far suggest the contrary) to add: "or harmful in response to the body's reaction to a circumstance indicating a risk of potential dysfunction".

5.4 Nordenfelt's critique of evolutionary approaches to disorder

Nordenfelt (2003) argues that disorder is primarily a value concept. In order to reject the contrary view that factual evolutionary considerations are primary in determination of health and disorder, he takes some pains to show that the HD analysis cannot offer an adequate account of disorder judgments. I consider here his main objections, but cannot enter into a discussion of his own quite different view.

5.4.1 Nordenfelt's four arguments against the HD analysis

Nordenfelt offers four main arguments in support of his claim that the evolutionary concept of natural function is plagued with difficulties.

(1) The objection based on degrees of natural function

Nordenfelt argues:

> What was a proper pulse of the human heart a million years ago may not—intuitively speaking—be a proper degree today, but one of the old appropriate degrees must determine the degree of the natural function.
>
> (Nordenfelt 2003, p. 53)

Nordenfelt is suggesting that, if evolutionary history determines function, then function and performance may be divergent in a novel circumstance where the originally selected feature may no longer be adaptive. I think Nordenfelt is here confusing normal variation in functioning with malfunction. Normal variations sometimes include adaptations to circumstances that are biologically designed capacities of the organs.

For example, the heart is designed to speed up and slow down depending on the demands placed on it for getting oxygen to the body's cells. Regarding the pulse rate, an overriding function—at a higher level than the pulse because the pulse varies depending on circumstances—is the distribution of oxygen to the body's cells. The heart is designed to respond to such needs with different rates of beating, which remarkably it can do without the electrical symphony of heart tissue contraction being thrown out of synchrony.

So, when one walks or runs, the heart beats faster than when one is sedentary. If our lives now are more sedentary, then, yes, the proper functioning of the heart would involve a slower average pulse rate than it did in primitive times when activity was more constant. But if this is so, it is completely normal given the way the heart is biologically designed. Or, we know that lungs and blood vessels and hearts tend to adapt to the amount and pressure of oxygen in the atmosphere over time, as in adaptation to the lesser oxygen at higher altitudes. There are thus levels of function, of which none are more "real" than the others.

(2) The "free rider" argument

Consider an organ that has no function either because it is vestigial or because it results as a side effect of other evolutionary processes. One example might be the appendix. Now, suppose there is an infection in this organ:

> The argument from free riders says that several organs are present in existing organisms today that have never been selected. According to the definition of natural function, such organs cannot have natural functions. They cannot be disordered.
>
> (Nordenfelt 2003, p. 53)

Nordenfelt uses the example of the appendix, which as also used by Murphy and Woolfolk (2000). I think his discussion tends to run together a feature that never had a function and is a side effect ("spandrel" in Gould's terminology) of other selected features, and a feature that no longer has a function, but it hardly matters. The point is that it is generally agreed that the appendix currently has no function and that even the function it did once have is no longer possible for it to perform.

The argument, then, is that without a function to go wrong, the appendix cannot be disordered according to the HD analysis, yet of course there are disorders of the appendix, such as appendicitis, that involve acute inflammation. How can such inflammation be a disorder according to the HD analysis? It cannot, argues Nordenfelt.

Nordenfelt anticipates an obvious response, given that appendicitis can lead to toxins migrating to other parts of the body and ultimately death. The response is that the inflammation has consequences for the functions of other parts of the body, and this interference with other functions accounts for the "dysfunction" in appendicitis. But, Nordenfelt answers, we can fix the example by imagining a focal infection that did not have any such implications for the functioning of other features of the individual and affected only the appendix.

Nordenfelt claims that if the organ itself were hurt or infected without affecting other mechanisms, then there is no dysfunction. (He sees that I cannot hold that it is the painfulness of the condition that makes it a disorder, because pain is a normal, biologically designed response to such an infection in the body.)

This claim is obviously mistaken. It is based on arbitrarily focusing on the absence of one high-level function of the overall organ, but the HD analysis says nothing about the level at which a dysfunction must be occurring, just that the harm must be due to a dysfunction. The objection ignores the many levels of dysfunction occurring in the appendix's tissue. Surely in the infected, inflamed tissue, various functions are failing in mechanisms operating at the tissue level, cellular level, and subcellular level. The tissue that is infected has functioning parts, and they are failing to perform their functions, yielding the inflammation as a response to cope with the infection or other failure.

(3) The argument from dying species.

The argument from the dying species says that the members of a dying population must have some natural functions. Since the species has survived so far—for instance, for a hundred million years—some causes must exist for this survival, in terms of the working of the organs of the members of the species. But in the present environment, these same functions have become disadvantageous and are killing the population. Nordenfelt suggests it would be odd to base a theory of the health of this species on a set of functions that are killing them:

> On the causal interpretation of natural function, the members of the dying species have natural functions: the ones that have, in the past, made them survive. But these functions are virtually killing the remaining individuals; they contribute to the extinction of the species. Should we then conclude that the features that are killing

these individuals represent the natural order and thereby the healthy working of these organisms?

(Nordenfelt 2003, p. 47)

Nordenfelt admits that a catastrophe, for example a meteor falling and destroying a species (because, say, it could not fly), cannot be translated into the ill-health of the species! But what about a process over a thousand years during which a species does not have time to adapt to changed circumstances?

Well, we don't have to look to bizarre counterfactuals about the creatures in the Galapagos Islands to test this claim. In our own society there are features of ours that were likely naturally selected, such as a taste for fat and sugar, a fight-of-flight response to stressful situations, shyness about speaking to groups of strangers, and sexual desire not limited to our spouses, that cause us much trouble. In the new environment we have created for ourselves (plentiful food availability, high stress, mass communication, lots of interaction with other people's spouses) these features, which may previously have helped us survive, cause us problems and in some cases if not controlled do lead to disorders. But nobody thinks that they are disorders! They are considered natural traits that are in conflict with the changing environment. Indeed, it would be frightening if society could redefine us as disordered when our natural qualities failed to match social changes, for then human failure to adapt would be blamed on disorders within the individual rather than on an interaction of human nature with social changes.

Let's take a simple analog of Nordenfelt's dying species. Imagine a moth species whose natural coloration is white because it is biologically adapted to be camouflaged against the white bark on the trees in its environment to evade predators. Now, imagine that for either natural or environmental reasons (e.g., a new factory opens nearby and emits soot), the bark on the local trees rapidly turns brown, and the whiteness of the moth becomes a fatal signal to predators. The moths' coloration is killing them, and perhaps killing the whole species, due to the change in the environment. Yet one is not inclined to construe this misfortune as a vast epidemic of disorder among the moths in which there is a dysfunction of their coloration. Rather, the moths are simply unlucky.

On the other hand, it is true that in such an eventuality the theory of moth health will change in certain ways. Before, a white moth was considered a healthy moth, and if a moth was born with a mutation that made it brown and thus lacking in the crucial camouflage it needed for survival, it would have been considered disordered. But if there is no harm, then there is no disorder, and with the changed circumstances of the local bark's color, the brown coloration of the moth due to a mutation is no longer disadvantageous. So a white moth is still a healthy, though unlucky, moth, but if a moth in that species suffers a mutation that makes it brown, it might no longer be labeled as having a disorder. In summary, it seems most consistent with our intuitions to say that those Galapagos creatures Nordenfelt describes—where a changed environment makes their natural patterns negative—are unlucky, not disordered, and we would indeed adjust some judgments about disorder accordingly, but basically their features still have the functions they always did even if the environment makes it impossible for them to accomplish those functions and in fact those functions are now harmful.

(4) The exaptation objection

Nordenfelt raises the objection that some useful features of the organism are not selected features but side effects of selected features, and if such side effects fail, the resulting disorder is a failure of a non-selected feature, thus not an evolutionary dysfunction. For example, acalculia, dyslexia, and amusia involve disorders that may be presumed to be failures of exaptations, for the respective capacities are of too recent origin to have been evolutionarily reshaped and are likely to be exaptations of existing brain structures.

My straightforward answer to this criticism has been given at length in earlier publications (see especially Wakefield 1999a). Dyslexia, for example, is a harm (inability to read) caused by a dysfunction; the dysfunction is at a lower level, and is presumed to be some brain dysfunction. The inability to read is a harm, not a dysfunction, thus need not be an evolutionary failure to allow the condition to qualify as a disorder under the HD analysis. I thought that extended discussion would have laid this misguided objection to rest, but no such luck.

Nordenfelt understands the power of my answer and acknowledges that referring to an underlying dysfunction that causes the harm does seem to be a winning strategy for responding to the exaptation objection to the HD analysis. However, he then goes on to argue that, just because it is such a winning strategy, my theory is therefore untestable! He writes:

> How effective is Wakefield's strategy? He appears always to have a way out. If something exists, for instance a mental ability, that is considered to be a valuable trait in human beings today but which, from an evolutionary point of view, cannot be anything else than an exaptation, then Wakefield sees to it that this case can still be dealt with within his theory of health. The trait is seen as a positive effect of some basic natural function to be found on the organic level. A deficiency with regard to the mental trait can then still be regarded as a disorder, given the combined theory that requires the existence of a dysfunction in the evolutionary sense and some harm caused by the dysfunction in question.
>
> But such a move does not impress; the theory then becomes irrefutable. Assume that we find a counterexample to Wakefield's theory. We find a typical exaptation and we agree that defects with regard to this exaptation are generally regarded as disorders

or diseases. Wakefield can then always answer: The defect in the exaptation is only a harm caused by some underlying dysfunction in the evolutionary sense; the exaptation should not in itself be considered to be a function.

But we cannot let Wakefield escape using this simple formula. It leaves his theory completely untestable. Even if Wakefield cannot tell what exactly is the dysfunction in a particular case, we must at least require that he indicate on which level of abstraction we will identify his functions...

(Nordenfelt 2003, p. 50)

Nordenfelt is claiming that in these cases the HD analysis cannot be tested because I can just create deeper dysfunctions out of thin air to explain the source of a non-evolutionary harm, so what appears to be a test is really spurious. This is an incorrect assessment of the situation—the fact that my theory is unfalsified does not mean it is untestable! If harmful exaptations were considered disorders even when there was no reason to think there is a dysfunction underlying its harm, that would offer disconfirmatory evidence. Contrary to Nordenfelt's description, I cannot manufacture evidence that people judge the condition to be a disorder when and only when they also suspect an underlying dysfunction. That conclusion is based on observation about how people actually think about the condition supported by reading of the literature, not an armchair out-of-the-blue speculation of mine.

So, contrary to Nordenfelt's claim, it is easy to test such theories: one simply examines whether exaptations (or their failures, whatever is the harmful condition) are judged disordered/non-disordered on the basis of the degree to which it is believed that there is/is not a dysfunction responsible. To take a simple example, illiteracy and dyslexia are both failures of the reading exaptation: the use of our brain mechanisms for the socially approved purpose of reading. Illiteracy is not thought to be a disorder because it is thought to result from normal processes, such as lack of opportunity or motivation to learn. Dyslexia is considered a disorder because, on the basis of a variety of circumstantial evidence, experts infer that there is a brain dysfunction underlying the individual's incapacity to learn to read despite adequate opportunity and motivation.

I did not manufacture this distinction. Good reading disorder experts begin the diagnostic process by going through a list of alternatives to a genuine dysfunction, such as lack of opportunity to learn, lack of motivation, lack of general intelligence, unfamiliarity with the language in question, and so on, and eliminating them and attempting to support the neurological hypothesis from the details of the condition before rendering a diagnosis of dyslexia. The professional learning disabilities literature contains many discussions of precisely this sort of issue, elaborating the basis for imparting a dysfunction. In contrast to the fit of the data of classificatory judgments to the HD analysis, a values or intentional capacity view of disorder of the kind held by Nordenfelt or for that matter a Cummins-style "systems effects" account of function or Boorse's (1975, 1976, 1977) "species-typical contribution to a goal" cybernetic-style account of function may not so easily be able to explain why the incapacity to read would not be a disorder whatever the cause, at least not without some *ad hoc* adjustments. In this case and many others, as I have shown above, the HD analysis offers the greatest explanatory power of any current view of the concept of disorder when tested against actual diagnostic judgments and the beliefs leading to those judgments.

References

Aristotle (Jonathan Barnes, ed.) (1971) *The Complete Works of Aristotle: The Revised Oxford Translation*. Princeton University Press, Princeton, NJ.

Aristotle (trans. Hugh Tredennick) (1989) *Aristotle in 23 Volumes*, volumes 17 and 18. Harvard University Press, Cambridge, MA.

Boorse, C. (1975) On the distinction between disease and illness. *Philosophy and Public Affairs*, 5, 49–68.

Boorse, C. (1976) Wright on functions. *The Philosophical Review*, 85, 70–86.

Boorse, C. (1977) Health as a theoretical concept. *Philosophy of Science*, 44, 542–73.

Campbell, G. (2003) *Lucretius on Creation and Evolution: A Commentary on* De Rerum Natura *Book Five, Lines 772–1104*. Oxford University Press, Oxford.

Clouser, K.D., Culver, C.M., and Gert, B. (1981) Malady: a new treatment of disease. *The Hastings Center Report*, 11, 29–37.

Clouser, K.D., Culver, C.M., and Gert, B. (1997) Malady. In J.M. Humber and R.F. Almeder (eds), *What is Disease?* Humana Press, Totowa, NJ, pp. 175–217.

Culver, C.M. and Gert, B. (1982) *Philosophy in Medicine: Conceptual and Ethical Issues in Medicine and Psychiatry*. Oxford University Press, New York.

Cummins, R. (1975) Functional analysis. *Journal of Philosophy*, 72, 741–65.

Darwin, C. (1964) *On the Origin of Species: A Facsimile of the First Edition*. Harvard University Press, Cambridge, MA.

Gert, B. and Culver, C.M. (2004) Defining mental disorder. In J. Radden (ed.), *The Philosophy of Psychiatry: A Companion*. Oxford University Press, New York, pp. 415–25.

Lennox, J.G. (1993) Darwin *was* a teleologist. *Biology and Philosophy*, 8, 409–21.

Lilienfeld, S.O. and Marino, L. (1999) Essentialism revisited: Evolutionary theory and the concept of mental disorder. *Journal of Abnormal Psychology*, 108, 400–11.

Lucretius, T.C. (W.H.D. Rouse trans., M.F. Smith rev. trans.) (1992) *De rerum natura* (On the nature of things) (revised 2nd edition with revision). Harvard University Press (Loeb Classical Library), Cambridge, MA.

McNeil, D.G. Jr. (2009) US Says Older People Appear Safer From New Flu Strain. *New York Times*, May 21, 2009, p. A30.

Murphy, D. and Woolfolk, R.L. (2000) The harmful dysfunction analysis of mental disorder. *Philosophy, Psychiatry, and Psychology*, 7 (4), 241–52.

Nordenfelt, L. (2003) On the evolutionary concept of health: Health as natural function. In L. Nordenfelt and P.-E. Liss (eds), *Dimensions of Health and Health Promotion*. Rodopi, Amsterdam, pp. 37–53.

Perring, C. (2009) http://christianperring.blogspot.com/2008_11_01_archive.html (accessed 6 November 2009).

Tengland, P.-A. (2001) *Mental Health: A Philosophical Analysis*. Kluwer, Dordrecht.

Wakefield, J.C. (1992a) The concept of mental disorder: On the boundary between biological facts and social values. *American Psychologist*, **47**, 373–88.

Wakefield, J.C. (1992b) Disorder as harmful dysfunction: A conceptual critique of DSM-III-R's definition of mental disorder. *Psychological Review*, **99**, 232–47.

Wakefield, J.C. (1993) Limits of operationalization: A critique of Spitzer and Endicott's (1978) proposed operational criteria for mental disorder. *Journal of Abnormal Psychology*, **102**, 160–72.

Wakefield, J.C. (1999a) Evolutionary versus prototype analyses of the concept of disorder. *Journal of Abnormal Psychology*, **108**, 374–99.

Wakefield, J.C. (1999b) Disorder as a black box essentialist concept. *Journal of Abnormal Psychology*, **108**, 465–72.

Chapter 6

Evolutionary foundations for psychiatric diagnosis: making DSM-V valid[1]

Randolph M. Nesse and Eric D. Jackson

The third and fourth editions of the *Diagnostic and Statistical Manual of Mental Disorders* (DSM) have brought much-needed reliability to psychiatric diagnosis. However, as is often the case, progress comes at a price. In this chapter, we support Wakefield's argument that DSM-III and DSM-IV typically ignore one of the most fundamental distinctions in medicine— the distinction between symptoms and the situations or diseases that cause them. In the case of emotional disorders, such as mood and anxiety disorders, this mistake is particularly deplorable, because many emotions are responses that evolved because they are protective in untoward circumstances. Here we suggest that an evolutionary perspective can advance the nosology of emotional disorders in several ways. First, this perspective confirms that the normality of an emotion depends necessarily on the context. Furthermore, it notes that variations in brain mechanisms that make a person susceptible to anxiety or depression are only sometimes diseases; more often they may have the same causal significance as variations in brain mechanisms that make a person especially prone to cough or fever during a cold. An evolutionary perspective also indicates that biologically normal responses may be aversive and even harmful to individuals. Finally, it suggests the importance of a detailed and evolutionarily informed analysis of the motivational structure of every patient's life.

---

[1] Adapted from Nesse, R.M. and Jackson, E.D. (2006) *Clinical Neuropsychiatry* 3 (2) 121–131.

Hundreds of researchers and clinicians have collaborated for the past three decades to revise the diagnostic criteria for mental disorders (Wilson 1993). The products of their labors are the source of widespread dissatisfaction and apparently irresolvable debates (Beutler and Malik 2002; Horwitz 2002). Clinicians often ignore the official diagnostic system. Researchers find themselves constrained by categories with no theoretical foundation and questionable reliability that include heterogeneous patients who show vast comorbidity. Nonprofessionals and experts look at prevalence rates of 50% and ask if there is a scientific justification for defining what is pathological. Even the architects of the system suggest the need for fundamentally new perspectives:

> Science strives for simplicity of explanation. Descriptive models tend to be piecemeal and complicated. We are at the epicycle stage of psychiatry where astronomy was before Copernicus and biology before Darwin. Our inelegant and complex current descriptive system will undoubtedly be replaced by explanatory knowledge that ties together the loose ends. Disparate observations will crystallize into simpler, more elegant models that will enable us not only to understand psychiatric illness more fully but also to alleviate the suffering of our patients more effectively.
>
> (Frances and Egger 2003)

Such extensive dissatisfaction after Herculean efforts suggests that persisting in the same path will not solve the problem. This chapter argues that evolutionary behavioral biology is a crucial but neglected scientific foundation for psychiatric nosology. Posing evolutionary questions about why we all are so vulnerable to negative emotions highlights a fundamental misunderstanding at the heart of the *Diagnostic and Statistical Manual of Mental Disorders* (DSM) that has kept psychiatric diagnoses artificially different from those in the rest of medicine. In the rest of medicine, symptoms such as pain and cough are carefully distinguished from disease such as appendicitis and pneumonia. In psychiatry, we are often trying to craft diagnoses based on symptoms, with predictable frustration. An evolutionary understanding of emotions reveals why the quest for simple criteria for emotional disorders is so frustrating, and where we can look for solutions.

6.1 Diagnosis and its discontents

Our core argument is simple. Negative emotions can be normal and useful in certain situations, so, except in the extreme cases, distinguishing normal and abnormal emotions requires close attention to the situation. The logic is that of the medical model. Consider pain. Pain is normal when its severity matches the amount of tissue damage. Pain is pathological when it is disproportionate to the cause. Decisions about normality and pathology depend on the situation.

The logical response to this argument would be to modify diagnostic criteria to take situations into careful account. As Jerome Wakefield has suggested, for instance, the grief exclusion for depression could be expanded to include other dire circumstances that can cause normal symptoms of depression (Wakefield et al. 2007). Instead, the DSM-V Committee is now apparently considering eliminating the grief exclusion! Instead of simply being aghast at such obliviousness, we should try to understand this response; it can help us understand the problem: the absence of any theoretical foundation for validity, with a resulting huge over-emphasis on reliability. This is understandable on two counts.

First, allowing exclusions for situations such as having a child with cancer, or loss of a marriage or job, would decrease reliability. Who is to say if a particular situation is severe enough to account for the symptoms? Reliability would decrease, and that would be fatal to many studies whose results are already on the border of significance because of the limited reliability of current criteria.

The second issue is more profound. We have no scientific foundation for establishing the validity of criteria for diagnosing emotional disorders. The foundation is being constructed by those working to describe how emotions evolved, how they give advantages, and how selection shaped the mechanisms that regulate them. This work is, however, just getting under way, and it is revealing the inherent difficulties of diagnosing disorders that result from a dysregulation of protective responses. The challenge is hard enough for physical responses such as pain, fever, and fatigue. For emotional responses, the appropriate intensity depends not only on the objective situation, but on how the individual appraises the meaning of the situation for his or her ability to reach personal goals. Emotions arise not from events; they arise from an individual's motivational structure, that is, from the interaction of an objective external situation with an individual's goals, strategies, and subjective assessments of ability to reach these goals and strategies.

Such complex causes, different in each case, make it very difficult to formulate reliable diagnostic criteria. If the decision about whether symptoms are normal or abnormal depends on a decision about the severity of the life situation, subjective judgment is unavoidable. The obvious solution is to ignore the situation and focus entirely on the severity and duration of symptoms. If this strategy was used in internal medicine, "cough disorder" would be diagnosed whenever the frequency, duration, and severity of a cough exceeded defined thresholds, irrespective of the cause of the cough. The problem is, of course, that life situations cannot be measured as objectively as a pulmonary infiltrate. Change will eventually come as researchers discover that their findings become stronger when they differentiate subpopulations according to how disproportionate symptoms are to the situation.

This transition will take time. It will be facilitated by creating methods to measure variables that are hard to measure, such as the size of the gap between a person's resources and aspirations, the extent to which the problem is an objective inability to get crucial resources, the scale of the individual's aspirations, and the extent of distorted negative thinking. But it will also be sped by neuroscientists and other psychiatric researchers who recognize the opportunity to ground their work in behavioral biology. Perhaps this chapter, and others like it, will fire the curiosity of some researchers to explore our growing knowledge about how evolutionary behavioral biology can inform psychiatric research.

Although this chapter emphasizes the utility of evolutionary principles for classifying emotional disorders, the same principles are also useful for classifying other psychiatric disorders. For instance, behavioral disorders such as addiction or eating disorders make much more sense in an evolutionary framework. Personality disorders can be organized based on the strategies people use to influence other people. Even psychoses and neurological conditions are illuminated by evolutionary considerations of the selection forces that maintain the frequency of predisposing genes and how they interact with novel aspects of the modern environment. Here, however, the focus is on the emotions and the categories that describe their disorders.

Much has been written about how an evolutionary approach can help distinguish pathological from nonpathological conditions (Wakefield 1992; McGuire and Troisi 1998; Clark 1999; Troisi and McGuire 2002). Wakefield's concept of "harmful dysfunction" brings a biological foundation to the question of where normal stops and pathology begins (see Chapter 5). Dysfunction grounds diagnosis in the selective advantages of normally operating brain mechanisms (but see Chapters 7 and 8). By also requiring a condition to be "harmful", Wakefield's approach acknowledges that what is good for our genes is not necessarily good for our selves, and what is good in one culture may not be good in another. In more recent work, Horwitz and Wakefield make a powerful case for basing psychiatric diagnosis on an evolutionary understanding of emotions. They point out that if depression symptoms can be normal in bereavement, they very likely can also be normal in other situations, therefore careful consideration of the situation is essential to any scientific nosology for depression (Horwitz and Wakefield 2007).

6.2 From clinical diagnosis to the DSM

The history of mental illness taxonomy began with highly speculative informal categories originating thousands of years ago. At turns biological, phenomenological, and moral in orientation, such informal systems prevailed well into the nineteenth century, when Emil Kraepelin took the first steps toward modern, systematic classification in collaboration with his colleague Allen R. Diefendorf (Kihlstrom 2002). In the USA formal classification systems for mental disorders were first adopted not by clinicians, but by the federal government because of its need to track asylum populations accurately. This encouraged the American Medico-Psychological Organization (AMPA) to publish the first standardized psychiatric nosology, the *Statistical Manual for the Use of Institutions for the Insane* (*Statistical Manual*) in 1918. The absence of the word "diagnosis" in the title accurately represents the marginal utility of the manual to the era's mental health practitioners.

The *Statistical Manual* was revised for the last time in 1942, just as the USA entered World War II (Grob 1991; Houts 2002). Military practitioners found the statistical categories woefully inadequate to describe battlefield psychological casualties. Dr George Raines, then head of the American Psychiatric Association (APA) Committee on Statistics and Nomenclature, noted in the introduction to the first edition of DSM that "only about 10% of the total cases seen [in World War II] fell into any of the categories ordinarily seen in public mental hospitals" (American Psychiatric Association 1952, p. vi). Such dissatisfactions led the APA to replace the *Statistical Manual* with a new standardized nosology in 1952: the first edition of the *Diagnostic and Statistical Manual of Mental Disorders* (DSM-I).

Although more useful to practitioners, DSM-I and its revision, DSM-II (American Psychiatric Association 1968), were unsatisfactory for research. Prior to the formulation of the Research Diagnostic Criteria and the publication of the third edition of the *Diagnostic and Statistical Manual* (DSM-III) (American Psychiatric Association 1980), research reports were hard to compare because the subjects in one study of "depression" might have quite different conditions from those in another. Critics pointed to such inconsistencies to argue that psychiatry was unscientific or even that mental illnesses were not diseases at all (Szasz 1974). Examples of malingering were published in *Science* as evidence for the subjectivity of psychiatric diagnosis (Rosenhan 1973). At about the same time, the utility of psychotropic drugs was being widely recognized and insurance companies began paying only for the treatment of specific medical disorders. These several crises combined to create a consensus that psychiatry should become more like the rest of medicine (Houts 2002; Jackson unpublished). Operationalizing diagnostic criteria was the obvious place to start.

The committee charged with creating the DSM-III quickly found there would be no agreement on a theoretical foundation for psychiatric nomenclature (Wilson 1993). Psychoanalysts remained powerful, and their views of mental disorders were fundamentally at odds with "biological" psychiatrists, who emphasized the brain origins of mental disorders. To get past this impasse, the DSM created diagnostic categories avowedly without theoretical foundation. The goal was a system derived empirically from clinical observations of observable signs, symptoms, and the disease course. Building on criteria from the International Coding Diagnoses and the Research Diagnostic Criteria group at Washington University (Feighner et al. 1972), the DSM-III and DSM-IV attempted to create categories defined by observable data (American Psychiatric Association 1994).

The inauguration of operationalized diagnoses transformed psychiatry (Guze 1992; Wilson 1993; Jackson unpublished). Indeed, the history of medicine contains few transitions so sudden and complete (Shorter 1997). Prior to the DSM-III, psychiatrists' diagnostic categories were theoretically based and used to complement highly valued narrative explanations for how an individual came to have his or her particular constellation of symptoms. Clinicians crafted idiographic explanations for a particular individual's problems in much the same way that historians explain the origins of a war or economic collapse in a particular country. Nomothetic (universally applicable) principles were incorporated into such explanations, but different clinicians used different principles. For instance, psychoanalysts emphasized the ubiquitous importance of defenses against Oedipal wishes, while behaviorists emphasized the reinforcement history. Arriving at a diagnostic formulation was an occasion for deep thought, sophisticated discussion, theoretical battles, and frequent flights of fancy. Two diagnosticians often arrived at plausible formulations with little in common and no way to decide between them. Reliability was low. Such diagnoses were nearly worthless for research.

Current criteria are nearly the polar opposite of their predecessors. Individualized explanations for symptom constellations have been replaced by categories defined by the presence or absence of specific signs and symptoms. For instance, a diagnosis of major depression applies to anyone who has had at least five of nine symptoms for at least 2 weeks, at least one of which is depressed mood or lack of pleasure (American Psychiatric Association 1994). Precipitating events are not taken into account, with the exception of bereavement in the past 2 months. Whether symptoms arise during a relaxing vacation or a stay in intensive care is irrelevant. Such exclusion of life context is mindless, but it does sidestep the serious problem of how to measure the kind and severity of precipitants. If criteria for depression required assessing the severity of recent life events, complexity would increase and reliability would plummet.

The quest for criteria that yield reliable diagnoses is well justified. If different clinicians examining a patient arrive at different diagnoses, the system is not at all that useful (Goodwin *et al.* 1979). Explicit criteria made possible standardized interviews that further enhanced reliability (Spitzer *et al.* 1992). Versions useable by lay interviewers have made extensive epidemiology possible for the first time, not just in the USA, but in over 39 countries where the same questions are administered using the same instrument translated into different languages (Kessler and Ustun 2004). This is real progress, and the data are useful for public health planning as well as research.

In short, the DSM has been essential for most recent progress in psychiatry. Treatment trials now target groups of well-defined patients and the results can be applied to other similarly defined groups. Research studies can measure genes, neurotransmitters, or brain structures in well-characterized groups of patients as compared to controls. Reliable diagnostic criteria have advanced psychiatric research more than any individual research project could.

6.3 **The price of progress**

Given such dramatic progress, why such dissatisfaction with the DSM approach to diagnosis? Many objections are based on the tangible factors outlined above—high comorbidity, heterogeneity within groups, and questionable reliability (Beutler and Malik 2002; Phillips *et al.* 2003; Watson 2005). However, larger issues are even more important.

First, the distinction between normal and abnormal remains fundamentally arbitrary. For cancer, pneumonia, rheumatoid arthritis, and pinworms you either have the condition or you don't. A zone of rarity separates the condition from normal (Kendell 1975). Most emotional disorders offer no such clean demarcation, leading some to suggest that diagnoses should be dimensional instead of categorical. However, to communicate, humans tend to use words that refer to categories or essences. People demand to know the boundary that separates pathology from normality. Dimensions are not diagnoses. Even high blood pressure is defined by a specific cut-off.

Second, the DSM diagnoses are often presented as products of clinical observation unconnected explicitly to any theory of human behavior. However, this presentation is not quite correct. Because explicit theories are excluded, the DSM criteria tacitly foster thinking about mental disorders as if they are diseases. This makes them fit easily into neuroscience models that seek to identify brain abnormalities correlated with each disorder.

Third, is the problem of how to incorporate context (Faust and Miner 1986). The DSM approach relegates much of what we know about the effects of life events to "stress", as if stress hormones mediated most adverse effects of social experience. Clinicians understand the far more complex relationships between life events and psychological structures (Brown and Harris 1978; Monroe *et al.*

2001). Many see the need to adjust the diagnostic threshold depending on the situation, lowering it for apparently unprovoked symptoms, increasing it in extreme life situations. However, with the exception of bereavement, the DSM criteria ignore context (Wakefield and First 2003). For instance, the diagnosis of panic disorder is applied whenever someone has symptoms for a month after recurrent unexpected episodes that include four out of ten possible panic symptoms. It makes no difference whether the patient had the onset in a grocery store or in a prison camp. The reason for this rigidity is that attempts to include context would require difficult-to-define objective criteria for levels of provocation.

At the same time, almost everyone recognizes the need to consider the circumstances in order to judge whether an emotion is normal or not. Following the debut of the multi-axial diagnostic system in DSM-III, it appeared that the editors of DSM had at least partially recognized the need to integrate life circumstances and context. Severity of psychosocial stress (Axis IV) and level of adaptive functioning (Axis V) were added to enrich the clinical context of the individual (Klerman 1984). However, the inclusion of environmental circumstances in separate axes excludes important contextual information from their important role in making Axis I diagnoses.

The DSM gives us categories for emotional disorders, but says nothing about what these disorders are. Are they diseases? Disorders? Are some merely responses to life circumstances? Is the cause located mainly in brain differences, in cognitive habits, or in exposure to environmental events? Almost everyone pays lip service to the bio-psycho-social model, but few are willing to get into the complexity of how individual differences interact with situations, events, and cognitions to give rise to symptoms that have evolutionary significance (Gilbert 1989).

Thinking about patients as DSM diagnoses instead of people impoverishes clinical understanding (Faust and Miner 1986). For instance, a resident recently concluded a case presentation by saying, "The diagnosis is major depression so I prescribed an SSRI." When asked why this person was depressed now, the resident replied "Well, we think depression is caused mostly by genetic factors, but also by stress," omitting any mention of why this particular patient was depressed now. When pressed to do so, he explained that there was a family history of depression and the patient had been abused in childhood, was in a bad marriage, and had recently lost his job after a drink-driving conviction. The resident clearly imagined that his job was to place his patient in the category "major depression" and to prescribe a treatment that was usually effective for someone in that category. He had not even tried to figure out whether the person had had a previous satisfactory and stable life adaptation, whether the alcohol use initiated the marital problems or came later, and whether or not the person was capable of maintaining good relationships. In short, like many young clinicians, the resident viewed DSM criteria as if they described specific diseases with specific consistent causes. He assumed that the diagnosis contained all that he needed to know to arrive at a treatment plan.

The same physician would never undertake such a crude approach to diagnosing and treating cough or pain. If a patient presented with a severe cough and fever, he would not be satisfied with a diagnosis of "cough disorder", he would instead consider all the possible causes of cough, and would not prescribe treatment until arriving at the best possible understanding of why this person had this cough now. Is it chronic obstructive pulmonary disease (COPD), or pneumonia, or congestive heart failure, or COPD and congestive heart failure complicated by pneumonia? The physician would find out whether the individual was especially vulnerable because of immunosuppression or steroid use, if there were exposures to infectious agents, and if the person had allergies. General physicians recognize that cough is not a disease, it is a response to a disease. Likewise, while pain can be abnormal, physicians recognize that pain is usually a response to pathology, not a disorder in itself. Psychiatrists sometimes think of anxiety as a potentially useful response to a danger, but other emotions such as depression and jealousy are usually thought of as abnormalities instead of being recognized as potentially useful responses to untoward situations.

6.4 The basic fault

The flaw in the DSM approach to emotional disorders is fundamental: the DSM fails to distinguish protective responses from diseases. This flaw is by no means new; the DSM merely extends the Kraepelinian tradition. Kraepelin excluded etiology and anatomic considerations from mental disorder classification because reliable information was not accessible except in the case of obvious injuries and post-mortem assessment of neural lesions (Kihlstrom 2002). In his 1904 textbook, Kraepelin recognized the limits of a nosology based on symptoms, but he also noted that diagnostic systems based on a comprehensive knowledge of symptoms or pathological anatomy or etiology should provide "uniform and standard classifications" that mapped well onto one another, no matter what the starting point was (Kraepelin and Dierdorf 1907).

It is a short leap from this to equating the outcome of exhaustive identification of symptomatology with the exhaustive identification of etiology; if all nosologies carve up the pie identically, then any one system should work as well as any other. This explanation is especially appealing if some systems are inaccessible, as neural systems were in the late 1800s. However, assuming that symptomatic categories will match etiological categories comes at the high price of blurring the directional relationship between cause and effect, leading to two kinds of errors. First, categories based on symptom constellations may contain subgroups that arise from fundamentally different causes. Second, such categories fail to distinguish symptoms that arise from pathological causes from those that are aroused by normally functioning systems. The former is an error of failing to distinguish distinct disorders (e.g., yellow fever vs spotted fever), while the latter fails to distinguish disorders from the symptoms of disorders (e.g., mistaking fever or cough for disorders, when they are actually protective responses to the disorder of pneumonia).

The rest of medicine long ago replaced symptomatic diagnoses such as "cough disorder" with etiologically based diagnoses such as pneumonia or lung cancer (Kihlstrom 2002). The rest of medicine recognizes cough, fever, pain, nausea, fatigue, diarrhea, vomiting, and inflammation as responses to diseases, not diseases themselves. These responses are aversive, and they can be dangerous, disabling, and even fatal. High fever can cause convulsions and diarrhea causes thousands of deaths each year. Nonetheless, fever, diarrhea, and other defenses are the body's adaptive responses to problems, not usually diseases themselves. They give important clues to the diagnosis, but they are themselves diagnoses only in special circumstances.

One circumstance is when the cause cannot be found. For instance, "fever of unknown origin" is a stand-in for a diagnosis when no reason for a fever can be identified. The other circumstance is when the system that regulates the response is presumed to be abnormal, as is the case in chronic pain syndromes. Chronic fatigue is likewise usually thought to arise from an abnormal regulation system. When every other possible cause has been eliminated, even fever or pain may be attributed to an abnormal regulation mechanism.

The error of failing to distinguish defenses from diseases needs a name. Most simply it can be called "the fallacy of mistaking defenses for diseases". It could be called "the DSM fallacy" because the DSM so resolutely ignores this basic medical distinction. The DSM takes great pains to define when symptoms are severe enough to justify a diagnosis, but it mostly ignores the more fundamental distinction between symptoms and the problems that arouse them.

This argument is based on the supposition that negative emotions are protective reactions akin to pain and fever. The next section reviews reasons to think this is correct. However, major differences between physical protective responses and emotional responses make the correspondence hard to see at first. The situations that arouse fever and cough are observable changes in specific tissues. Most arise from diseases or injuries. The situations that arouse negative emotions are also adverse, but few are specific diseases with identifiable tissue pathology. Many are injuries to social resources such as relationships or social status, which are less tangible despite their importance to function and Darwinian fitness. Some situations, such as exclusion from a group, directly arouse negative emotion. Other connections between situations and emotions are far less direct, such as the anxiety that follows a subtle vocal inflection that suggests new distance in a previously close relationship.

Fever and cough indicate the presence of an infection or some other disadvantageous abnormal state. Anxiety and sadness arise from states that are disadvantageous, but generally not abnormal. This apparently major difference can be turned on its head by noting that the infections that arouse fever and cough are not exactly diseases, they are just conflicts with pathogens of the sort that our bodies manage constantly. The symptoms are aspects of the body's well-established plan for dealing with infections. Both physical and emotional responses are useful only in certain situations. For physical responses these situations are more tangible and more likely to be abnormalities. For emotional responses, the etiology is not usually a disease process. To avoid confronting the complex social situations that arouse negative emotions, psychiatry has defined extremes of negative emotions as disorders. The result is a major emphasis on individual differences in "vulnerability" to negative emotions and a relative neglect of causes in the environment.

6.5 Evolution and emotions

The proper foundation for understanding emotional disorders is an evolutionary understanding of why the emotions exist at all (Nesse 1990; Tooby and Cosmides 1990; Nesse 1998; Nesse and Ellsworth 2009). The same logic is at the heart of pathophysiology. To understand the kidney, we first try to understand what it is for. Armed with this knowledge, we can understand how the nephron works and why it is the way it is. Such evolutionary functional understanding is so intrinsic to physiology that it is easy to overlook that it includes two separate kinds of knowledge, one an evolutionary explanation for why a trait exists at all, the other a proximate explanation for the details of the trait's structure and how it works (Mayr 1961).

It is tempting to posit functions for emotions that are just as straightforward as functions for abdominal organs, but this is a mistake. The abdominal organs are always present and constantly useful, while emotional states are aroused only in certain situations and they are useful only in those situations. Panic, for instance, may be life-saving when serious danger is present, otherwise it is worse than useless. The correct way to analyze the utility of an emotional state is to define the situations in which it is useful and the adaptive challenges posed by those situations. In the face of life-threatening danger, rapid breathing oxygenates the blood, muscle tension increases strength, and insulin allows glucose to flow into muscles. Emotions have utility for communication, motivation, and for adjusting physiology and behavior, but there is no need to consider which of these is primary. All are part of a special coordinated state that gives an advantage in a certain situation (Nesse 1990; Tooby and Cosmides 1990). For instance, sweating, rapid heartbeat, muscle tension, and a wish to escape are all useful when confronted by dangers that demand fight or flight, and they serve a variety of related functions. Emotions are like computer programs that adjust multiple aspects of the organism to cope with the exigencies of situations that have recurred over evolutionary time. Organisms with such abilities to adjust have an advantage over those that make no adjustments.

Emotions are positive or negative for the simple reason that special states are useful only in situations that pose opportunities or threats. Positive or negative subjective experience is but one aspect of an emotional state that includes changes in arousal, motivation, physiology, memory, and action endencies (Plutchik 2003). Negative emotions are naturally associated with untoward situations, so it is easy to incorrectly conclude that they are themselves problems. This "clinician's illusion" is a serious impediment to understanding and treating emotional problems (Nesse 2005; Nesse and Ellsworth 2009).

It would be grand if all who treat emotional disorders could take several courses about emotions or at least read one good textbook, such as Plutchik's (2003), but some of the debates in emotions research would likely be more distracting than illuminating (Ekman and Davidson 1994). For instance, arguments continue about whether emotions are best viewed as dimensions or as a few distinct basic kinds with combinations. An evolutionary approach offers a possible resolution by tracing the phylogeny of various emotions over evolutionary time as they have been gradually but only partially differentiated from one another in order to cope with diverse kinds of situations (Nesse 2004).

This view has profound implications for psychiatric diagnosis and the comorbidity of emotional disorders. For instance, instead of attempting to determine whether the various anxiety disorders are fundamentally the same or fundamentally different, it suggests that anxiety has been partially differentiated into subtypes shaped to cope with a variety of different kinds of dangers. We should, therefore, not expect to be able to differentiate subtypes of anxiety sharply; the boundaries between them are blurred (Marks and Nesse 1994). Similarly, the profound overlap between anxiety, sadness, low mood, and depression arises because they are responses to related kinds of danger. Anxiety is aroused by situations that pose threats of possible future loss. Sadness is aroused by loss. Low mood is aroused by the expectation that one will be unable to reach an important goal. The decreased motivation encourages seeking another strategy or, if nothing works, disengaging from pursuit of the goal. If efforts persist nonetheless, ordinary low mood is likely to escalate to clinical depression.

There is no room here for a detailed consideration of the full spectrum of emotions, to say nothing of the extensive research and writing about them (Barlow 1991; Izard 1992; Oatley and Johnson-Laird 1995; Lewis and Haviland-Jones 2000; Fessler 2003; Fessler and Haley 2003). Instead, consider a list of some common situations and the emotions they arouse:

- opportunity → desire, excitement
- success → joy, happiness
- failure → disappointment
- threat of damage → fear
- threat of social loss → anxiety
- loss → sadness
- failure to make progress towards an important goal → low mood
- inability to get or protect an essential resource → despair
- betrayal → anger
- contamination → disgust.

The list could be greatly extended, but the relationship among different emotions becomes clearer if they are organized into groups that correspond to the two main classes of situations individuals need to cope with (Nesse 1990, 2004). The first is goal pursuit and the problem of what to do when, and with how much effort and persistence. Living is a sequence of episodes in which organisms attempt to reach goals and avoid losses. Table 6.1 summarizes the emotions that arise in the situations associated with goal pursuit. It presumes that a somewhat consistent set of brain mechanisms has regulated the pursuit

Table 6.1 Emotions shaped to deal with the situations that arise during goal pursuit

Situation	Before	During	Obstacle	After success	After failure
Opportunity					
Social	Excitement	Engagement	Frustration	Joy	Disappointment
Physical	Desire	Flow	Anger	Happiness	
		Interest	Despair	Pleasure	
Threat					
Social	Anxiety	Confidence	Dread	Relief	Sadness
Physical	Fear	Coping	Despair		Pain

Table 6.2 Emotions shaped to deal with the situations that arise in relationships

	Other cooperates	**Other defects**
You cooperate	Trust	Before: suspicion
	Friendship, love	After: anger
You defect	Before: anxiety	Rejection
	After: guilt	Disgust

of diverse goals in different organisms over hundreds of millions of years. For any particular species, these global emotions gradually become somewhat specialized to cope with particular kinds of goals. For instance, when faced with the possibility of losing a mate most humans experience not just generic anxiety, but the complex emotion of jealousy. The regulation of these emotions is further specialized by life experience.

The other group contains emotions shaped to deal with the situations that repeatedly arise in managing social relationships. As most readers will know, evolutionists and economists often model the trading of favors as a prisoner's dilemma in which the maximum net outcome emerges from repeated mutual cooperation, but on any given move, a player who defects gets a big gain at the expense of the other player (Axelrod and Hamilton 1981). We and others have argued that these situations are so ubiquitous that they have shaped specific emotions: trust and friendship after repeated successful exchanges, suspicion and anger before and after the other defects, and anxiety and guilt before and after the self defects (see Table 6.2) (Ketelaar and Clore 1997).

These tables are not intended to be exhaustive. For instance, surprise is a more general emotion aroused by situations that give rise to unexpected outcomes. Disgust probably evolved to protect us from contaminated materials, but it seems to have been co-opted for use in the mechanisms that keep us away from those who are judged morally unclean. All of the above emotions deserve extended explanations that are available elsewhere (Nesse 1990; Plutchik 2003). They are summarized briefly here as a prelude to addressing the question of emotional disorders.

6.6 **Emotional disorders**

An evolutionary perspective on emotions has several implications for a nosology of psychiatric disorders.

1. Emotional disorders should be recognized as distinct from other mental disorders. They are, like chronic pain, abnormalities of the regulation of useful responses and thus very different from disorders such as psychoses that are abnormal in any amount and any situation (Watson 2005).

In DSM-II they were better unified, but they have since each been pulled out as separate disorders.

2. Because emotions adjust the organism to cope with certain kinds of situations, the normality of an emotional state cannot be assessed without information about the situation (except for certain extreme emotional states that will be abnormal no matter what the situation).

3. The word "disorder" implies an abnormality of the mechanisms that regulate emotions, for instance panic in safe situations. Such abnormal expressions of emotions must be carefully distinguished from emotions that arise from normal mechanisms but nonetheless cause distress or impaired function, such as depressive symptoms arising from a fruitless job search (Wakefield 1992).

4. Two global classes of abnormalities are possible for each emotion:
   a. Too much: too quickly aroused, too intense, too long, or aroused by nonspecific cues.
   b. Too little: too slowly aroused, too mild, too short, or aroused only by excessively specific cues.

5. Emotions researchers now recognize that emotions arise not from directly apprehended cues, but from an appraisal of what the new information means for an individual's ability to reach personal goals (Ellsworth 1991), a perspective that encourages attention to the life of the individual.

6. Negative emotions are just as useful as positive emotions. It is essential to avoid the clinician's illusion that makes all negative emotions seem abnormal and all positive emotions seem normal. No one comes to the clinic complaining of too little anxiety or an inability to feel sad, but this is just an artifact of our limited imagination and the absence of a scientific foundation for diagnosis of emotional disorders. People with these disorders exist, they just are not complaining or coming for treatment. Instead, they show up in the emergency room or jail or unemployment lines.

7. The mechanisms that regulate expression of emotions are governed by the smoke detector principle: inexpensive defenses are often subject to false alarms that are perfectly normal (Nesse 2005).

8. What is useful for our genes is not necessarily useful for our selves. Much normal emotion, especially negative emotion, may not be worthwhile for individuals at all, but only for their genes, and sometimes only for their genes in kin.

9. It is also important to recognize that some emotions may have been shaped in the Paleolithic, which may render them useless or even harmful in the

modern environment, even though they arise from normal mechanisms. For instance, expressing normal anger towards one's boss is likely to be maladaptive in a modern bureaucracy.

10. The distinction between negative and positive emotions intersects the distinction between abnormalities of excess and deficit to define four broad classes of emotional disorders, two of which have been neglected because they do not lead to subjective complaints. See Table 6.3 for details.

These and related principles provide a foundation for a scientific nosology for emotional disorders. An improved diagnostic system based on them will seem senseless to those who do not understand the behavioral biology of emotions.

This framework encourages systematic consideration of disorders of excess and deficiency for every emotion, not just anxiety and depression. The vast majority of treatment is for anxiety and depression, of course. They are usually called affects instead of emotions, to reflect their more enduring presence and the difficulty of connecting them to a very specific situation, but the conclusions are the same nonetheless.

An emphasis on the evolved utility of negative emotions should not lead to the conclusion that they are always useful, nor should it distract attention from the huge genetic variation in emotional predispositions. Some people rarely experience guilt while others feel constantly that they have somehow transgressed. Some people rarely worry, others worry constantly. Some people have never experienced romantic love, others fall madly in love with remarkable regularity. This variation poses a major problem for any attempt to determine what emotional experiences are normal.

Part of the answer is in how natural selection shapes the systems that regulate behavior. About half of the variation among individuals in most emotional traits arises from genetic differences. Why hasn't natural selection shaped a much more narrow range of responsiveness that we can recognize as "normal"? It is because humans have evolved in varying physical and social environments, so variations for a substantial range around the mean may not have a consistent effect on fitness. The resulting variation in personality traits is so large as to sometimes make us wonder if we are even justified in talking about human nature (see also Chapter 7).

Table 6.3 Categories of emotional disorders

	Excess	Deficit
Positive emotions	Mania, erotomania	Lack of joy, love, interest
Negative emotions	The usual emotional disorders: anxiety, depression, etc.	Deficits of anxiety, low mood, jealousy, etc.

We can now return to the DSM approach to diagnosis and the problem of taking context into account. The criteria for some disorders have built-in exclusions that generally ensure that anyone who meets criteria does indeed have a disorder. For instance, the criteria for panic disorder refer to "unexpected attacks", which excludes panic in life-threatening situations. Panic disorder is a reliably pathological condition in which the threshold for panic is so low that attacks emerge spontaneously. What an evolutionary perspective adds is recognition that panic is a normal response that is expressed too readily in panic disorder (Nesse 1987). This simple fact is useful in psychotherapy. Patients who have spent months fearing they have heart disease or a brain tumor often can be helped to recognize that their symptoms would indeed be useful in extreme danger and that they are experiencing mere false alarms.

Jealousy is a more complicated example. In the face of threats to a mate's fidelity, jealousy is normal and its absence is abnormal (Buss 2000). However, in many instances jealousy seems to be pathological. In many such cases, later evidence reveals the emotional response was an accurate indication of what was actually happening. In others, jealousy is aroused in someone who is depressed or who otherwise feels that his or her partner could do better with someone else. Then there is the psychoanalytic observation that jealousy can arise from projecting illicit desires onto an innocent partner. The important point here is that different cases of pathological jealousy may have different origins, but differentiating and understanding them requires knowing the situations in which the emotion is useful.

The overwhelmingly common disorder is, of course, depression. Increasingly, patients receive a diagnosis after a brief interview with a general physician, who prescribes antidepressants and advises a return visit in a month. Such perfunctory treatment is often justified by noting that the patient has met criteria for a pathological condition, major depression, whose presumed etiology is a deficiency of brain neurotransmitters. Drug treatment seems indicated and has been proven somewhat effective, so why not get on with it? This sequence completely ignores any possible utility of low mood, to say nothing of the causes of an individual's depression.

An evolutionary approach recognizes that low mood is useful to disengage effort from enterprises that are failing (Price and Sloman 1987; McGuire et al. 1997; Nesse 2000; Wrosch et al. 2003; Nettle 2004; Nesse 2009). If the person persists in useless efforts, the low mood escalates to full depression. It sounds easy to recommend giving up a fruitless pursuit until you realize that the goal may be getting a child off drugs, finding a job, or ending an affair. Treating depression without a careful examination of a patient's motivational structure is like treating a cough without first trying to find its cause (Nesse 2005, 2009).

6.7 The importance of analyzing motivational structure

The most useful contribution evolution makes to classifying, diagnosing, and treating emotional disorders may be the framework it offers for analyzing the motivational structure of an individual's life. Emotions arise from perceived problems and opportunities in the motivational structure. Like other organisms, humans must allocate three kinds of effort to get resources in six different areas. Somatic effort yields personal resources and material resources. Reproductive effort yields mates and offspring. Social effort yields allies and status. Many people seem to imagine that there is some normal way to live without compromises, but an evolutionary perspective reminds us all that every human action is an investment in getting one kind of resource at the expense of others. More time working out means less time working. More time impressing potential mates means less time for childcare. More time seeking status means less time for everything else.

The motivation regulation system seems to be designed, sensibly enough, to focus effort where it is most needed, that is, wherever it will yield the greatest payoffs of reproduction-limiting resources for the least investment. It would be so nice if our minds settled comfortably to a focus on what we have, but after any satisfaction the mind turns quickly to solving the next problem (Nesse 2004). Many tasks are simply enterprises that work well, such as a job or a marriage. What then is a life problem? A life problem is a difficulty in getting or keeping some important resource. People describe their problems in such diverse ways it is at first amazing to see how easily they all fit into a behavioral biological framework. The foundation for any therapy, especially psychotherapy, is a detailed understanding of what resources and sources of resources the person has, what he or she wants, how he or she is going about reaching these goals, and what the expectations are for success or failure. Many depressed people seem to have nothing major lacking in their lives, but as we get to know them, we find that they are striving to get love from a cold mother, sex from an uninterested spouse, or praise from a competitive boss. Or, they are trying to be truly good at all times, or to be the world's best in some status competition, achievements that are rare, and always temporary.

Good clinicians intuitively grasp motivational structures and the exigencies that give rise to an individual's problems. An evolutionary perspective and knowledge about emotions can help nearly every therapist to do this even better. There is the risk, of course, of using such insights to make crude suggestions. A patient who visited the emergency room attributed his depression to his wife's disinterest in sex. He was told, "Well, you will have to leave her or put up with it, those are your choices." Better therapists know that people have good reasons for why they live in the way they do. They examine their patients to see if symptoms arise from bipolar disorder or some other distinctive condition, but they recognize that diagnoses are no substitute for a deep understanding of a person's life.

6.8 Towards an evolutionary foundation for psychiatric nosology

The crucial missing ingredient for a truly medical nosology for emotional disorders is a functional understanding of the emotions and their regulation that is comparable to the functional understanding that physiology provides for the rest of medicine. Brain mechanisms are an essential part of this missing knowledge, but they are no more complete in themselves than the anatomy and mechanisms of the kidneys are for understanding the causes of renal pathology. Understanding the adaptive utility of a system is just as important for emotional as for physiological systems. Evolution provides the missing functional perspective for understanding the emotions and their disorders.

Many readers may agree with much of the above argument and yet find themselves asking, "Yes, the problems are large and clear, but how can we craft a DSM-V that avoids them?" A straightforward approach is to classify emotional disorders in the same way medical disorders are classified, based on the etiological factors that give rise to them. As already noted, medical symptoms are usually aroused by fairly specific tissue-changing pathologies, while emotional symptoms are most often aroused by untoward social situations that are much less susceptible to neat classification. I think it is likely that finding reliable and valid categories for emotional disorders has been difficult because they are not distinct diseases with specific causes. They arise from interactions between neural and cognitive diatheses interacting with inherently subjective appraisals of complex situations.

Sometimes, as in bipolar disorder or panic disorder, nearly all the variance is in genetic individual differences. In most cases, however, the circumstances giving rise to the emotion also play a major role. Although such situations are diverse, they can be categorized nearly as neatly as the aversive emotions. Here are a few examples of some of the causal situations that clinicians recognize intuitively:

- unrequited love (inability to give up a hopeless romantic goal)
- unable to find an intimate partner
- unable to leave an unsatisfactory intimate relationship
- unable to find a job anywhere near as high status as one's parents
- unable to leave an unsatisfactory job
- personality disorder that disrupts adaptation in multiple domains
- being blackmailed

- unable to help a child in trouble
- health problems that prevent functioning in crucial roles
- an affair that threatens major relationships
- partner may be having an affair
- partner is ill or disabled.

If someone is shivering, we do not look to the brain center that mediates shivering for an explanation, we instead look at the temperature, clothing, possible infection, etc. There is variation, both innate and acquired, in how readily different people shiver, but this is only part of the picture. We don't know what proportion of patients in our clinics have disorders of emotion regulation, and what proportion have basically normal mechanisms interacting with untoward circumstances. We need to know. Axis IV calls attention to life events. But because these events are carved out from consideration in reaching an Axis I diagnosis, diagnosis in psychiatry remains fundamentally different from that in the rest of medicine. General physicians no longer diagnose "cough disorder", they use different diagnoses depending on the etiology. DSM-V should incorporate life events and life situations into the main diagnostic categories, where their role as elicitors of emotions will be clearer.

Even stating the problem as distinguishing between individual differences or environmental effects is a mistake. Every emotional disorder arises from interactions among an individual's brain mechanisms, cognitive patterns, and his or her appraisals of the significance of information for reaching personal goals. The first variable is influenced by genes, early experiences, drugs, and other direct influences on brain mechanisms. The cognitive appraisal is influenced by personal and cultural experiences as well as individual idiosyncrasies from many sources. The events that arouse emotions arise from complex socio-cultural contexts, but also from the social network that grows around an individual, which is influenced by all the other variables. This is complicated. There is no getting around the complexity without excluding important factors or causal links.

We began with the supposition that continuing consternation about psychiatric nosology suggests that we are missing something basic. We are trying to categorize emotional disorders without a foundation of the understanding of the emotions and their origins and functions. This foundation illuminates many of the problems encountered by nosologies for emotional disorders. Unfortunately, however, it does not offer a simple solution. Instead, it shows that extreme emotional states arise not from one source, but from interactions of individual brain differences with complex life circumstances interpreted by diverse cognitive appraisals and psychological defenses. The categories of cleanly differentiated well-defined emotional disorders that we have been seeking do not exist (Nesse and Ellsworth 2009).

One could conclude from this that the DSM-IV approach to emotional disorders is about the best that can be done. This would be like nineteenth century physicians being satisfied with the diagnosis of "fever" because they can measure it reliably even though they don't know its causes. Instead, we need to proceed in the same way general physicians approach symptoms. They consider all possible causes in a differential diagnosis, then they investigate to find the etiology in any particular case. For emotional disorders, we must investigate the motivational structures of individuals in the same kind of detail that has been lavished on brain mechanisms.

To escape from abstractions, consider three cases.

Case 1

This 35-year-old woman has moderate depression and anxiety with intense anger and jealousy.

Situation: She learned her spouse is having an affair and wants to leave him, but she has no income and would have to give up her friends and her art career.

Person and vulnerability factors: Somewhat emotional in general, she has a slight tendency towards negative affect, but no enduring abnormal regulation of emotions in general and no family history of mental disorders.

Etiology: Her emotions are normal responses to her life situation.

Case 2

This 35-year-old woman has moderate depression and anxiety with intense anger and jealousy.

Situation: She suspects her spouse is having an affair but has no evidence of this. He denies it and tries to reassure her.

Person and vulnerability factors: She has always believed men will prefer other women and has been pathologically jealous in most of her relationships. She attributes this to her father leaving her mother when she was 5 years old. No family history of emotional disorder.

Etiology: Personality problem likely related to early life events; intense jealousy and other emotions are secondary.

Case 3

This 35-year-old woman has moderate depression and anxiety with intense anger and jealousy.

Situation: She accuses her spouse of having affairs, but only after she has been without sleep for several days, often while drinking.

Person and vulnerability factors: Strong family history of bipolar disorder.

Etiology: Genes causing bipolar disorder, complicated by alcoholism, relationship problems, and extreme jealousy.

These cases illustrate what most clinicians know: the same clinical conditions can arise from fundamentally different causes. It therefore makes no sense to view these emotions as a specific disorder. In case 1 the symptoms arise from an untoward situation, in case 2 from a personality disorder, in case 3 from bipolar disorder. Every clinician will think of more realistic and complex cases, for example a man with bipolar tendencies, chronic relationship difficulties, a low threshold for jealousy and anger, who drinks heavily and is having an affair.

The implications for the DSM-V are substantial, but not simple. Detailed consideration of the opportunity by the DSM-V Committees is indicated. It is essential to recognize that an evolutionary foundation is fully compatible with other biological and medical approaches. The DSM-IV has encouraged much useful work on the problem of why some people have tendencies to excessive anxiety and depression, and the brain mechanisms that mediate affects. Evolution puts this knowledge in perspective by emphasizing that these affects can be normal, their regulation mechanisms were shaped by natural selection, and there are likely good evolutionary reasons why these mechanisms are so vulnerable to failure. It also highlights the need to look for disorders of regulation for all emotions, especially the neglected disorders characterized by deficient negative or excessive positive affect.

Some people think that the utility of negative emotions means that they should not be treated. This is a serious mistake. We have much to learn from general medicine, where both the utility and the harm caused by responses such as pain and diarrhea is well recognized, and where relief of suffering by blocking defensive responses is a routine goal of clinical work, whether the symptom is being aroused normally or arises from a faulty mechanism. Campaigns to convince the public and practitioners that depression and anxiety are brain diseases have motivated much useful research and have decreased stigma, but they are biologically naïve. An evolutionary approach supports a more medical model in which clinicians recognize many symptoms as defenses shaped by natural selection that are aroused by more primary causes, and others arising from defects in the systems that regulate defenses. The clinician tries to identify and remove the factors arousing the symptoms when possible. When that is not possible, a good psychiatrist tries to relieve suffering, often by using drugs to block normal responses. If that is not possible, then the clinician tries to relieve suffering, even if that means using drugs to block normal defensive responses. Evolutionary biology offers a biological foundation for a genuinely medical model for understanding and diagnosing emotional disorders.

References

American Psychiatric Association (1952) *Diagnostic and Statistical Manual of Mental Disorders.* American Psychiatric Association, Washington, DC.

American Psychiatric Association (1968) *Diagnostic and Statistical Manual of Mental Disorders.* (2nd edition). American Psychiatric Association, Washington, DC.

American Psychiatric Association (1980) *Diagnostic and Statistical Manual of Mental Disorders.* (3rd edition). American Psychiatric Association, Washington, DC.

American Psychiatric Association (1994) *Diagnostic and Statistical Manual of Mental Disorders.* (4th edition). American Psychiatric Association, Washington, DC.

Axelrod, R. and Hamilton, W.D. (1981) The evolution of cooperation. *Science,* **211**, 1390–96.

Barlow, D.H. (1991) Disorders of emotions: Clarification, elaboration, and future directions. *Psychological Inquiry,* **2** (1), 97–105.

Beutler, L.E. and Malik, M.L. (2002) *Rethinking the DSM: A Psychological Perspective.* American Psychological Association, Washington, DC.

Brown, G.W. and Harris, T. (1978) *Social Origins of Depression.* The Free Press, New York.

Buss, D.M. (2000) *The Dangerous Passion: Why Jealousy is as Necessary as Love and Sex.* Free Press, New York.

Clark, L.A. (1999) Special section on the concept of disorder. *Journal of Abnormal Psychology,* **108**, 371–472.

Ekman, P. and Davidson, R.J. (1994) *The Nature of Emotion: Fundamental Questions.* Oxford University Press, New York.

Ellsworth, P. (1991) Some implications for cognitive appraisal theories of emotion. In K.T. Strongman, *International Review of Studies of Emotion.* Wiley, Chichester, pp. 143–61.

Faust, D. and Miner, R.A. (1986) The empiricist and his new clothes: DSM-III in perspective. *American Journal of Psychiatry,* **143** (8), 962–7.

Feighner, J.P., Robins, E., Guze, S.B., Woodruff, R.A., Jr., Winokur, G., and Munoz, R. (1972) Diagnostic criteria for use in psychiatric research. *Archives of General Psychiatry,* **26** (1), 57–63.

Fessler, D.M. and Haley, K. (2003) The strategy of affect: emotions in human cooperation. In P. Hammerstein (ed.), *Genetic and Cultural Evolution of Cooperation*: Dahlem Workshop Report 29. MIT Press, Cambridge, MA, pp. 7–36.

Frances, A.J.F. and Egger, H.L. (2003) Whither psychiatric diagnosis. *Australian and New Zealand Journal of Psychiatry,* **33**, 161–5.

Gilbert, P. (1989) *Human Nature and Suffering.* Lawrence Erlbaum, Hove.

Goodwin, D.W. and Guze, S.B., (1979) *Psychiatric Diagnosis.* Oxford University Press, New York.

Grob, G. (1991) Origins of DSM-I: A study in appearance and reality. *American Journal of Psychiatry,* **148**, 421–31.

Guze, S.B. (1992) *Why Psychiatry is a Branch of Medicine.* Oxford University Press, New York.

Horwitz, A.V. (2002) *Creating Mental Illness.* University of Chicago Press, Chicago.

Horwitz, A.V. and Wakefield, J.C. (2007) *The Loss of Sadness: How Psychiatry Transformed Normal Sorrow into Depressive Disorder.* Oxford University Press, New York.

Houts, A.C. (2002) Discovery, invention, and the expansion of the modern *Diagnostic and Statistical Manuals of Mental Disorders*. In L. Beutler and M. Malik (eds), *Rethinking the DSM: A Psychological Perspective.* American Psychological Association, Washington, DC, pp. 17–65.

Izard, C.E. (1992) Basic emotions, relations among emotions, and emotion-cognition relations. *Psychological Review*, **99** (3) 561–5.

Kendell, R.E. (1975) The concept of disease and it implications for psychiatry. *British Journal of Psychiatry*, **127**, 305–15.

Kessler, R.C. and Ustun, T.B. (2004) The World Mental Health (WMH) Survey Initiative Version of the World Health Organization (WHO) Composite International Diagnostic Interview (CIDI). *International Journal of Methods in Psychiatric Research*, 13 (2), 93–121.

Ketelaar, T. and Clore, G.L. (1997) Emotion and reason: distinguishing proximate effects and ultimate functions. In G. Matthews (ed.), *Personality, Emotion, and Cognitive Science*. Elsevier Science Publishers, Amsterdam, pp. 355–96.

Kihlstrom, J.F. (2002) To honor Kraepelin. . .: From symptoms to pathology in the diagnosis of mental illness. In L.E. Beutler and M.L. Malik (eds), *Rethinking the DSM: A Psychological Perspective*. American Psychological Association, Washington, DC, pp. 279–303.

Klerman, G.L. (1984) The advantages of DSM-III. *American Journal of Psychiatry*, **141** (4), 539–42.

Kraepelin, E. and Dierdorf, A.R. (1907) *Clinical Psychiatry: A Textbook for Students and Teachers*. (7th edition). Macmillan, New York.

Lewis, M. and Haviland-Jones, J.M. (2000) *Handbook of Emotions*. Guilford Press, New York.

Marks, I.M. and Nesse, R.M. (1994) Fear and fitness: An evolutionary analysis of anxiety disorders. *Ethology and Sociobiology*, **15**, 5–6, 247–61.

Mayr, E. (1961) Cause and effect in biology. *Science*, **134** (3489), 1501–6.

McGuire, M.T. and Troisi, A. (1998) *Darwinian Psychiatry*. Oxford University Press, Oxford.

McGuire, M.T., Troisi, A., and Raleigh, M.M. et al. (1997). Depression in evolutionary context. In S. Baron-Cohen (ed.), *The Maladapted Mind*. Psychology Press, East Sussex, pp. 255–82.

Monroe, S.M., Harkness, K., Monroe, S.M., Harkness, K., Simons, A.D., and Thase, M.E. (2001) Life stress and the symptoms of major depression. *Journal of Nervous and Mental Disorders*, **189** (3) 168–75.

Nesse, R.M. (1987) An evolutionary perspective on panic disorder and agoraphobia. *Ethology and Sociobiology*, **8**, 73S–83S.

Nesse, R.M. (1990) Evolutionary explanations of emotions. *Human Nature*, **1** (3), 261–89.

Nesse, R.M. (1998) Emotional disorders in evolutionary perspective. *British Journal of Medical Psychology*, **71** (4), 397–416.

Nesse, R.M. (2000) Is depression an adaptation? *Archives of General Psychiatry*, **57**, 14–20.

Nesse, R.M. (2004) Natural selection and the elusiveness of happiness. *Philosophical transactions of the Royal Society of London. Series B, Biological Sciences*, **359** (1449), 1333–47.

Nesse, R.M. (2005) Natural selection and the regulation of defenses: a signal detection analysis of the smoke detector principle. *Evolution and Human Behavior*, **26**, 88–105.

Nesse, R.M. (2009) Explaining depression: Neuroscience is not enough, evolution is essential. In C. M. Pariente, R.M. Nesse, D.J. Nutt and L. Wolpert (eds), *Understanding Depression: A Translational Approach*. Oxford University Press, Oxford, pp. 17–36.

Nesse, R.M. and Ellsworth, P.C. (2009) Evolution, emotions, and emotional disorders. *American Psychologist*, **64** (2), 129–39.

Nettle, D. (2004) Evolutionary origins of depression: a review and reformulation. *Journal of Affective Disorders*, **81** (2), 91–102.

Oatley, K. and Johnson-Laird, P.N. (1996) The communicative theory of emotions: Empirical tests, mental models, and implications for social action. In L.L. Martin and A. Tesser (eds), *Striving and Feeling: Interactions among Goals, Affect, and Self-regulation*. Erlbaum, Hillsdale, NJ, pp. 363–93.

Phillips, K.A., First, M.B., and Pincus, H.A. (2003). *Advancing DSM: Dilemmas in Psychiatric Diagnosis*. American Psychiatric Association, Washington, DC.

Plutchik, R. (2003) *Emotions and Life: Perspectives from Psychology, Biology, and Evolution*. American Psychological Association, Washington, DC.

Price, J.S. and Sloman, L. (1987) Depression as yielding behavior: An animal model based on Schelderup-Ebbe's pecking order. *Ethology and Sociobiology*, **8**, 85s–98s.

Rosenhan, D.L. (1973) On being sane in insane places. *Science*, **179**, (4070), 250–8.

Shorter, E. (1997) *A History of Psychiatry: From the Era of the Asylum to the Age of Prozac*. John Wiley & Sons, New York.

Spitzer, R.L., Williams, J.B., Gibbon, M., and First, M.B. (1992). The structured clinical interview for DSM-III-R (SCID). I: History, rationale, and description. *Archives of General Psychiatry*, **49** (8), 624–9.

Szasz, T.S. (1974) *The Myth of Mental Illness: Foundations of a Theory of Personal Conduct*. Harper & Row, New York.

Tooby, J. and Cosmides, L. (1990) The past explains the present: Emotional adaptations and the structure of ancestral environments. *Ethology and Sociobiology*, **11** (4/5), 375–424.

Troisi, A. and McGuire, M. (2002) Darwinian psychiatry and the concept of mental disorder. *Neuroendocrinology Letters*, **23** (Suppl 4), 31–8.

Wakefield, J.C. (1992) Disorder as harmful dysfunction: a conceptual critique of DSM-III-R's definition of mental disorder. *Psychological Review*, **99** (2), 232–47.

Wakefield, J.C. and First, M. (2003) Clarifying the distinction between disorder and non-disorder: Confronting the overdiagnosis ("false positives") problem in DSM-V. In K.A. Phillips, M.B. First and H.A. Pincus (eds), *Advancing DSM: Dilemmas in Psychiatric Diagnosis*. American Psychiatric Press, Washington, DC, pp. 23–56.

Wakefield, J.C., Schmitz, M.F., First, M.B., and Horwitz, A.V. (2007) Extending the bereavement exclusion for major depression to other losses: evidence from the National Comorbidity Survey. *Archives of General Psychiatry*, **64**, 433–40.

Watson, D. (2005) Rethinking the mood and anxiety disorders: a quantitative hierarchical model for DSM-V. *Journal of Abnormal Psychology*, **114** (4), 522–36.

Wilson, M. (1993) DSM-III and the transformation of American psychiatry: a history. *American Journal of Psychiatry*, **150** (3), 399–410.

Wrosch, C., Scheier, M.F., Wrosch, C., Scheier, M.F., Miller, G.E., Schulz, R., and Carver, C.S. (2003) Adaptive self-regulation of unattainable goals: Goal disengagement, goal reengagement, and subjective well-being. *Personality and Social Psychology Bulletin*, **29** (12), 1494–1508.

Chapter 7

Normality, disorder, and evolved function: the case of depression

Daniel Nettle

About suffering they were never wrong,
The Old Masters; how well, they understood
Its human position; how it takes place
While someone else is eating or opening a window or
just walking dully along;

W.H. Auden, *Musée des Beaux Arts*

Following the work of Emil Kraepelin, many philosophers and psychiatrists firmly believe that it is possible to distinguish neatly between normality and disorder. In this chapter, I scrutinize this belief while focusing on the difference between (normal) low mood and depression. My main argument is that evolutionary thinking *cannot* guide us in discriminating between low mood and depression, even though it helps us understand the many functions of depressive symptoms. Evolutionary considerations cannot help us here because determining whether one's mood system is dysfunctional requires an assessment of the proportionality of a mood reaction, which in turn requires a deep idiographic understanding of one's cognitive and ecological context. And yet such practical difficulties pale before the observation that there is substantial individual variation in the threshold of activation for the human mood system, which eventually implies that the individual is the ultimate benchmark on which to distinguish between health and disease. But why should we even bother to find (or draw) a fine line between health and disease, or between function and dysfunction? One of the arguments behind our obsession with this issue, I argue, is that it is held to have important implications for psychiatric practice. After all, don't we need a way of distinguishing normality from disorder in order to determine who needs treatment, and who doesn't? In a final section, I note that even if we would be able to discriminate (mental) function from (mental) dysfunction, it would be of limited use, because it would lead us to treat people who seek no treatment, and to deny treatment to people who want to be treated. On this view, subjective distress is a better guide in allocating aid.

7.1 Introduction

Psychiatry has a great deal of business with depression. In the World Health Organization's *Global Burden of Disease* report of 1996, major depression is listed as the single greatest cause of medically relevant problems in mid-life for the contemporary human population. Manic-depressive illness, which of course involves depression, additionally appears at number six in the list, and four others in the top ten (alcohol misuse, self-inflicted injuries, violence, and schizophrenia) are conditions in which depressed mood plays an important part (see Andreasen 2001, p. 5). Thus, understandably, much diagnostic and research effort within medicine and allied sciences goes into identifying and explaining depressive conditions.

Prevailing diagnostic practice identifies a countable number of discrete disorders of mood, which are held to be categorically distinct from one another, and from normal mood fluctuations. This chapter examines the basis of such distinctions, and particularly the distinction between major depressive disorder (MDD) and normal low mood. Broadly speaking, one can take a *realist* approach to such distinctions (that is, claim there is a discrete difference in the world between normal and abnormal moods, which would exist regardless of our detecting it) or a *conventionalist* approach, in which the normal/disorder boundary is placed by human activity onto a world that contains no natural break. Realist claims about the MDD/normality distinction could be mounted on the basis of inductive evidence, for example by showing that there are two nonoverlapping sets of mood symptoms observed in the human population, one rarer than the other and associated with different outcomes. In section 7.2 I suggest that this claim fails evidential tests. An alternative realist approach would be to appeal to evolved functions (and hence an evolutionarily grounded criterion for dysfunction). Here, the difference between normal low mood and MDD would be to do with whether or not mood mechanisms were fulfilling their evolved function in a particular case. In sections 7.3 and 7.4 I argue that although evolutionary thinking does elucidate the functions of low mood, it does not at present aid in distinguishing "normal" low mood from depressive disorder. I conclude that, at the very least, it is empirically difficult to distinguish normal low mood from depressive disorder, and, more tentatively, that it may be impossible in principle because all biological populations contain inter-individual variation in reaction norms.

In section 7.5 I argue that most currently treated MDD may not be, or at least cannot be shown to be, disorder *sensu stricto*. However, this does not mean that it is not a treatable condition, nor that treatment should be withheld. Rather, the need for medical support depends on the sufferer's distress and their perception that their low mood represents a problem for them. To the extent that discrete categorical distinctions are made, they are purely conventional. This means that they will change over time and between societies. I also suggest that they may not always be particularly helpful for either researchers or patients.

7.2 Inductive evidence for a categorical depression/normality distinction

One way to argue for the reality of a categorical normality/disorder distinction would be to show that, in the distribution of symptoms across persons, there is a point of rarity, with a small group of people on one side of the dividing line experiencing poorer health outcomes than the majority non-ill group. This would make the diagnosis of MDD conceptually similar to that of many non-psychiatric disorders, and open the way for the eventual discovery of a specific lesion or discrete pathophysiological marker responsible for the symptom set. The search for discrete syndromes and subsequently for underlying lesions has been a recurrent preoccupation of psychiatric research since Kraepelin, and informed the creation of (the third and fourth edition of) the American Psychiatric Association's *Diagnostic and Statistical Manual of Mental Disorders* (DSM) (see also Chapter 6).

However, administration of self-report surveys of current depressive symptoms in general populations reveals very few people with no symptoms at all (although this depends on how the questionnaire is framed), but, more importantly, yields absolutely continuous distributions from fewest symptoms to most. For example, the short General Health Questionnaire (GHQ-12) asks respondents to report on their experience of 12 symptoms, such as lack of enjoyment of normal day-to-day activities, feeling unhappy and depressed, feeling worthless, and inability to concentrate, within the last few weeks (see Kalliath *et al.* 2008). The GHQ-12 was administered to a representative sample of the British population (the National Child Development Study cohort, n = 11,281) at age 42, and, as Fig. 7.1 shows, there is a right-skewed but continuous distribution of symptoms, with the vast majority of individuals experiencing at least some. The developers of instruments such as the GHQ or the widely used Beck Depression Inventory often provide suggested threshold scores for clinical concern. However, these reflect a conventional cut-off point rather than a real point of rarity. As for outcomes, depressive symptoms are associated with poor long-term health, but the relationship is graded, with a smooth proportionality between negative symptoms and outcomes across the whole continuum from sub-clinical symptoms, to mild diagnosed disorder, to severe diagnosed disorder, rather than there being a step-function at a boundary point (Neeleman *et al.* 2002; Kessler *et al.* 2003).

DSM-IV (American Psychiatric Association 1994) recognizes the need for clinicians to distinguish MDD from sub-clinical low mood. Diagnosis of MDD requires (i) the presence of at least five symptoms from a list (this includes feeling low or sad, anhedonia, weight loss or gain, insomnia or hypersomnia, psychomotor agitation or retardation, fatigue, feelings of worthlessness or excessive guilt, diminished ability to think or concentrate, and suicidal ideation), (ii) that these symptoms cause clinically significant distress or impairment, and (iii) that they be present for at least 2 weeks. There are a number of points to make about these criteria. Under criterion (i), several of the symptoms listed include opposite alternatives (e.g., insomnia or hypersomnia, agitation or retardation) and, as shown in Fig. 7.1, these kinds of symptoms are common and continuously distributed across the population. There is also an element of arbitrariness. Is there really something qualitatively different about experiencing

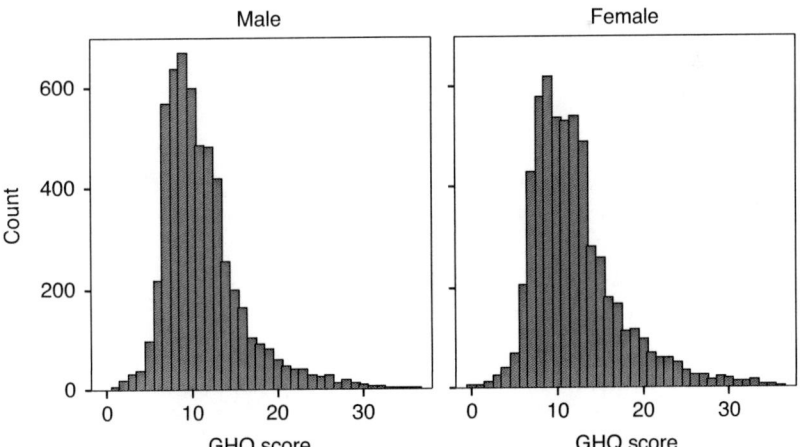

Fig. 7.1 Distribution of scores on a self-report measure of affective symptoms (the short form of the General Health Questionnaire; higher scores indicate more symptoms reported; "count" means the number of individuals in the sample with that symptom score), by 11,281 British 42-year-olds from the National Child Development Study cohort.

five symptoms as opposed to four? Criterion (ii) is a strict circularity, since presumably the criterion for clinical significance is that some illness be present, and if the criterion for the presence of illness is clinically significant impairment, then neither concept can be defined independently of the other. Interestingly, this criterion was added to the 4th edition of the DSM in response to concerns that DSM-III's criteria were too inclusive, since in community samples those criteria led to around 30% of all mid-life adults being deemed to have some disorder or another (Kessler et al. 2003). As for criterion (iii), we again have issues of arbitrariness to deal with. Is a mood episode of 13 days' duration really qualitatively different from one lasting a day longer? One would suspect the answer to be no, and indeed, Kendler and Gardner (1998) examined the evidence for discontinuities in the distribution of mood fluctuations and found no real evidence of anything but a continuum. Using data from repeated interviews of female twins, they asked how current symptoms predicted future mood difficulties, and also how they predicted mood problems in the other twin. If there really is a categorical distinction between a mental illness, depression, and normal mood variation, and if that distinction lies where DSM-IV identifies it, then having five symptoms in the present ought to predict mood difficulties in the future or in the other twin much more strongly than having four symptoms does, and having symptoms lasting more than 2 weeks ought to predict future problems or problems in the other twin much more strongly than having symptoms of shorter duration. The researchers in fact identified gradient relationships: the more depressive symptoms people had in the present, the more likely they were to have depressive problems in the future, and the more likely their twins were to have depressive problems. These risks increased in smooth, linear fashion with both the number and the duration of current mood symptoms. It didn't make any difference whether current symptoms fell into DSM-IV's zone of "disorder" or not.

As well as inter-individual variation in mood, there is intra-individual variation in mood across time. DSM-IV recognizes that some of this fluctuation should not be considered pathological, but as normal functioning. For example, it specifically excludes from the MDD category sets of otherwise qualifying symptoms than can be related directly to bereavement. In other words, the same symptoms that would constitute illness at other times are considered normal functioning in the aftermath of the death of a loved one. The question of arbitrariness arises again: why bereavement, but not divorce, or the loss of a job? In terms of course or sequelae, there is no empirical justification for separating negative emotional responses to bereavement to negative emotional responses to other types of loss (Wakefield et al. 2007). Overall, then, there seems to be little pattern in the data on variation in human mood that would allow an inductive localization of a normal/abnormal boundary. Instead, we see only the continuum of human affective experience, from buoyancy to despair.

7.3 Evolved functions, dysfunctions, and depression

Purely inductive considerations, then, do not much illuminate where the normality/disorder boundary should be drawn for MDD. This section examines whether evolutionary thinking can clarify the matter. There are actually two sets of questions here, which I will tackle in turn. First, does evolutionary thinking help us understand why low mood exists and has the features that it does? Second, does evolutionary thinking help us localize the boundary between normal low mood and disorder? The first and second questions are linked because of Wakefield's (1999) influential definition of disorder as *harmful dysfunction*. Wakefield argued that the function of a mechanism is the recurrent effect of that mechanism on the organism which caused that mechanism to be selected or to be retained in its current form (for a full discussion of this notion of function, see Chapter 5). Thus, the function of a heart is to pump blood, since hearts in their current form were selected exactly because they had the recurrent effect of causing blood to circulate. Wakefield uses this evolutionary notion of function to in turn identify dysfunctions, which are failures of mechanisms to fulfill their biological functions. Thus, a heart that is not pumping blood is a dysfunctional one. Wakefield argues that disorder requires both dysfunction to be present and a judgment to be made that the dysfunctional is harmful. The latter judgment may vary according to local conceptions of what is harmful and what is tolerable, and thus Wakefield's position is a hybrid of realism (about function and dysfunction) and conventionalism (about what is harmful, or sufficient harm to warrant medical attention). Nonetheless, for Wakefield, a necessary condition for the existence of a disorder is that a mechanism be dysfunctional. This seems to open the door for a principled distinction between normal low mood, which would be cases where the low mood is fulfilling its selected function, and disordered mood, which would be where the system has gone wrong. However, to make this distinction work, it is first necessary to identify what the functions of low mood might be.

The fact that fluctuations of mood in response to circumstances are a universal, reliably developing feature of human minds strongly suggests some adaptive function, and the fact that low mood is unpleasant is no evidence of it representing dysfunction (see also Chapter 6). The capacity for physical pain, for example, is adaptive in that it protects the soma, but the experience of pain is unpleasant. In fact, physical pain is adaptive not in spite of, but exactly because of, its unpleasantness. A number of evolutionarily minded thinkers have suggested that low mood, like physical pain, is adaptive (Price 1992; Nesse 2000;

Allen and Badcock 2003; Andrews and Thomson 2009). Since my focus in this chapter is on the issue of disorder, I will provide only a partial review of these proposals here. The reader is directed to the references cited for more detail.

Low mood is prototypically triggered by loss or by a failure to progress towards important goals, and all the proposals discussed below are based around the idea that low mood is a conditional response to adverse situations. The proposals vary somewhat in what benefit they see low mood as having, given that the situation is adverse. Some key types of benefit are the following:

1. *Motivational disengagement.* Low mood, with its associated desire to withdraw and loss of motivation, serves to disengage from a goal that has become unattainable or is unlikely to be fruitful given current circumstances (Nesse 2000). This proposal is supported by clinical observations and by experimental evidence of the role of mood in goal engagement and disengagement in healthy volunteers (Klinger 1975; Carver and Scheier 1990).

2. *Risk regulation.* Allen and Badcock (2003) suggest that mood is involved in the regulation of risk-taking (in the behavioral ecological sense of preferences for behaviors whose outcome is variable). By this argument, when the individual is in a poor state, she can ill afford to take on novel endeavors that could be fruitful, but might go wrong, because she cannot currently bear the consequences of further depletion in her position. By contrast, an individual for whom things are currently going well can bear the cost of a failure, and can thus orient themselves towards riskier, but potentially more rewarding, novel behaviors. The proximate mechanisms for low mood reducing exposure to risk would be symptoms such as anhedonia (which makes the benefit of rewards seem smaller), pessimism (which makes novel ventures seem less likely to be worthwhile), and fatigue (which inhibits novel behaviors). Nettle (2009) provides a formal evolutionary model of this idea, but some of the best evidence for the general principle comes from studies of the effects of either experimentally induced or naturally occurring low mood on decisions in gambling tasks (Yuen and Lee 2003; Grable and Roszkowski 2008).

3. *Cognitive problem solving.* Andrews and Thomson (2009) marshall an array of evidence to suggest that the cognitive changes typical of low mood (a slower, more analytical thinking style with attention to the details of the problem, and a tendency to ruminate on the preeminent issue), although not generally optimal, are adaptive in the specific situation that the individual faces a complex, difficult to solve, fitness-relevant life problem. Apparently contrary elements of low mood, such as poor concentration, make sense once we realize that it is concentration on anything *other* than the triggering life problem that is impaired. Apparently unrelated symptoms, such as fatigue and anhedonia, may in fact function to divert all available resources

to the analysis of the triggering problem and to shut down distractions. The case for this function revolves around clinical observations on rumination, and also experimental evidence from psychology that whilst low mood impairs performance on some tasks, it improves performance on others.

4. *Social signaling.* Low mood often involves crying and seeking aid, and it seems a reasonable proposal that these behaviors function to elicit support from key allies in times of need (Watson and Andrews 2002; Keller and Nesse 2006). Similarly, low mood is associated with submissive and non-competitive behaviors, and thus reduces the likelihood of costly antagonistic interactions at a time when the person is not in a position to compete for status (Price 1992).

5. *Learning.* Many of the above functions could be fulfilled without low mood having a subjectively unpleasant component. One possible reason that defense states such as pain and low mood have a negative valence at the phenomenal level is that part of their function is to promote decisions that avoid such situations in future, and by being unpleasant they enter into processes of learning and decision-making with an aversive or deterrent character.

It is not necessary to choose any one of these as *the* function of low mood. Emotions (and presumably moods) are *suites* of coordinated body and brain changes. For example, fear involves changes in heart rate, immune activity, cognition, and vigilance. Each of these changes has a separate function, but they are coordinated in a suite because they all tend to be useful in a similar class of situations. Similarly, low mood is a suite consisting of a number of different elements, each of which may confer a different benefit, but all of which tend to be useful in situations of loss or failure. Perhaps the strongest evidence for this "suite" view comes from recent findings by Keller and Nesse that different types of precipitating life situations lead to different profiles of low mood symptoms (Keller and Nesse 2005, 2006). Thus, in a sense, the relevant evolutionary question is not "What is the function of low mood?", but rather "What is the function of rumination?", "What is the function of anhedonia?", and so on. I will not further consider the answers to these questions here. Rather, what is relevant is that there are cogent hypotheses concerning the functions of low mood, that they lead to testable predictions, and that empirical work is being done. Here, we need to examine what those evolutionary accounts have to say about the functional/dysfunctional boundary.

Early evolutionary work on depression did not explicitly seek to distinguish normal low mood from pathological depression, and sometimes used "low mood" and "depression" interchangeably (e.g., Nesse 2000). Indeed, one possible implication of evolutionary work is that much of what we currently medicalize as disorder may represent the proper functioning of evolved mechanisms. Andrews and Thomson (2009) give this position its fullest expression to date, explicitly relating their functional account to the view that much of what we currently diagnose and treat as disorder, especially in the USA, may in fact be "normal" sadness (Horwitz and Wakefield 2007). Note that Andrews and Thomson do not exclude the possibility that there may be cases where low mood mechanisms are dysfunctional. They merely point out the lack of empirical evidence for a sharp distinction between normal and disordered low mood.

Other evolutionary writers have sought to maintain a distinction between normal low mood, for which their adaptive explanations are appropriate, and clinical disorder, which represents dysfunction. For example, Allen and Badcock state that:

> mild (and predominantly transient) depressive states of the type that are experienced by most persons from time to time reflect a behavioural adaptation that has been shaped by natural selection. The more extreme states … associated with clinical disorders are, like many other diseases, perhaps best understood as pathologies that reflect divergence from the normal functioning of adaptive mechanisms.
>
> (Allen and Badcock 2003, p. 888)

In similar vein, Nesse and Stearns suggest that:

> "while there is no doubt that much anxiety and depression is pathological, the capacities for anxiety and depression were shaped by natural selection … the problem [in clinical cases] is dysregulation of a response that can be normal and useful."
>
> (Nesse and Stearns 2008, p. 40, for very similar comments see Wolpert 2008; Nettle 2009)

These distinctions all defend the commonsense intuition that intractable or severe bouts of depression (such as those that lead to suicide) might be pathological, even whilst accepting that psychological pain in general has a function. The distinction seems reasonable. It would be hard to defend the claim that the mechanisms which regulate mood *never* malfunction; this would make them different from any other known biological system. Unfortunately, however, as Andrews and Thomson (2009) point out, these writers (myself included) tend to neither produce actual evidence for the dysfunctional nature of clinical depression, nor provide a principle for distinguishing between low mood and disorder in particular cases.

If the low mood system is working correctly, then it should produce mood shifts whose severity and duration are appropriate to the triggering life event. One potential principle for identifying dysfunction, then, would invoke some kind of proportionality between precipitators and symptoms. The DSM-IV implicitly accepts this view in part. This is evident in the bereavement exclusion for otherwise qualifying depressive symptoms. It is also evident in the inclusion of the category of "adjustment disorder with depressed mood" as distinct from MDD itself. Adjustment disorders are indicated where the individual has had a recognizable life stressor within 3 months of the onset of symptoms, and the symptoms terminate within 6 months of the resolution of the stressor. However, it is defined as a disorder using the criterion that the magnitude of the affective reaction is "in excess" of what would be expected given the nature of the stressor.

The proportionality criterion is superficially appealing. It works well at distinguishing some clear cases. For example, mood changes that occur when the person reports that life is going very well, but they have had a bang on the head, would obviously be a candidate for disorder, whereas more severe symptoms consequent on divorce might not be. A person who suddenly becomes suicidally hopeless but reports no adverse personal circumstances would appear a strong candidate for disorder, whereas a person who is equally distressed after years of abuse is probably not. However, there are many difficulties with the proportionality criterion. For one, the absolute magnitude of the distress becomes irrelevant. Extreme distress may be entirely functional in the evolutionary sense, if the person's situation is indeed bad. Similarly, constant rumination, anxiety, and weeping in a woman with no job and financial problems might be signs that the system is working perfectly well. These points are conceptually clear, but scant consolation for the physician dealing with a distraught woman across the table and proposing no treatment because her mood system is in great shape. A second difficulty is that a clinician cannot really know in full the life context of a patient. The same magnitude of symptoms could be produced by a healthy mood system in a very bad situation or a disordered mood system in a relatively good one, and it is very hard to tell how bad someone's situation is without a deep idiographic understanding of their goals, opportunities, and ecological context. Even assuming such an understanding can be achieved, the proportionality criterion would require that different people's negative life events be translated into a common currency, and the person's symptoms be compared to population norms for how many or how severe symptoms the person *should* have given their situation. Needless to say, this is difficult to imagine working in practice, and may be problematic conceptually too (see section 7.4). Given this, it is inevitable that diagnostic criteria will be based on simple and rather arbitrary conventional cut-off points, and we are therefore back with the view that the boundaries of MDD are conventional rather than real.

Another issue with the dysfunction-based approaches to identifying depression is that the design features of mood itself generate an asymmetry. Low mood symptoms are supposed to be unpleasant, and the subject is supposed to want to avoid them (see function 5, above). That is, their aversive character is one aspect of how they achieve their function. It follows that people who are experiencing low mood symptoms, regardless of whether they are functional or not, will want them to go away, whereas people whose mood system is dysfunctional and does not produce *enough* low mood given the situation will feel fine and not show up in clinics, regardless of how much harm their insouciance is doing to their success in life. Such people must exist, but have never been systematically studied, since by definition they do not feel distressed. We have only fragmentary indications of the problems they might face, for example from studies showing more accidents when young and reduced educational success amongst those with low neuroticism (McKenzie *et al.* 2000; Lee *et al.* 2006; Chapter 6).

Overall, then, a physician who attempted to follow a criterion based on evolved function might well be in conflict with her patients, who would demand treatment to be apportioned (mainly) on the basis of how bad things felt, not on the basis of whether the mechanisms were maximizing current reproductive success or fulfilling their ancestral function. I return to this issue in concluding the chapter.

7.4 **The challenge of individual variation**

The prospects for discriminating function from dysfunction may be worse than suggested so far when we consider in more detail the way that selection is likely to act on emotional systems. Research in evolutionary biology generally finds that although core design features of adaptations are species-typical, quantitative characters exhibit heritable variation (Nettle 2006). For example, all guppies have predator avoidance mechanisms, but there is heritable inter-individual variation in how close they have to be to the predator before these mechanisms are triggered (Dugatkin 1992; O'Steen *et al.* 2002). Why does such variation exist? In part, for a polygenic character such as a threshold, there is simply a high effective mutation rate due to the number of genes involved. Stabilizing natural selection can keep the central tendency of the population at the optimum, and can act to reduce the variation, but can never reduce it quite to zero because of the constant input of new mutations (Keller and Miller 2006). There is also a second reason. The optimal threshold from a fitness point of view may not be the same for every individual in every part of the habitat. This follows from simple reverse engineering. The theoretical optimal threshold for an anti-predation mechanism, for example, can be shown using signal detection theory to depend on the cost of false positives (detecting a predator who is not in fact there), the cost of false negatives (failing to detect a predator who is in fact there), and the actual frequency of predators in the environment (Haselton and Nettle 2006). In other words, if you live in an environment that is in fact more dangerous, the best threshold for detecting danger to have is lower than it would be if you lived where it was safer. For the guppies, the consequence of this is that individuals living in the upper tracts of streams, where piscivorous predators can seldom stray, have higher thresholds for anti-predator behavior than those living downstream, and the introduction of predators produces a measurable evolutionary response in terms of antipredator thresholds (O'Steen *et al.* 2002).

Human beings vary significantly on the personality dimension of neuroticism, which amounts to variation in the threshold of activation of negative emotion systems such as anxiety, sadness, and stress (Nettle 2004, 2006). Just as in the guppy case, this may reflect the effects of locally specific selection in the immediate past. Perhaps my ancestors lived in more dangerous, socially precarious parts of the habitat than yours. If this persisted for a few generations, it could be enough to have produced differential selection. How does this apply to the current topic? Cosmides and Tooby (1999) argue, correctly, that the normality/disorder threshold is related to what would have been adaptive in the environment of evolutionary adaptedness (EEA). They also correctly point out that the EEA is not any particular time or place, but a statistical composite of past selection pressures. The implication of this that they do not follow through is that no two individuals (siblings excepted) have exactly the same EEA. If my grandparents lived a tough life on the streets and yours lived in a secure castle, then our EEAs were not the same. For most purposes, such small differences will be insignificant; what our EEAs share will dwarf the ways they differ. However, small personality differences between two individuals can quite plausibly, as in the guppy case, reflect small differences in lineage history.

These considerations are relevant to the normality/disorder boundary in the following way. Individuals high in neuroticism will require only a small circumstantial trigger to evoke a given level of depressive symptoms, whereas individuals low in neuroticism will require a much larger stimulus to produce the same outcome (Fig. 7.2; see Caspi *et al.* (2003) for some compelling empirical evidence of genotype-circumstance interactions for this kind). Most of the burden of clinical depression falls on individuals with heritably high neuroticism, which is why depression is so recurrent and familial (see Nettle 2004; Lahey 2009). For these individuals, should we define as disordered all mood change that would be disproportionate to the eliciting circumstances for the *average* human or for a human with that person's genotype? Or, given that early experience is likely to calibrate the development of the stress system, that person's genotype and a comparable ontogenetic history? One seems soon forced to the conclusion that a mood reaction is dysfunctional if it is out of proportion to the way *that individual* would normally react. In practice, many people will make judgments about their depressive conditions by comparing their current experiences to their own past, and so this has some face validity, but it hardly serves to demarcate a clear, universally applicable natural kind that is depressive disorder.

Sex differences in depressive symptoms constitute a milder form of this problem. Women in community samples reliably report more depressive

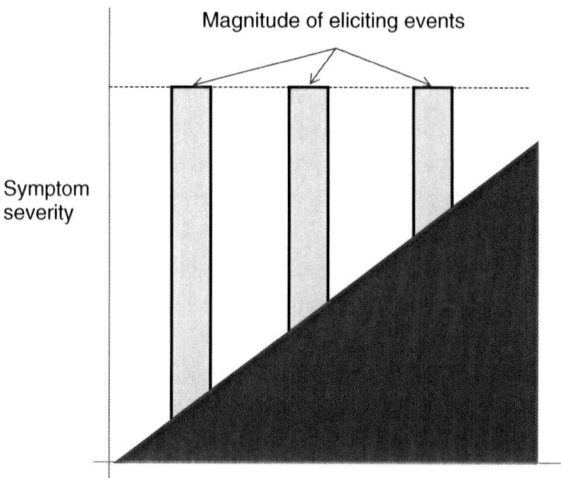

Fig. 7.2 The same level of depressive symptoms could be produced by different combinations of personal emotional reactivity (neuroticism) and magnitude of life events. This is problematic for any attempt to discriminate a taxon of "disproportionate, disordered" mood responses from "normal, functional" ones.

symptoms than men. (This effect can be discerned from Fig. 7.1, where there are considerably more women than men with scores in the range of 12–22, although, interestingly, no more at the very highest extremes). This is a result of larger emotional reactions to life events, and this might plausibly be related to a slightly different history of selective pressures. Does this mean that women's moods are more often disordered than men's or does it mean that the normal function of their mood systems is different? This is just one instance of the more pervasive problem of intra-specific variation in EEA, and its implications for identifying dysfunction.

7.5 Disorder versus complaint as the basis for identifying depression

In view of the issues discussed above, we have to accept the following conclusions.
i. Current normality/disorder distinctions for MDD, rather than carving nature at its joints, represent the arbitrary splitting of a continuum.
ii. The distinction between mood changes that are fulfilling the evolved functions of the mood system and mood changes that are dysfunctions is difficult to make without a profound idiographic understanding of each individual's life circumstances, and may be difficult even then.
iii. The evolved-function criterion is made even more problematic by individual differences in the threshold of activation for negative emotion

systems, which may reflect individual differences in selective history. Such individual differences require us to qualify normality statements as "normal for person X" or "normal for a person of X's temperament", rather than being able to apply them universally.

iv. Allocating treatment based on whether the mood system can be demonstrated to be dysfunctional would in principle lead to treating many people who seek no treatment (because their mood system is dysfunctionally unresponsive), and denying treatment to many people who want it (because they are in terrible distress).

One response to these conclusions would be to side with the many recent calls to abandon the idea that entities like depression should be viewed as discrete "diseases" (Bentall 2003, 2006; Charlton 2009). In the remainder of this chapter, I discuss what the implications of this bold move might be. Although my focus is on MDD, all of the points that follow (and many made earlier in the chapter) could be and have been made in respect of other psychiatric categories, such as schizophrenia (Bentall 2003; Adriaens 2008).

One concern that people might have about abandoning the category of depression is that such categories, arbitrary as they are at the boundaries, serve a useful function in communication amongst specialists, treatment choice, and research. This is a reasonable concern, and it is likely that some arbitrary cut-off points will always be retained as clinical heuristics. However, the problem is that these heuristics become reified, and such reification can actually impair rather than promote research and understanding. What is currently diagnosed as MDD is heterogeneous in terms of triggering situation, symptom profile, and treatment response (Keller *et al.* 2007; Charlton 2009), which may be obscured by the use of a single categorical construct. The categorical model, in its implication of watertight distinctions between disorders, also underestimates the problems common to all psychiatric difficulties, such as the problem of comorbidity, and leads to fruitless proliferation of intermediate-category disorders at the boundaries (e.g., schizoaffective disorder at the boundary between schizophrenia and bipolar disorder, bipolar II, depressive mixed states, and agitated depression at the boundary between MDD and bipolar I disorder, adjustment disorder with depressed mood at the boundary between MDD and normality, and so on). Accepting that there are no discrete disease states, but merely different collections of symptoms of different severities, and talking about and treating the symptoms the individual has, could help clinical practice as much as harm it.

A second concern might be that abandoning the idea that depression is a discrete disease means that people will not receive treatment for their problems. This need not be so if we are prepared to break the link between medical treatment and the presence of disorder. As well as disorders in the sense of harmful dysfunctions, there are classes of medical problems known variously as treatable conditions (Cosmides and Tooby 1999), life difficulties (Bolton 2000), or complaints (Bentall 2006), which warrant treatment simply on the basis of the distress they cause, even though they are not demonstrable dysfunctions. To accept these conditions as medical is to acknowledge that psychiatric diagnosis is in fact based on values, such as the need to reduce human suffering, rather than only on natural kinds of mental functions or dysfunctions. We can't clearly tell when people's mood systems are disordered, but we can clearly tell when they are having life difficulties as a result of their moods. We can tell this because they can tell us that this is the case. Fortunately, the case of depression, which necessarily involves subjective distress, is less difficult in this regard than cases of psychiatric conditions where the person does not feel he or she has a problem.

In this view, medical conditions are not (all) natural kinds, and the normality/abnormality boundary is conventional and somewhat indeterminate (Bolton 2000), although the continuum of emotional experience along which the cut-off is placed maybe universal and objective. This means that temporal change in the prevalence of depressive conditions is very hard to interpret, since improving physical health and aspirations over time may mean that people's tolerances for distress are changing. Thus, it is unclear that the increasing prevalence of treated depression in recent decades represents a change in human experience, rather than a change in values in terms of what people are prepared to tolerate, how they frame their experiences, and what labeling frameworks are available to them and their doctors (Charlton 2009). This does not, of course, mean that there has not been any increase in distress, only that it is very hard to interpret the meaning of changes in the prevalence of diagnosed mental disorder.

A related issue is that if antidepressant medication is offered when there is no actual dysfunction present, it may actually be suppressing the functioning of an adaptive response. This seems a cause for concern: pharmaceutical blocking of diarrhea in shigellosis slows recovery (Dupont and Hornick 1973) and acute anemia often represents the innate immune systems attempt to combat infection by depriving pathogens of iron, in which case iron supplementation is harmful (Ong *et al.* 2006). Might antidepressant medication, for example, prevent a person from learning to avoid situations that are bad for them or prevent them ruminating sufficiently on an important life goal? (Note that this danger is not limited to pharmaceutical interventions, since effective cognitive therapy could in principle be equally effective at disabling adaptive functions). This important question has scarcely been addressed empirically, since all measures of treatment effectiveness for depression medications focus on abatement of distress and prevention of recurrence, not on the life decisions that people make. However, there are a number of relevant points to be made. First, antidepressant medications do not completely disable mood systems. If they did, they would be as dangerous for well-being as surgically induced total analgesia. Second, defenses designed to detect dangers of large effect operate according to the "smoke-detector principle" (Nesse 2005). That is, as a consequence of the history of selection acting on them, they produce many more instances of over-response than under-response, suggesting that the effects of their partial suppression may be relatively innocuous (although, as mentioned, total suppression would be dangerous). Third, the contemporary affluent environment tends to be less dangerous in many respects than ancestral contexts. For example, social ostracism was generally fatal to our ancestors, whereas it is merely unpleasant and disruptive now; angering others often led to homicide in the past, whereas it does so very rarely now; and temporary resource shortages are buffered by redistributive mechanisms operating at a larger scale now than was true in the past. Thus, we may have a tendency to over-estimate the magnitude of consequences of certain types of problem. A useful comparison may be analgesics, which have been widely available without prescription for decades and are doubtless used by people in many cases where pain is not dysfunctional. Physical pain mechanisms also operate according to the smoke-detector principle, and the disabling created by the drugs is also partial. To my knowledge there is no evidence of people doing damage to their somas as a consequence over-using analgesics in such a way as to eliminate pain but also eliminate pain's protective functions. We must hope that the same holds true of antidepressant therapies.

7.6 Conclusion

In conclusion, evolutionary thinking has produced plausible hypotheses for the functions of different low mood symptoms, hypotheses which generate new predictions and deeper understanding of psychiatric phenomena. It seems a reasonable view that mild or transient depressive symptoms usually reflect the proper functioning of evolved adaptations, but that severe or inexplicable bouts may sometimes reflect the dysfunction of these same mechanisms. However, epidemiological evidence does not yield a clear demarcation between normality and disorder for mood phenomena, and the evolutionary literature does not much aid in delineating the boundary either. We may have to accept that much of what we currently identify as clinical depression cannot be shown to be dysfunction, and moreover that the clear presence of dysfunction cannot be used as the criterion for applying or withholding medical attention.

References

Adriaens, P.R. (2008) Debunking evolutionary psychiatry's schizophrenia paradox. *Medical Hypotheses*, **70**, 1215–22.

Allen, N.B. and Badcock, P.B.T. (2003) The social risk hypothesis of depressed mood: Evolutionary, psychosocial and neurobiological perspectives. *Psychological Bulletin*, **129**, 887–913.

American Psychiatric Association (1994) *Diagnostic and Statistical Manual of Mental Disorders* (4th edition). American Psychiatric Association, Washington, DC.

Andreasen, N.C. (2001) *Brave New Brain. Conquering Mental Illness in the Era of the Genome.* Oxford University Press, New York.

Andrews, P.W. and Thomson, J.A. (2009) The bright side of being blue: Depression as an adaptation for analyzing complex problems. *Psychological Review*, 116, 620–54.

Bentall, R.P. (2003) *Madness Explained.* Penguin, London.

Bentall, R.P. (2006) Madness explained: Why we must reject the Kraepelinian paradigm and replace it with a "complaint-oriented" approach to understanding mental illness. *Medical Hypotheses*, 66, 220–33.

Bolton, D. (2000) Alternatives to disorder. *Philosophy, Psychiatry and Psychology*, 7, 141–53.

Carver, C.S. and Scheier, M.F. (1990) Origins and function of positive and negative affect: A control-process view. *Psychological Review*, 94, 319–40.

Caspi, A., Sugden, K., Moffitt, T.E., Taylor, A., Craig, I.W., Harrington, H., McClay, J., Mill, J., Martin, J., Braithwaite, A. and Poulton, R. (2003) Influence of life stress on depression: Moderation by a polymorphism in the 5-HTT gene. *Science*, 301, 386–89.

Charlton, B.G. (2009) A model for self-treatment of four-types of symptomatic "depression" using non-prescription agents: Neuroticism (anxiety and emotional instability), malaise (fatigue and painful symptoms), demotivation (anhedonia) and seasonal affective disorder (SAD). *Medical Hypotheses*, 72, 1–7.

Cosmides, L. and Tooby, J. (1999) Toward an evolutionary taxonomy of treatable conditions. *Journal of Abnormal Psychology*, 108, 453–64.

Dugatkin, L.A. (1992) Tendency to inspect predators predicts mortality risk in the guppy (*Poecilia reticulata*). *Behavioral Ecology*, 3, 124–7.

Dupont, H.L. and Hornick, R.B. (1973) Adverse effect of Lomotil therapy in shigellosis. *Journal of the American Medical Association*, 226, 1525–8.

Grable, J.E. and Roszkowski, M.J. (2008) The influence of mood on the willingness to take financial risks. *Journal of Risk Research*, 11, 905–23.

Haselton, M.G. and Nettle, D. (2006) The paranoid optimist: An integrative evolutionary model of cognitive biases. *Personality and Social Psychology Review*, 10, 47–66.

Horwitz, A.V. and Wakefield, J.C. (2007) *The Loss of Sadness: How Psychiatry Transformed Normal Sadness into Depressive Disorder.* Oxford University Press, New York.

Kalliath, T.J., O'Driscoll, M.P., and Brough, P. (2008) A confirmatory factor analysis of the General Health Questionnaire-12. *Stress & Health*, 20, 11–20.

Keller, M.C. and Miller, G.F. (2006) Resolving the paradox of common, harmful, heritable mental disorders: Which evolutionary genetic models work best? *Behavioral and Brain Sciences*, 29, 385–452.

Keller, M.C. and Nesse, R.M. (2005) Is low mood an adaptation? Evidence for subtypes with symptoms that match precipitants. *Journal of Affective Disorders*, 86, 27–35.

Keller, M.C. and Nesse, R.M. (2006) The evolutionary significance of depressive symptoms: Different adverse situations lead to different depressive symptom patterns. *Journal of Personality and Social Psychology*, 91, 316–30.

Keller, M.C., Neale, M.C., and Kendler, K.S. (2007) Different negative life events are associated with different patterns of depressive symptoms. *American Journal of Psychiatry*, 164, 1524–9.

Kendler, K.S. and Gardner, C.O. (1998) Boundaries of major depression: An evaluation of DSM-IV criteria. *American Journal of Psychiatry*, 155, 172–7.

Kessler, R.C., Merikangas, K.R., Berglund, P., Eaton, W.W., Koretz, D.S., and Walters, E.E. (2003) Mild disorders should not be eliminated from the DSM-V. *Archives of General Psychiatry*, **60**, 1117–22.

Klinger, E. (1975) Consequences of commitment to and disengagement from incentives. *Psychological Review*, **82**, 1–25.

Lahey, B.B. (2009) The public health significance of neuroticism. *American Psychologist*, **64**, 241–56.

Lee, W.E., Wadsworth, M.E.J. and Hotopf, A. (2006) The protective role of trait anxiety: A longitudinal cohort study. *Psychological Medicine*, **36**, 345–51.

McKenzie, J., Taghavi-Knosary, M. and Tindell, G. (2000) Neuroticism and academic achievement; the Furneaux factor as a measure of academic rigour. *Personality and Individual Differences*, **29**, 3–11.

Neeleman, J., Sytema, S. and Wadsworth, M. (2002) Propensity to psychiatric and somatic ill-health: evidence from a birth cohort. *Psychological Medicine*, **32**, 793–803.

Nesse, R.M. (2000) Is depression an adaptation? *Archives of General Psychiatry*, **57**, 14–20.

Nesse, R.M. (2005) Natural selection and the regulation of defenses: A signal detection analysis of the smoke-detector problem. *Evolution and Human Behavior*, **26**, 88–105.

Nesse, R.M. and Stearns, S.C. (2008) The great opportunity: Evolutionary applications to medicine and public health. *Evolutionary Applications*, **1**, 28–48.

Nettle, D. (2004) Evolutionary origins of depression: A review and reformulation. *Journal of Affective Disorders*, **81**, 91–102.

Nettle, D. (2006) The evolution of personality variation in humans and other animals. *American Psychologist*, **61**, 622–31.

Nettle, D. (2009) An evolutionary model of low mood states. *Journal of Theoretical Biology*, **257**, 100–3.

Ong, S.T., Ho, J.Z.S., Ho, B. and Ding, J.L. (2006) Iron-withholding strategy in innate immunity. *Immunobiology*, **211**, 295–314.

O'Steen, S., Cullum, A.J. and Bennett, A.F. (2002) Rapid evolution of escape ability in Trinidadian guppies (*Poecilia reticulata*). *Evolution*, **56**, 776–84.

Price, J. (1992) The adaptive function of mood change. *British Journal of Medical Psychology*, **71**, 465–77.

Wakefield, J.C. (1999) Evolutionary versus prototype analyses of the concept of disorder. *Journal of Abnormal Psychology*, **108**, 374–99.

Wakefield, J.C., Schmitz, M.F., First, M.B. and Horwitz, A.V. (2007) Extending the bereavement exclusion for major depression to other losses. Evidence from the National Comorbidity Survey. *Archives of General Psychiatry*, **64**, 433–40.

Watson, P.J. and Andrews, P.W. (2002) Towards a revised evolutionary account of depression: The social navigation hypothesis. *Journal of Affective Disorders*, **72**, 1–14.

Wolpert, L. (2008) Depression in an evolutionary context. *Philosophy, Ethics and Humanities in Medicine*, **3**, 8–10.

Yuen, K.S.L. and Lee, T.M.C. (2003) Could mood state affect risk-taking decisions? *Journal of Affective Disorders*, **75**, 11–18.

Chapter 8

Function, dysfunction, and adaptation?

Kelly Roe and Dominic Murphy

Jerome Wakefield has argued that mental disorders are harmful dysfunctions. In claiming to capture people's intuitions, however, Wakefield argues that people think of mental disorders along the lines of a two-stage model. Now, the two-stage model sees psychiatry as a branch of medicine, in that it rests on a scientific account of the normal function of the human mind/brain. If psychiatry is continuous with medicine and physiology in this way, its analysis of function and malfunction should reflect that continuity. In this chapter we argue that the way the relevant biomedical sciences determine function does not presume a selectionist concept of function. We argue that the relevant accounts of function are those drawn from mechanistic explanation rather than historical explanation; the life sciences ask all sorts of questions, but the questions which medicine asks are not those which a selectionist account of function can answer. The chapter contrasts Wakefield's (and others') selectionist (or historical) view with a causal (or mechanistic) theory of functions - the systemic capacity view, which sees the function of a system as its contribution to the maintenance of the larger system in which it is embedded. The authors argue that the systemic capacity view provides a better account of how functions are understood and functional claims tested in medicine and physiology. However, they end by wondering whether there is in fact any role for science to play in determining what the overall functional state of an organism is, and hence whether science can draw the line between health and illness in the way that the two-stage system takes for granted.

8.1 Introduction

The medical model of psychiatry sees mental disorder as a bio-medical disorder. The difference in subject matter between psychiatry and another branch of medicine is supposed to reflect the fact that the disorder is mental rather than, say, cardiovascular, but when we call a bodily system disordered we mean exactly the same thing throughout medicine. In this chapter we attend to the puzzle of what "dysfunction" means by looking at rival conceptions of function in the context of the bio-medical conception of disorder. Throughout, our concerns will be dictated by what we call (following Murphy 2006) the two-stage view of the concept of mental disorder, which distinguishes facts about dysfunction from normative judgements about the adverse consequences of dysfunction.

The problem of offering a naturalistic or scientifically respectable account of biological function and dysfunction has long been a concern in the philosophy of biology. The debate on dysfunction within psychiatry and philosophy of psychiatry has usually gone on without looking at this wider philosophical literature, but we will rely on it. After introducing the "two-stage view" and saying something about the picture of psychiatry we presuppose, we will discuss two accounts of biological function and assess the prospects of incorporating them into psychiatry. We will argue that the *systemic capacity view* of biological function and dysfunction seems better suited than the *selectionist view* to capture what bio-medical scientists take themselves to be doing. We will then consider the objection that neither account can explain dysfunction. Finally, we ask whether biological dysfunction is in fact necessary for psychiatric and bio-medical disorder. We do not offer a comprehensive overview of the function debate; we only care about the features of it that seem relevant to psychiatry.

8.2 The two-stage view

The two-stage view is the most popular account of psychiatric disorder among theorists who deny that ascriptions of mental illness are entirely normative. It was introduced by Wakefield (1992), who borrowed heavily from earlier work by Boorse (1975, 1976a). Two-stage theorists hold that there are two individually necessary and jointly sufficient conditions for disorder. First, there is a biological dysfunction. Second, the dysfunction must result in harm to the individual and/or society, as judged by prevailing social norms. "Harm" is uncontroversially a normative notion, but psychiatric dysfunction is assumed to be a matter for medicine to establish, just as it would establish that an esophagus is dysfunctional.

The two-stage view aims for a middle ground between (i) a scientism that says psychiatry has no role for values at all and (ii) a constructivist claim that our judgments that a person is disordered depend entirely on their having violated some norm. The view supposedly respects both the role of science in psychiatry and that of social norms. However, the two-stage view faces two sets of conceptual problems. First, there are the difficulties involved in justifying the intuition that science plays a role in the discovery of objective facts about disorder. In the final section of this chapter we will discuss what that role is. Second, we have the intuition that norms have a role to play in whether an individual is harmed by his or her dysfunction. This is thought to have normative implications along the lines of rights and duties to treatment. Now, unpacking the notion of harm is at least as problematic as unpacking the notions of function and dysfunction (e.g., see De Block 2008). Perhaps one kind of harm is simply physical injury, but the relevant concept of harm involves judgments about the quality of someone's life. These judgments need to be sensitive to both the individual's own needs and goals, and the ideas about well-being that feature in the wider society. Here we put "harm" aside to ask what the best way to think about dysfunction might be.

Boorse (1975) distinguished "disease" from "illness". Disease is failure to function as designed, meaning that some biological system does not conform to the "species-typical design" of humans. Illness is a matter of judgment that a disease is undesirable, entitles one to special treatment, or excuses bad behavior. Boorse proposed an account of functions as properties of systems designed to contribute to survival and reproduction (1976a, p. 62–63). Wakefield advocates a selectionist view of function, according to which the function of a psychological system is what natural selection designed it to do (see Chapter 5).

So the question we take up has to do with function talk within a medical context. We assume for the sake of argument that psychiatry is a branch of medicine in a strong sense (Murphy 2009), so that mental illnesses are caused by distinctive cognitive neurophysiological pathologies. An example of this view is Nancy Andreasen's (2001, p. 172–6) argument that an explanation of mental illness will ultimately cite destructive processes in brain systems, just as bodily diseases are explained by such processes in other organs. The process at issue need not be entirely endogenous: it can mediate the effects of cultural forces or other environmental risk factors. Nor does the cause of disorder have to completely destroy a brain system: it may be enough to put the system into a stable but chronically dysregulated state. The way to understand the scientific part of the two-stage view, then, is that there are, in psychiatry, phenomena that fit the conception of disease as a destructive process that predominates in biomedicine generally. Our question is, what does it mean to say that psychiatry is concerned with departures from the natural functions of the human mind/brain? Put differently, we aim to understand the idea of dysfunction in bio-medical contexts. In the next section, we will consider both evolutionary and systemic capacity views of function with an eye to their prospects of accounting for dysfunction as a naturalistic grounding for bio-medical norms.

8.3 Theories of function

The modern literature on function stems from two seminal papers. Wright (1973) argued that ascriptions of function to a structure are causal-historical. His analysis applies to any structure that participates in a selection process and thus is not explicitly evolutionary. Millikan (1984) and Neander (1991) built an evolutionary analysis of function on Wright's foundations. Cummins (1975) was the other key paper. Cummins's concept of function was not historical but causal. He understood the function of an entity to be the causal contribution it makes to the operation of the overall system(s) that includes it. This is the underlying idea behind the "systemic capacity" analysis of function, which we will defend in this chapter. According to Cummins, a component may have a function even if the component was not "designed", therefore parts with no selection history can be ascribed a function.

Now if psychiatry is a branch of medicine, its function concepts should be continuous with those of medicine and physiology more generally. In this section we argue that the way the relevant sciences determine function does not presume a selectionist concept of function. We will argue that the relevant accounts of function are those which find a home in mechanistic explanation (the systemic capacity view) rather than historical explanation (the selectionist view). The life sciences ask all sorts of questions, but the questions which medicine asks are not those which a selectionist account of function can answer.

8.3.1 The selectionist view

Evolutionary views of function involve causal-historical explanations of traits that we will call *selectionist*. Wakefield's argument is that we look to science for our best theory of how human traits come about, and science says that functions are fixed by the historical process of evolution by natural selection. The heart is a standard example. Millikan (1984) said that the heart is a pump because it is the heart's pumping that causally contributes to the successful reproduction of organisms with hearts: if x is a member of a biological category it is not because of "the actual constitution, powers, or dispositions" of x, but because of the "proper function" of x (Millikan 1984, p.17). X's proper function depends on the history of x's lineage, which explains x's being supposed to do whatever it does. The point is quite subtle because the relevant history consists of correlations obtained between ancestors of x having a certain character and their having been able to perform x's function. So the structure of a heart explains why it pumps, but it does not count as a heart in virtue of having that structure.

The selectionist account of function seems to offer two big benefits. First, it promises to give a definite specification of the function of an organic system and hence a clear criterion for calling it dysfunctional. Second, it seems to offer a scientifically unproblematic way to say what a system ought to be like. If you are worried about the accusation that function talk is normative you can embrace natural selection. Teleological notions are commonly associated with the pre-Darwinian view that the biological realm provides evidence of conscious design by a supernatural creator. The point about evolutionary views is that they assuage this metaphysical concern by showing how norms are part of nature. We are not going to suggest that there is something wrong with the Darwinian picture of natural order. But we will suggest that an evolutionary concept of function and dysfunction is a poor bet for psychiatry.

We begin with a question about inquiry. On the face of it, the evolutionary view confronts a simple epistemic problem. How do we ascertain whether the relevant history is actually obtained? And even if a certain history is obtained, what is the proper assignment of evolutionary function and dysfunction? We are not saying that it is impossible to test claims that something is an adaptation, but examples of successful tests, we claim, look nothing like the tests we see in medicine or the parts of biology adjacent to medicine.

Cain and Sheppard (1950) provide an example of a test of a selectionist function. They demonstrated that the appearance of the shells of the grove snail (*Cepaea nemoralis*), which had been believed to be a neutral trait ascribed to random factors, actually varied adaptively. For example, darker and more uniform shell patterns predominated in darker and more uniform habitats like deep woodland. Populations of snails that were more heavily predated by song thrushes had a higher incidence of more conspicuous patterns. Cain and Sheppard persuaded everybody that *C. nemoralis* patterns were adaptive camouflage, and therefore one can appeal (defeasibly) to current utility to fix function. One can also measure reproductive success over several generations to see if the variant one believes to be adaptive is in fact spreading in the population in response to selection pressures. This has been done, for example, by Peter and Rosemary Grant, who spent more than 30 years tracking the responses to selection pressure of Darwin's finches (Grant 2000).

It is even more difficult to test an ascription of evolutionary function in humans. Remember, we need an account of function that will license judgments that some system is dysfunctional. One analysis would be that traits that are selected against (are becoming less prevalent) in the face of the success of other variants are dysfunctional. To do this, we need to show that humans with that trait are reproductively less successful. That would take generations, and no branch of medicine or physiology tests claims about the functional architecture of humans by showing how possession of a trait correlates with reproductive success. The sorts of tests that would be needed to demonstrate that are almost impossible to do on humans, given the difficulty of establishing over several generations that a part of our biology does in fact enhance reproductive success relative to the competition. Of course an unhealthy heart can kill you, but we do not decide that your heart is healthy by looking at the number of offspring you have. The situation is even worse for mental illness. We have little systematic evidence on whether alcoholics or depressives or psychopaths have fewer children than control subjects. (Psychopaths, for instance, tend to have irregular, promiscuous sexual lives, but that doesn't mean that they don't reproduce successfully.)

At this point, somebody with a commitment to the evolutionary picture may object that in fact the relevant test is that of current fitness. We can work out what a trait does for an organism in the current environment and thereby show that it is adaptive. However, there are problems with this strategy too. Lloyd (2005, p.166–7) notes that to assume, based on current evidence, that mammalian fur has always functioned to help with thermoregulation we have to also buy into the claim that selection pressures have remained constant in the past. To do this we really need independent evidence from other historical sciences, and the relevant assumption about the stability of past pressures is in any case very hard to make in human biology. There seem to have been substantial changes in our environment and our relations to it, even over small time scales, like the last few thousand years.

But don't medicine and physiology routinely try to work out what a system contributes to the overall functioning of the organism? Yes, but that doesn't mean that in doing so we are trying to establish that a biological component has a selectionist function. For example, take Hubel and Wiesel's famous program of mapping the receptive fields of cells in the visual cortex and then establishing further visual information-processing channels in the brain. That program, and the research on the neurobiology of vision inspired by it, depended on a set of engineering assumptions about the way the brain is organized to process information. It did not test assumptions about the selective advantage and history of the components of the visual brain. Most physiological research is based on establishing the components, and the functional relations between components, in biological systems. It is not aimed at uncovering evolutionary relationships. It may be that the facts uncovered in physiology are evidence for evolutionary relationships, and of course all biological systems have an evolutionary history, but when we determine what normal function is, in medicine, we do not even try to establish what something's selectionist function is.

Schaffner (1993) argued that although medicine might use teleological talk in its attempts to develop mechanistic explanations, that talk is just heuristic. It focuses our attention on entities that are useful to the organism. Schaffner suggested that as we learn more about the role a structure plays in the overall functioning of an organism, the need for functional ascriptions drops out. It is replaced by the vocabulary of mechanistic explanation: the causal relationship of parts that jointly produce phenomena of explanatory interest. Functional explanations that draw on evolutionary considerations are, he claimed "necessary, but empirically weak to the point of becoming almost metaphysical" (Schaffner 1993, p. 389–90).

In our view, biomedical ascription of function to a system makes no claims about adaptedness or selective history. It requires only that we can identify the role played by a system in the overall economy of the organism. How is dysfunction determined? By the use of a biomedical concept of normality that is an idealized description of a component of a biological system in an unperturbed state that may never be attained in actual systems. It does not rest on the failure of a biological part to function as its ancestors did, but by its failure to be close enough to the causal contribution of the analogous part in the idealized overall system.

Wachbroit (1994, p. 588) argues persuasively that when medicine or physiology says that an organ is "normal", the relevant conception of normality "is similar to the role pure states or ideal entities play in physical theories". Such an idealization represents actual organs or systems in unperturbed states (see also Ereshefsky 2009). To understand a real case we add information to develop a model that resembles actual hearts (Wachbroit 1994, p. 589). For instance, Gross (1921) was able to establish post mortem that anastomotic communication between main arteries increases over a typical lifespan, thereby establishing that we need to model younger and older hearts differently. The point of such idealizations is not to represent the statistically average heart, but to describe hearts in a way that allows departures from the ideal to be recognized and to serve as template from which more realistic models can be built. In general, physiological theories are families of such idealizations, and bodily systems are understood as functional parts of larger systems, typed unhistorically. In so far as psychiatry is a branch of medicine, the concept of function it needs will resemble the unhistorical concepts of physiology and bio-medicine. Evolutionary considerations are just beside the point.[1]

---

[1] Furthermore, a trait's evolutionary function (and hence its dysfunction) might be quite different from what psychiatrists think of its function and dysfunction.

Here is one last argument against reading the selectionist view of function into psychiatry. Suppose we discovered that schizophrenia, bipolar disorder, depression, and psychosis were evolutionary adaptive strategies: forms of behavior evolved to further the interests of the sufferer. (Theories of this sort are reviewed by Murphy 2005.) Would we then re-think their status as mental disorders or would we be led to reject the evolutionary dysfunction view as providing an adequate account of dysfunction in psychiatry?

Horwitz and Wakefield (2007) argue that the human mind includes a system that has evolved by natural selection to respond to loss. They argue that the intuitive distinction between normal sadness and morbid depression tracks the workings of their hypothetical system, which explains why we become sad in situations where sadness seems like the right response. Major depression occurs when the system kicks in for no reason or produces excessive responses to trivial misfortunes, whereas normal sadness occurs in response to serious misfortune, for the system they hypothesize is "biologically designed to produce such responses at appropriate times" (Horwitz and Wakefield 2007, p. 25). They suppose that there is a system misfiring in cases of major depression because they are committed to an evolutionary view of function that underpins their two-stage picture. Their conjecture, accordingly, is that since depression is obviously a matter of the mind going wrong, it must be a matter of some system failing to function as natural selection has designed it. It follows that if whatever underlies depression is not an adaptation failing to function as designed, they would have to say that depression is not a mental disorder.

It is not impossible that we should be led to conclude that what we had seen as a mental illness is actually not one, and there are historical examples of that happening. (Psychiatrists famously changed their minds about the pathological nature of homosexuality in the 1970s.) However, overturning our judgments about depression would be a real revolution. Depression and psychosis are the two paradigmatic kinds of mental disorder. A theory which said, based on evolutionary considerations, that depressed people or psychotics are really healthy, would almost certainly be rejected. That is a conceptual revisionism that we would be very unlikely to accept. It would be like finding out that cancer is not a disease. Nothing is settled for ever in science, but it is very hard to imagine us siding with the revisionist, rather than simply saying that they had uncovered some interesting causal explanations of the nature of a disease. Or, if we did accept it, it is unclear whether the concept of mental disorder would survive, since its very utility would be called into question by the removal of its exemplary instances. The facts might also go the other way and point to some condition being a mental illness even though nobody had ever thought it was. An apparently normal or beneficial part of our everyday psychology might reflect a failure of an ancient adaptation to do its job: suppose that giving to charity is caused by the failure of an adaptation designed to make us altruistic towards members of our immediate communities. In that case, charitable giving would be a candidate mental illness. Although commonsense is not definitive, we would need very good grounds indeed to accept an analysis that raises the possibility that giving to charity is pathological and paranoid delusions are not, and we do not think that the grounds to accept such revisionism are present. Actual scientific explanations of disease, psychiatric or not, are accepted routinely in the absence of evolutionary considerations. Suppose that a good explanation of the underlying neurobiology of depression were discovered. Psychiatrists would dislocate their shoulders from all the patting themselves on the back that would ensue in the light of what would be seen as a scientific triumph, and nobody would care whether there was an evolutionary rationale, or testable Darwinian hypothesis, for the system that was uncovered. Testing in biomedicine does not work that way, as we have tried to show, and our ordinary intuitions do not demand an evolutionary rationale either.

8.3.2 The systemic capacity view

Cummins (1975) claimed that when we say that the function of the heartbeat is to circulate blood through the organism we are not accounting for adaptation but *explaining circulation*. We begin by assuming that the circulatory system is what needs to be explained, and we identify a system as having a function in the context of explaining it. We may explain the *advantage* of the heartbeat by identifying the activity it facilitates. This is different from explaining the existence of the heartbeat. Cummins (1975, p. 746) thought that we could only say why a system exists by appealing to the intentions of a designer. We suppose that it can often be done for organisms by appealing to evolutionary history. But we agree with Cummins that explaining what a systemic component presently does is different from explaining why it is there. The life sciences ask lots of questions, and trim their accounts of function to fit.

Cummins argued that the basic explanatory use of function talk in the life sciences derives from a particular analytic strategy in which the biologically significant capacities of a whole organism are explained by breaking down the organism's biology into a number of "systems", the circulatory system, the digestive system, the nervous system, and so on, each of which has its characteristic capacities. These capacities are in turn analyzed into the capacities of their component organs and structures. We can reiterate the systemic capacity framework down through levels of physiology, explaining the workings of the circulatory system, the heart, certain kinds of tissue, certain kinds of cell, and so on. Much mechanistic research in biology exemplifies this approach.

Many theorists have argued that there are biological contexts in which Cummins's ahistorical analysis is broadly correct (Godfrey-Smith 1998; Kitcher 1998; Craver 2007; Bechtel 2008). *Function* in biochemistry and physiology typically refers to the contribution a structure makes to the overall organismic system containing it and therefore differs from functional talk in evolutionary disciplines. As Kitcher (1998, p. 266) already noted, "philosophical analyses reveal unresolved ambiguities in biological practice", and philosophical analysis should respect, rather than try to reform, the differing scientific usages. The view of function within physiology is, in outline, the conception of systemic functional analysis we introduced above and aim to defend as the right characterization of function talk in medicine, and hence in psychiatry.

The systemic capacity conception of function uses an analytic strategy to explain how something is able to perform a function by treating functions as dispositions of a component of a larger system. On the systemic capacity account functions are assigned to components in virtue of the role that they play in the production of a phenomenon, typically the output in some greater system. Once one has the relevant system then one proceeds to analyze the system into components and assign functions to the components in virtue of the role they play with respect to the production of the phenomena that one wants to explain.

One longstanding objection to the use of a systemic capacity account of function is that it robs us of the power to say when a system is dysfunctional. This is held to be a problem because assignments of systemic capacity function look to be relative to the interests of the researcher rather than a feature of the world, as the two-stage view requires. Davies (2001) is one recent theorist who reviews this objection. He claims that assignment of function to components is doubly relative. First, which components are relevant is going to partly depend on what phenomena the researcher is interested in. So, if we are mostly interested in the noises that the body makes, it is the sound of the heartbeat which is the relevant property of the heart. This makes functional analysis mind-dependent. There is no such thing as the natural functions of a system, just whatever the investigator finds interesting. Cummins (1975, p. 763–4) saw this objection coming. He argued there would be little point in applying a functional analysis because the disposition of the body to make a noise is just a process of the same type as the disposition of the heart to make a noise. Functional talk would not be interesting in such cases. But this is a rather pragmatic, rather than principled, reason for ruling out the sound the body makes as the disposition of interest. Davies argues that a better defense can be made by restricting functional ascriptions to hierarchically organized systems in which lower level capacities realize upper level ones. The noise of a heart is part of the overall noise of the body, but it is at the same level, so it does not count as a functional subpart of the noise system. That gives us a characterization of function independent of our explanatory interests.

Cummins and Davies have both maintained that systems (and remember, a system from one perspective may be a subsystem of a more encompassing unit) must consist of two distinct levels. There is the level of the phenomena and the system that produces them and there is the lower level of the components and their functions. Once we hit a level at which the outputs are basic, where the "system" cannot be analyzed into further components, then we have reached the end of the systemic capacity chain of explanation and moved into a different science, which analyses the physical composition of the system. But where does it peak? Is there a principled topmost level of analysis such that all components at lower levels are functioning to realize that topmost disposition? The two-stage view seems to require that major organ systems can be assessed as functional and as making a contribution to overall health, independently of what anyone thinks. If mind-independence is established via participation in a hierarchy of natural levels, we want to know whether a human being is a level in the hierarchy. If the overall person is a level in the relevant hierarchy, we have a level of analysis of which the major organ systems are the components.

Our current question, then, is whether there is a level at which each individual counts as a functional system with the major sub-systems—respiration, circulation, and so on—as the proper parts of that system. For the two-stage view to work there needs to be a mind-independent level of analysis corresponding to the individual human. This is necessary because if there is not such a mind-independent whole-organism level we face the danger that whatever aspects of overall human flourishing we attend to in psychiatry (or elsewhere in medicine) will just be assessed relative to the interests of the investigator. If that is the case, then the evaluation of one's social cognition or self-esteem or mood cannot be conducted in terms of its principled contribution to an overall level of function, but just in terms of the particular output of interest. Bluntly, you could show how thoughts of suicide play a role in explaining self-destructive behavior if you are interested in self-destructive behavior. However, the two-stage view is interested in establishing not just that self-destructive behavior exists and is contrary to our view of a good human life, but in showing how it rests on dysfunction. So for the two-stage picture to apply there must be a natural functional level of the human person, with a set of outputs whose character is mind-INdependent. If this cannot be specified, then we can say that biological subsystems make a contribution to overall functioning, but not in the mind-independent way that the two-stage picture needs. We will be left with an account of overall human function that is a reflection of our norms, and identify dysfunction in that light. And that is just the situation that the two-stage view is designed to keep us out of.

The problem we face is that of trying to establish what the major organ systems of the body (and, for psychiatry, the major components of our psychology) are actually for. Boorse's answer (1976b) is that they are there to help us survive and reproduce. Griffiths (2006, p. 2) agrees: the "causal functions which are of primary interest to biologists are those which contribute to an organism's capacity to survive and reproduce." We are not convinced this is right. Boorse offers no real argument for his point; he just asserts that in physiology the relevant system for functional analysis is the organism and the relevant goals are those of its survival and reproduction (1976b, p. 84). Griffiths (2006, p. 2) does have an argument. He says biology makes use of a causal conception of function to identify "adaptive traits, which increase the fitness of organisms that possess them relative to other types." Adaptive traits may not yet be adaptations (i.e., we may not explain them as products of evolution by natural selection) but they are adaptive in so far as they explain survival value; the functional analysis explains the survival value of a particular piece of behavior or physiological process. However, there are other questions about the role that biological systems play. As well as questions of survival value, we can ask questions that simply aim to find out how a system does what it does in the context of the superordinate system.

There is an idea almost as old as natural selection that may help to answer these questions. It is Claude Bernard's 1927 [1865] suggestion that major systems in the human body seek to maintain stable internal homeostatic states, in his favorite example by regulating the chemistry of the blood. Bernard argued that organisms can only explore and transform the external environment if they have sufficient internal stability. In this view the answer to the question "What is the function of the major physiological systems?" is "To keep the internal environment stable." We suggest that homeostasis, not survival value, is what guides physiological answers to questions about causal explanations of biological systems.

Physiological mechanisms are mutually integrated into an organism-level system that depends on the external environment but is not integrated into it the same way. The overall physiological system can move through different environments, for instance, in a way that its components cannot; they must remain in situ to do their job. Similarly, what goes on within the cell is integrated relative to the relationship between the cell and its surrounds. At every level of the hierarchy the internal environment is relatively independent of the external one, and more fully integrated internally than the higher level is integrated with its surrounds. If mechanisms co-operate to maintain a constant internal environment, the system can compensate for changes in the external environment.

This relative uncoupling of the overall system from its range of possible environments is well defended for psychology by Rupert (2009); in psychiatric contexts we can follow Rupert in treating the organism as a relatively enduring cognitive system. The specific problem for psychiatry is to identify failures of cognitive parts that constitute the mechanism's underlying intelligent agency.

We have argued that the problem of relativity in the systemic capacity view of function can be solved in medicine and physiology by appeal to a hierarchy of integrated systems. A further, distinct, problem concerns not the relativity but the indeterminacy of functional ascriptions. Millikan (2002, section 6) points out that we might ascribe a broader or less broad function to a trait; after all, if my ear canals can keep me upright in a gravitational field of 1G they can also work under forces of 0.9G, 0.8G and so on. If we regard these as different functions then we have an infinity of them. She proposes to solve this problem by introducing "the descriptive generality requirement", which states that functions (of whatever type) should be described according to the most general principles available. We agree, and pause only to note that the principles will typically derive from the even broader principles guiding the idealization of the system.

So we have a way to explain function, this time in terms of the contribution a system makes to the operation of a wider, integrated set of systems that maintain themselves in equilibrium. This view has a lot in common with the causal conceptions of function that philosophers have worked within since Cummins. It detaches function from natural selection and sees it working in ecological and physiological time rather than evolutionary time, to make contributions to the maintenance of a living system that are distinct from questions of survival value and reproductive potency. In the next section we will see whether the systemic capacity view of function can also account for dysfunction, and then in the final section we will discuss whether this conception of function licenses the attribution of mind-independent norms to systems of psychiatric interest.

8.4 Dysfunction

Something clearly needs to be said about dysfunction. Any theory must be able to meet the problem of distinguishing the functional from the dysfunctional. Some theorists might not insist that a theory of function has the resources to account for dysfunction, but it is clear that any account of function that purports to be relevant for general medicine or psychiatry has to meet the dysfunction challenge. In this section, we will evaluate both the selectionist and systemic capacity views on dysfunction. If either view can cope with the challenge and the competing view cannot, we have a powerful argument in favor of the successful view. To do this, we will discuss Davies' argument against both selectionist and systemic accounts. We will argue that this argument fails, in both views, mainly because Davies cannot substantiate his key contention that the classification of entities in the life sciences is essentially functional.

Supporters of an evolutionary account of function (e.g., Neander 2002) often think that one of the virtues of their theory is the straightforward way in which an account of dysfunction (or malfunction) follows from the account. Their idea is that we can say when a system is malfunctioning by observing that it is not carrying out the job which natural selection designed it to perform.

In contrast, it is widely bruited that systemic accounts of function cannot deal with dysfunction at all. If the function of a system is relative to our explanatory interests, it seems that a putative malfunction can just be understood as a contribution to a different overall disposition of the system. We endorse Godfrey-Smith's (1993) response to this objection: a token component in a system is malfunctioning when it cannot play the role that lets other tokens of the same type feature in the explanation of the larger system.

Davies (2001, p. 212) denies this. He says that Godfrey-Smith's point works only if "incapacitated functions' tokens retain their membership in the functional types and not just the generic type" and he denies that there are any grounds for thinking that is true. That is, a dysfunctional heart just belongs to the generic type "circulatory device" because functional types are defined in terms of what they can do. Therefore, if a component cannot carry out its normal contribution to the overall system then it ceases to be a member of the type.

However, Davies also thinks that roughly the same argument works to deny that the selectionist can explain dysfunction. Evolutionary accounts, too, he argues, individuate types according to their functions. Since the functions are thought to be necessary and sufficient for membership in the type, it is thus impossible for an instance to both be a member of the type (possess the necessary and sufficient condition or function) and yet lack the function and hence dysfunction. So Davies (p. 210) maintains that instead of saying that a heart is malfunctioning, all the selectionist view lets us say is that the instance that does not pump is not a heart after all. It therefore doesn't have the function of pumping and thus it isn't malfunctioning—it just lacks the function that we wanted to assign to it. Dysfunctional hearts lack the defining capacity of historical success, and hence do not belong to the type at all. Davies claim seems correct to the extent that if having some function F really is both necessary and sufficient for F's being classified as a member of the functional kind K, then it follows that if F were to lack the necessary and sufficient condition for being a member of kind K then it would simply stop being a member rather than becoming a dysfunctioning member. By analogy, if we consider an instance of gold and we then apply a proton gun and remove one of the protons then the instance isn't a malfunctioning or abnormal instance of gold in virtue of having one fewer proton. Rather, it would no longer be an instance of gold.

Davies' argument, if correct, shows that no naturalistic concept of function can accommodate malfunction. He therefore concludes that there are no norms of performance in nature that a naturalist can embrace (p. 214). This seems to mean that medicine is built on purely nonnaturalistic or normative assumptions that people make about whether other people are flourishing. If that is right then it seems that the two-stage view is indefensible not just in psychiatry but in all of medicine.

However, Davies' argument can be refuted. It relies, as we just saw, on the assumption that both the selectionist and the systemic capacity theorist individuate biological types according to the function that they assign to the type. Insofar as types possess their function as a matter of necessity he seems correct that an instance of a type cannot malfunction. Davies maintains that the burden of proof is on theorists who think there can be norms of nature to provide an account of how to individuate kinds.

However, we follow the biological consensus and deny that biological components are essentially typed according to their function—a counterargument that seems to work for both the evolutionary view and the systemic capacity view. The point that traits are not typed by their function was first made by Amundson and Lauder (1994), and it has been endorsed widely for the best of reasons, that is, it appears to capture biological practice. So Griffiths (1997, p. 215–6), for instance, argues that if biology has kinds at all, they are either cladistic kinds (members of a shared lineage) or disjunctions thereof: it is common descent rather than function that determines that a heart is a heart. More recently, Griffiths (2006, section 6) has argued that homology determines that a heart is a heart. He argues that functional classifications in biology type organisms by analogy (or shared evolutionary purpose) and are logically dependent on classifications in terms of homology. An organ can still be a token of a type defined in terms of homology even if it is not currently functioning.

So Davies' objection appears to fail for contexts supplied by evolutionary questions, since it seems untrue to insist that traits are typed according to their function in biology generally. We will now argue that it is also untrue for systemic capacity accounts. Griffiths (2006) has argued that many disciplines in experimental biology type traits by homology. In medicine there is a long tradition of identifying systems anatomically before going on to investigate their physiology. In the neurosciences, for instance, which are plainly relevant to psychiatry, there is a century-old tradition of identifying brain regions (Brodmann's areas) based on the physical architecture of cells within the region and then going on to ask what the function of that region might be. We appeal to this kind of procedure to defend our strategy of arguing that organs have functions, rather than insisting that they are essentially tokens of functional types. We will not investigate the precise relationship between this tradition and the tradition of typing by homology, which also involves identifying biological components in terms of their anatomical structure and position and relationships to other organs, and not just solely as functional types. We can therefore say that a heart, even if it has lost the pumping capacity that hearts often have and hence is malfunctioning, retains its identity as a heart because it is still in the position characteristic of hearts and it retains some of the musculature and internal anatomy of a heart. Or consider a doctor conducting an autopsy. She can identify bodily organs as tokens of organ types in order to assess the degree of pathology in each case. Reasoning like this is perfectly satisfactory scientific practice: we mentioned earlier how Louis Gross' research in cardiology depended on generalizing from post-mortem hearts to living ones. However, every system in a corpse no longer possesses its function in the sense that Davies's argument relies on. In summary, sciences type components of biological systems in nonfunctional ways. Its function is a property of a biological unit, not its essence. Thus it is that Davies' challenge can be met for both evolutionary and systemic accounts.

8.5 Dysfunction and the role of science

We now turn to ask more general questions about the role of science in discovering disorder. The two-stage view says that scientific facts play a significant role in determining whether or not a condition is a disorder. The dysfunction criterion was initially introduced to help us determine which individuals were in fact disordered, in a way that avoids subjective, mind-dependent, or culturally relative judgments. For the two-stage view to work, the science of mental disorder (and, indeed, that of disease more generally) can't be just a particular application of a nonnormative neuroscience or molecular biology, but a distinct province of that wider science, one concerned with dysfunctions rather than just unusual cognitive or physiological processes. We will end this chapter by asking whether the two-stage view can be sustained with respect to the systemic account of function as we have sketched it, and whether it makes any difference if it can't be.

The scientific aspect of the two-stage view thus has the job of rebutting the skeptical claim that disorders are just violations of norms that currently prevail in a society. Because it must play this role in the two-stage picture, science must go beyond the role of simply determining what kinds of conditions there are, how they develop, and what interventions are effective for them. We need a definitive list of dysfunctions that justify our regarding a condition to be a disorder, not just knowledge of how the mind and body work in physiological contexts. In the remainder of this section we sketch a defense of a more modest role for science in a program that is basically normative.

Cooper (2005, 2007) and Murphy (2006) have drawn an analogy between the concept of mental disorder and that of weed. Weeds are not a scientifically relevant category of entities. We can perhaps say that a weed is a fast-growing species that negatively impacts on economically valuable crops, usually through competition for nutrients, sunlight, and space. What fixes the extension of "weed" (and similar concepts like "vermin" or "precious metal") is a set of contingent human interests that can change over time. There is nothing inherently dysfunctional about a weed; weeds are just species that we don't like because of certain interests that we have. Suppose that determining that a condition is a disorder is like determining that a plant is a weed. The judgment is determined by value judgments we have already made. So "weed" is not a technical term in ecology and the science of weeds is just the science of plants, put to special use. "Weed" rarely appears within publications in reputable ecological journals, but nonetheless there is real, explanatory mind-independent knowledge to be had about each sort of "weed". For those who are skeptical about the two-stage view, science does not uncover dysfunction in a way that is independent of our value judgments; science is directed by those value judgments.

We will consider cancer as an example. It is an obvious instance of a biomedical disorder. This example has been chosen to emphasize that our skeptic's questions aren't specific to psychiatric disorders. Theorists have often thought that the presence of normativity in psychiatry would undermine psychiatry's status as a branch of medicine, but we think that psychiatry is useful in helping us become clearer on the role and limits of science for medicine more generally.

Let's suppose that one wants to understand or explain cell development roughly along the lines of the systemic view. One way to do this is to construct an idealization of the development of cells of a given type. One kind of information that we would want our model to contain would be the causal information inferred when we make interventions that seem to have a fairly robust bearing on the future course of the cell. An alternative would be to attempt to model subtypes of cell development. There are a number of considerations that bear on whether we should "lump" or "split" phenomena in these ways. One consideration is whether we discover differences in the response to our manipulations that result in differential outcomes that seem important. The occurrence of fairly robust responses under intervention is thus one important consideration for our individuating importantly different kinds of phenomena. Medicine has one further refinement of this that we shall consider shortly.

Suppose we want to understand cancerous cell development. We have two fairly different ways of proceeding. One way is to initially proceed as before. We build a model of cell development in general or a particular kind of cell development. We can then proceed to model cancerous cell development by explaining what "break downs" occur in our model in order to explain cancer as a "biological malfunction" of the cell in the systemic sense. An alternative would model cancerous cell development on its own terms, in much the same way as we initially modeled the development of the noncancerous cell. Our model would be constructed on the basis of some idealization of the development of particular cancerous cells. Now it seems that different groups of scientists could proceed differently on this and we could well end up with two distinct models of cancer. According to the first model cancer would be a "biological dysfunction" whereas according to the second model cancer would be a distinctive pathway that cells can take. Of course, everyone thinks cancers are pathologies, but our skeptic is asking how we establish that by scientific investigation of cell development, as opposed to merely using our prior assumptions about what seems intuitively pathological and plugging a causal model into it.

Both models seem capable of capturing precisely the same causal information with respect to providing different points at which we can intervene to disrupt the process we have modeled. We can disrupt the course of cell development and we can disrupt the course of cancerous cell development. But surely cancer can't be both a biological dysfunction and a merely unusual kind of biological development at the same time! Given the set up we have imagined, what further scientifically discoverable fact is there that tells us whether or not cancer involves a dysfunction? How much is biological dysfunction an assumption of our modeling rather than something that is to be discovered by it?

To save the two-stage view, we need to answer the question we just asked: what fact is there that science can discover that discriminates between cancer as a dysfunction and cancer as an unusual developmental pathway? If there is no such fact then we must reconsider whether science is playing a foundational role in determining that conditions are disorders, as the two-stage view says. The skeptic's alternative is that science discovers important biological facts guided by prior normative judgments that something is a disorder. To save the two-stage view, we must uncover natural norms.

To reply to the skeptic we cannot stay at the level of the biological system, but we must move up to ask about the role of the system in the overall economy of the organism. The answer from the two-stage theorist who adopts our version of the systemic capacity is: look at what the system you are studying does for the organism. The reason why cancer is a dysfunction is that it drives the organism out of equilibrium and into a new state in which other systems stop being able to act as we usually explain them. This approach also requires a way of differentiating normal from abnormal development; basically, it defines normal development as the set of pathways that lead to the final, functional, adult form.

The systemic capacity theorist can use the idea of a natural hierarchy in the organism to defend the claim that disease perverts the functioning that is normal for an organ system. The textbook tells you what a healthy organ is like by reference to an abstraction—an idealized organ. This concept of normality is not justified by conceptual analysis. It draws its authority from its predictive and explanatory utility: we account for variation in actual hearts (a particular rhythm, say), by citing the textbook rhythmic pattern (which may be very unusual statistically) and identifying other patterns as arrhythmic. The role of the idealizations, as we have said, is to classify real systems according to their departure from the ideal, and the ideal must be justified by an appeal to overall organismic homeostasis.

Our skeptic just says that now the problem recurs. What justifies our idealized or assumed "normal" systems? Variation in biological traits is ubiquitous, so establishing whether or not a mechanism is functioning normally depends on whether an overall picture of normality for the organism can be adumbrated in a way that doesn't depend on our prior values. The skeptic just denies that can be done. The exponent of the two-stage view will say that it is possible.

We might think that disorders can be tied to a break between normal and abnormal functioning of an underlying mechanism, such as a failure of the kidneys to conserve electrolytes. Skeptics argue that while one way of construing the phenomenon is that the kidneys "fail to conserve", another is that they simply "don't conserve"; conserving electrolytes is not part of the model of what those kidneys are doing. The problem seems to recur at each level on the systemic analysis. Adding layers up (e.g., considering individuals as functional or homeostatic units in a social group) or down (the organ systems that comprise them) will not determine how we idealize or assume "normal" or "homeostatic" systems to be in a way that is independent of our values. No biological system can sustain a stable internal environment if its system for filtering waste has broken down. We may think that in that case we have a clear rationale for arguing that when a kidney does not filter waste then it is "failing". We are dealing with a "problem" for the overall system, not merely an alternative pathway for a component. But the skeptic can reply that lying behind this intuition is our (entirely reasonable) valuation of organismic integrity. It is in virtue of our valuing it that we are inclined to describe many component processes that threaten to disrupt it as "dysfunctional".

Distinguishing failures to flourish from functional systemic failures will always be a hard problem for psychiatry. For example, judgments of irrationality are central to many psychiatric diagnoses, and our standards of rational thought are not based on biological findings. They reflect standards derived from normative assessment (Murphy 2006, chapter 5). The possibility of psychiatric explanation employing the methods and models of physical medicine, then, depends on how much of our psychology is like the visual system, that is, decomposable into structures with a clear natural function that can be tied in to a biological hierarchy topping out in homeostasis. Some mental processes may lend themselves to such a treatment, but it is unclear how the notion of homeostasis even applies to most of our rational and emotional lives. We may also wonder whether the notion of homeostasis itself can be rendered nonnormatively. Accordingly, our skeptic's challenge is likely to be very hard for psychiatry to overcome. If this can't be done then the same problem arises in medicine more generally. Fortunately the skeptical position doesn't undermine the considerable role that science can play with respect to modeling traits of interest, individuating kinds of conditions, discovering their etiology and course, and developing more or less effective interventions for them. However, the skeptical view does threaten the two-stage view by arguing that the science is guided by judgments of disorder, instead of providing a foundation for them. The challenge to the naturalist is to establish, in medicine generally, not just psychiatry, the mind-independence of the functional hierarchy and the relevant notion of organism-level performance.

8.6 Conclusion

Since the rise of a mechanistic conception of nature in the seventeenth century, medicine, including psychiatry, has struggled to make sense of the apparent teleology of biological systems. Without a satisfactory account of function it is hard to see how we can have a satisfactory account of malfunction, which endangers any naturalistic perspective on disease. Many theorists have considered evolved, selected functions to be an attractive solution to this problem. In this chapter, we have argued against Darwinian accounts of function, as they do a poor job of accounting for medical practice and suffer from debilitating epistemic problems. In our view, psychiatry and medicine presume a systemic capacity view of function, in which functions are considered as components of mechanisms designed to keep a system in homeostasis.

Acknowledgments

We thank Paul Griffiths, Jack Justus, and a discerning anonymous referee for their comments.

References

Amundson, R. and Lauder, G.V. (1994) Function without purpose: the uses of causal role function in evolutionary biology. *Biology and Philosophy*, **9**, 443–69.

Andreasen, N.C. (2001) *Brave New Brain*. Oxford University Press, New York.

Bechtel, W. (2008) *Mental Mechanisms: Philosophical Perspectives on Cognitive Neuroscience*. Routledge, London.

Bernard, C. 1927 [1865] *Introduction to the Study of Experimental Medicine*, Macmillan, New York.

Boorse, C. (1975) On the distinction between disease and illness. *Philosophy and Public Affairs*, **5** (1): 49–68.

Boorse, C. (1976a) What a theory of mental health should be. *Journal for the Theory of Social Behavior*, **6**, 61–84.

Boorse, C. (1976b) Wright on functions. *Philosophical Review*, **85** (1), 70–86.

Cain, A.J. and Sheppard, P.M. (1950) Selection in the polymorphic land snail *Cepaea nemoralis*. *Heredity*, **4**, 275–94.

Cooper, R. (2005) *Classifying Madness: A Philosophical Examination of the diagnostic and statistical manual of mental disorders*. Springer, Dordrecht.

Cooper, R. (2007) *Psychiatry and the Philosophy of Science*. Acumen, Stocksfield.

Craver, C.F. (2007) *Explaining the Brain: Mechanisms and the Mosaic Unity of Neuroscience*. Oxford University Press, New York.

Cummins, R. (1975) Functional analysis. *Journal of Philosophy*, **72**, 741–64.

Davies, P.S. (2001) *Norms of Nature*. MIT Press, Cambridge, MA.

De Block, A. (2008) Why mental disorders are mental dysfunctions (and nothing more). Some Darwinian arguments. *Studies in the History and Philosophy of Science (Part C)*, **39**, 338–46.

Ereshefsky, M. (2009) Defining "health" and "disease". *Studies in the History and Philosophy of Biology and Biomedical Sciences*, **40**, 221–7.

Godfrey-Smith, P. (1993) Functions: consensus without unity. *Pacific Philosophical Quarterly*, **74**, 196–208.

Godfrey-Smith, P. (1998) Functions: consensus without unity. In D.L. Hull and M. Ruse (eds), *The Philosophy of Biology*. Oxford University Press, Oxford, pp. 280–92.

Grant, P.R. (2000) *Ecology and Evolution of Darwin's Finches*. Princeton University Press, Princeton, NJ.

Griffiths, P.E. (1997) *What Emotions Really Are*. University of Chicago Press, Chicago.

Griffiths, P.E (2006) Function, homology, and character individuation. *Philosophy of Science*, 73 (1): 1–25.

Gross, L. (1921) *The Blood Supply to the Heart in its Anatomical and Clinical Aspects*. Hoeber, New York.

Horwitz, A. and Wakefield, J. (2007) *The Loss of Sadness: How Psychiatry Transformed Normal Sadness into Depressive Disorder*. Oxford University Press, New York.

Kitcher, P. (1998) Function and design. In D.L. Hull and M. Ruse (ed.), *The Philosophy of Biology*. Oxford University Press, Oxford, pp. 258–79.

Lloyd, E.A. (2005) *The Case of the Female Orgasm: Bias in the Science of Evolution*. Harvard University Press, Cambridge, MA.

Millikan, R.G. (1984) *Language, Thought and Other Biological Categories*. MIT Press, Cambridge, MA.

Millikan, R.G. (2002) Biofunctions: two paradigms. In A. Ariew, R.Cummins, and M. Perlman (eds), *Functions*. Oxford University Press. Oxford, pp. 113–43.

Murphy, D. (2005) Can evolution explain insanity? *Biology and Philosophy*, 20, 745–66.

Murphy, D. (2006) *Psychiatry in the Scientific Image*. MIT Press, Cambridge, MA.

Murphy, D. (2009) Psychiatry and the concept of disease as pathology. In M. Broome and L. Bortolotti (eds), *Psychiatry as Cognitive Neuroscience: Philosophical Perspectives*. Oxford University Press, New York, pp. 103–117.

Neander, K. (1991) Functions as selected effects: The conceptual analyst's defense. *Philosophy of Science*, 58, 168–84.

Neander, K. (2002)Types of traits: Function, structure and homology in the classification of traits. In A. Ariew, R. Cummins, and M. Perlman (eds), *Functions: New Essays in the Philosophy of Biology and Psychology*. Oxford University Press, New York.

Rupert, R. (2009) *Cognitive Systems and the Extended Mind*. Oxford University Press, New York.

Schaffner, K.F. (1993) *Discovery and Explanation in Biology and Medicine*. University of Chicago Press, Chicago.

Wachbroit, R. (1994) Normality as a biological concept. *Philosophy of Science*, 61, 579–91.

Wakefield, J.C. (1992) The concept of mental disorder: On the boundary between biological facts and social values. *American Psychologist*, 47, 373–88.

Wright, L. (1973) Functions. *Philosophical Review*, 82, 139–168.

Part 3
Psychopathology, evolution, and human nature

Chapter 9

Mirroring the mind: on empathy and autism

Farah Focquaert and Johan Braeckman

From an evolutionary point of view, the prevalence of disorders that are characterized by (extremely) low-empathy, such as autism spectrum conditions and psychopathy, is hard to grasp. Empathy affects almost every aspect of our social world and (extremely) low empathizing skills are devastating to an individual's social life. The following questions loom large: Why are autism spectrum conditions so common, and why are they more prevalent in men? And in general, why does a 'negative' trait like low empathy even manifest itself at all? Keller and Miller's mutational load model, that describes mental disorders in terms of an 'overload' of harmful mutations, is put forward to explain more extreme instances of autism. Less extreme instances of autism spectrum disorders are explained by the presence of maladaptive extremes of low empathy in combination with compensatory systemizing mechanisms. Based upon existing evolutionary hypotheses and current (neuro) cognitive findings, we propose that high empathy, with a high threshold to 'shut down', is more likely to have evolved during female evolutionary history. Low(er) empathy, with a low threshold to 'shut down', is more likely to have evolved during male evolutionary history. Differential selection pressures on empathy may therefore explain the male/female ratio in autism spectrum conditions, and possibly, a difference in empathy between males and females in general.

9.1 **Introduction**

9.1.1 **Empathy**

Hume (1711–76) wrote in his *Treatise of Human Nature* that "the minds of men are mirrors to one another" (1739–1740/2000, p. 236). According to Hume, this mirroring of minds is achieved through *sympathy*. Sympathy does not reflect a feeling of *compassion* or *pity* in Hume's view, but a process or a means of communication, of acquiring and experiencing the passions and sentiments of others. In his view, others' affect(ion)s are first known by their effects and the external signs that can be observed. It is only after these ideas get infused by sympathy that they get converted into impressions. According to Hume, these impressions produce comparable emotions in the spectator as "any original affection" would (p. 206). Darwin (1872) similarly argued that the recognition of certain emotions in others, through sympathy, induces a similar emotion in oneself. He gives the example of his 6-month-old child, who, on witnessing his nurse pretending to cry, instantly replied by exerting a melancholic expression, with the corners of his mouth strongly depressed. Skeptical about a child's ability to reason on this matter at such a young age, it seemed to Darwin that some "innate feeling" must have "told" the child that the nurse's crying expressed sadness, which through sympathy elicited sadness in the child himself (p. 359). It is these kinds of "mirroring" processes that are impaired in autism.

A lack of empathy or empathy-related abilities (e.g., emotional contagion) is central to autism spectrum conditions and underlies many of the social difficulties that individuals with autism spectrum conditions face. However, although empathy is pivotal for social behavior, we must not forget that it is only one aspect of an individual's personality. Focusing on an isolated personality trait ignores the importance of interaction between personality traits and the way in which an overall personality manifests itself within a given environment. An isolated negative trait may be less of a problem when approached within the context of other personality traits or sets of personality traits. It is important to keep in mind that statistically extreme traits of a particular personality dimension do not necessarily constitute a disorder. There is a difference between statistical abnormality and functional abnormality when addressing mental disorders (Wakefield 2008; see also Chapter 5). Indeed, although extremely low empathizing skills in themselves are devastating for one's social life, they might not be as detrimental in combination with other personality traits or within certain environments. A negative condition (e.g., sadness, low empathy, aggression) may or may not amount to a disorder. For example, whereas grief after losing a loved one is normal, experiencing intense enduring sadness that is not triggered by a similar loss may be considered a disorder (i.e., depression) (Wakefield 2007).

9.1.2 Autism spectrum conditions

Autism spectrum conditions, including autism, high-functioning autism, and Asperger syndrome, are neurodevelopmental "disorders" involving a triad of impairments: (i) social impairments (e.g., passivity in social interaction or the opposite, inappropriate and repetitive approaches to others), (ii) communicative impairments (e.g., impaired speech, language delay), and (iii) repetitive and restricted behaviors and interests (e.g., obsessive interest in a narrow topic such as dinosaurs or planets). Each element of the triad can occur in different degrees of severity and may manifest itself in varying ways in different individuals (Happé et al. 2006). Autism spectrum conditions are highly heritable. Concordance rates of 70% for autism and 90% for autism spectrum conditions are found in monozygotic twins compared to concordance rates of 5 and 10%, respectively, for dizygotic twins (Sebat et al. 2007). Prevalence rates are approximately 1% of the population for autism spectrum conditions, with a male/female ratio of 4:1 for autism and an even higher rate for the whole spectrum (Yeargin-Allsopp et al. 2003; Baird et al. 2006). Because autism spectrum conditions are often devastating to one's social life, their prevalence in modern society is puzzling from an evolutionary perspective, so are we facing an evolutionary paradox? Or rather, if "autism genes" exist, shouldn't they have been eliminated by natural selection?

Although several genetic studies on autism spectrum conditions have been done, the search for such "autism genes" remains elusive. Why has it proven so difficult to reliably identify susceptibility genes for autism spectrum conditions? And why is it more common in boys, but more severe in girls? Several candidate genes have been put forward, but specific results are rarely replicated across studies (Gupta and State 2007). Perhaps we are asking the wrong questions. Recent findings provide evidence for the heterogeneity of the autism spectrum at the cognitive, neurological, and genetic levels (Happé et al. 2006). Autism spectrum conditions may prove to be an "umbrella concept", as was argued by Adriaens (2007, p. 525) for schizophrenia, covering a variety of different (sets of) impairments instead of exemplifying a more or less uniform condition or natural kind. According to Morrow et al.:

> ...the genetic architecture of autism resembles that of mental retardation and epilepsy, with many syndromes, each individually rare, as well as other cases potentially reflecting complex interactions between inherited changes.
>
> (Morrow et al. 2008, p. 218)

Hence, cognitive, neurological, and genetic research on autism spectrum conditions has much to gain from focusing on one of the components of the triad or on specific *endophenotypes* related to a single component, rather than exclusively focusing on the triad as a whole. Specific (genetic) anomalies that are related to a given neuropsychiatric disorder may also manifest themselves in other disorders (Lupski 2008), providing further reasons to investigate specific traits related to autism spectrum conditions.

Baron-Cohen (2008) describes autism spectrum conditions as empathizing–systemizing (E–S) conditions, in which individuals with autism spectrum conditions show below average empathizing alongside normal or above average systemizing. In line with Baron-Cohen's (2003) E–S theory of sex differences, this chapter focuses on a specific impairment related to autism spectrum conditions that is also manifest in certain other neuropsychiatric disorders (e.g., psychopathy): (extremely) low empathy. It addresses instances of (extremely) low empathy in the population as a whole, as well as instances related to autism spectrum conditions. Empathy is a key mechanism underlying social behavior and (extremely) low empathy, as observed in individuals with autism spectrum conditions, falls largely within the social component of the triad (Baron-Cohen 2003). Empathy is part of an individual's emotional and cognitive make up and is very strongly correlated with agreeableness (Nettle 2007), which is one of the factors in the five-factor model of personality comprising broad personality trait domains found throughout the population to varying degrees. On average, women score higher than men on empathizing (Baron-Cohen 2003; Rose and Rudolph 2006). What we see in a disorder like Asperger syndrome is that empathizing is (extremely) low, leading to all sorts of problems within the social domain (e.g., social anxiety, impaired imitation skills, inability to detect deception, difficulty building and maintaining relationships with others).

9.1.3 Aim of this chapter

Drawing on Baron-Cohen's E–S theory of sex differences and the related "extreme male brain theory of autism", we aim to unveil part of the evolutionary puzzle surrounding the prevalence of autism spectrum conditions and (extremely) low empathy in particular. We start this chapter with a section on the E–S theory of sex differences and how this relates to autism spectrum conditions. Then we summarize a selection of important recent genetic findings on autism spectrum conditions and discuss a recent theory by Keller and Miller (2006a), which fits in nicely with these findings. In the third and final section, we argue that differential selection pressures on males versus females with regard to empathy might provide a (partial) explanation for the male/female ratio observed in autism spectrum conditions. In summary, we put forward a hypothesis concerning evolved sex differences in empathy combined with recent (evolutionary) genetic findings and theorizing, in order to explain the presence of (extremely) low empathy as well as the male/female ratio in autism spectrum conditions.

9.2 Autism spectrum conditions: a lack of "mirroring" and empathy

9.2.1 Empathizing–systemizing in autism spectrum conditions

The E–S theory of psychological sex differences (Baron-Cohen 2003; Baron-Cohen *et al.* 2005) claims that whereas the female brain is predominantly hard-wired for empathy, the male brain is predominantly hard-wired for understanding and building systems. Empathizing can be defined as:

> ...the drive to identify another's mental states and to respond to these with an appropriate emotion, in order to predict and to respond to the behavior of another person.
>
> (Baron-Cohen *et al.* 2005, p. 820)

Systemizing can be defined as:

> the drive to analyze a system in terms of the rules that govern the system, in order to predict the behavior of the system.
>
> (Baron-Cohen *et al.* 2005, p. 820)

According to Baron-Cohen (2003), both traits can be seen as adaptations originating in our evolutionary history, systemizing being an answer to physical selection pressures (e.g., using and making tools, hunting and tracking, social dominance) and empathizing an answer to social selection pressures (e.g., mothering, making friends, mindreading). According to the E–S theory individuals can possess one of three brain types: (i) an individual's level of empathy can be higher than his or her level of systemizing (type E), (ii) an individual's level of systemizing can be higher than his or her level of empathizing (type S) or (iii) an individual can have comparable levels of empathizing and systemizing skills (type B). Type S is more common in men, whereas type E is more common in women (Baron-Cohen *et al.* 2005). Individuals with extreme E > S cognitive patterns are deemed "system-blind", whereas individuals with extreme S > E cognitive patterns are deemed "mind-blind". The extreme male brain theory posits that autism represents an extreme of the male brain type S (Baron-Cohen *et al.* 2005).

Baron-Cohen (2006) claims that human brains have a specific systemizing mechanism that engages in the interpretation of nonagentive changes in the world that are at least to some extent lawful and have narrow variance (or limited degrees of freedom). This mechanism is set at different levels in different individuals. Individuals with autism spectrum conditions have their systemizing mechanism set too high, to the extent that this does not easily allow for the interpretation of agentive change. This is due to the complexity of the changes occurring in agentive systems, which involve many degrees of freedom and wide variance, as opposed to nonagentive systems. This type of reasoning requires an empathizing system. In the case of high-functioning autism or Asperger syndrome, for example, empathizing may take the form of "hacking" or applying systemizing to agentive systems, which does not correspond to empathizing-driven social behavior and may seem artificial. According to Baron-Cohen (2006), the more severe a person's autistic traits, the higher their systemizing level appears to be and the more consequences this has for other types of reasoning, such as language. Language acquisition seems easier if one has a bit more tolerance to variance and change. This might explain why individuals with high-functioning autism have a language delay, but are still able to acquire language, in contrast to severely autistic individuals, with their systemizing mechanism set at the highest level, who do not acquire language at all. Having a closer to average systemizing level could therefore result in less language delay, less obsessive behavior, less impaired social reasoning skills and less "stilted social behaviour, such as attempts at systemizing social behaviour" (Baron-Cohen 2006, p. 4). Although we agree with Baron-Cohen that systemizing the social world might appear somewhat artificial, we suggest that having a systemizing mechanism set at a very high level might be useful or even life-saving for an individual with (extremely) low empathizing. When empathizing skills are unavailable, systemizing skills might provide the only means available to deal with the social world. Instead of resulting in impaired social reasoning, systemizing skills might prevent a total lack of social skills.

9.2.2 A lack of "mirroring"

From a cognitive neuroscience perspective, "mirroring" neural activity implies shared activity in a given brain area (e.g., medial prefrontal cortex for introspection and theory of mind) or shared activity in single cells (e.g., mirror neurons in the inferior frontal gyrus) for both self-aspects and other-aspects of cognition. Current imaging studies (e.g., Carr et al. 2003) support the hypothesis that human mindreading relies on mirroring (Goldman 2006; Focquaert et al. 2008). Moreover, several functional magnetic resonance imaging (fMRI) studies on mindreading have shown that self-reported empathy correlates with mirroring-type neural activity (e.g., Singer et al. 2004, 2006). If empathy is related to mirroring-type neural activity in the brain, the possibility arises that individuals with below average or low empathy need to rely (more strongly) on other, nonmirroring (neural) mechanisms to mindread. This indeed appears to be the case for individuals with autism spectrum conditions.

Recent imaging studies indicate that autism spectrum conditions involve deficits in mirror neuron activity and anatomy, even specifically related to (face-based) mindreading (Dapretto *et al.* 2006; Hadjikhani *et al.* 2006). In line with these findings, a study by McIntosh *et al.* (2006) revealed that individuals with Asperger syndrome do not automatically mimic or "mirror" emotional face expressions in others compared to normal participants. Although they are able to mimic facial expressions (voluntary mimicry), they do not tend to do so automatically. Moreover, a recent transcranial magnetic stimulation study by Minio-Paluello *et al.* (2009) found no "embodiment of others' pain" (p. 58) in individuals with Asperger syndrome during observation of painful stimuli in others. According to the authors, their results indicate that "embodied empathic pain resonance effects are absent in AS participants" (p. 58). This is a very important finding because it rules out mirroring in individuals with Asperger syndrome during pain-related mindreading. Most likely, such mirroring mechanisms are unavailable to individuals with Asperger syndrome for all aspects of mindreading. Because individuals with Asperger syndrome can be said to possess a late-acquired theory of mind system that is the result of effortful learning (Frith and Happé 1999), there must be a different nonmirroring mechanism at work. The autobiography of Temple Grandin (Grandin 1995), a doctor in veterinary science with Asperger syndrome, is in line with a systemizing approach to social interactions. Over the years, she says she formed a mental library filled with different social situations and their specific rules of behavior to guide her own social behavior. She knows how to act appropriately at social settings, not by mirroring others' intentions, wishes, feelings, etc., but by comparing the situational aspects of the social setting at hand to previous occasions that resemble it. Bering (2002) similarly argues that in cases of high-functioning autism, it is as if their systemizing skills are "translated to problem solving in the area of social matters" by exploiting observable cues in others. It allows these individuals to "get by in the real world" (p. 14).

Moreover, a recent study by Rutherford and McIntosh (2007) suggests that individuals with autism spectrum conditions rely more heavily on a rule-based strategy when observing emotional faces. They propose that normal individuals process faces predominantly based upon configural information, whereas individuals with autism spectrum conditions might rely more strongly on specific rules about individual facial features to identify emotional faces (see also Chapter 3). For example, according to a rule-based strategy, "sad" is associated with down-turned corners of the mouth, narrowed eyes, and lowered eyebrows. Individuals with autism spectrum conditions might conclude that someone is sad based on the following rule "*if* the corners of the mouth are turned down, *then* the person is sad". In line with their hypothesis, a recent study by Lahaie *et al.* (2006) found evidence for enhanced processing of individual face parts in individuals with Asperger syndrome compared to normal controls. Due to the absence of embodied empathy in individuals with autism spectrum conditions, these individuals likely rely on their systemizing skills to understand social behavior (e.g., reading emotional faces).

Baron-Cohen has argued that autism spectrum *disorders*—the term disorders is commonly used when referring to autism and Asperger syndrome—should be called autism spectrum *conditions* because less severe forms are often not dysfunctional for the individual in question. In our view, although (extremely) low empathizing is a social handicap, the presence of high systemizing might provide relief, either because systemizing provides a compensatory mechanism that allows for suboptimal but sufficient social reasoning skills or because specific niches are present where high systemizers, despite their low empathizing, can thrive. Silicon Valley provides just one example of a specific niche where high systemizing skills are valued (De Block 2006). Indeed, although extremely low empathizing is likely due to "a failure of the mind to work as designed", it does not amount to a disorder if one's behavior, resulting from one's overall personality traits, is not judged to be negative or harmful (Wakefield 2007, p. 153; see also Chapter 5). Whether or not a dysfunction constitutes a disorder depends on the individual and the situation in question.

9.3 The genetics of autism spectrum conditions

9.3.1 The triad of impairments

Until recently, it was assumed that the three clusters of impairments that are manifest in autism spectrum conditions co-occur and that they are somehow related to each other, even though the nature of this ascribed relationship was not known (Mandy and Skuse 2008). Ronald *et al.* (2005) investigated whether social impairments and obsessive and repetitive behaviors are phenotypically and genetically related using a large community sample of twins. A relative independence of both traits was found. Results indicate that social and nonsocial phenotypic traits were very modestly correlated in the general population and at the extreme ends of the social and nonsocial distributions. Genetic analyses of the twin database reveals that both social and nonsocial behaviors are substantially heritable (62–76%), while showing only modest genetic overlap. Happé *et al.* (2006) report that, within this large population-based sample, several children presented themselves with isolated difficulties in only one component of the triad, and while children with one component of the triad have an increased risk of showing a second or third, these risks are relatively low. So, although the components of the triad manifest themselves together at above-chance rates, there is considerable evidence for "fractionation of the three aspects of the triad" (Happé et al. 2006, p. 1218). The hypothesis that most of the genetic effects that are related to autism are specific, meaning that they act on only one component of the triad, fits the family data. Some relatives of individuals with autism spectrum conditions show only isolated traits (i.e., in one component of the triad). Hence, it is likely that different susceptibility genes exist for each component of the triad or for different endophenotypes presenting themselves within these different behavioral components (e.g., impaired eye-to-eye gaze).

9.3.2 Susceptibility genes

Neuropsychiatric disorders typically display complex phenotypes with multiple genetic and environmental factors influencing their development. Hence, the search for susceptibility genes likely involves a quest for several genes contributing to a given neuropsychiatric disorder. Moreover, different genes or sets of genes may lead to the same neuropsychiatric phenotype in different individuals. During an individual's lifetime, his/her genome is passed from mother cells to daughter cells by self-replication. Because of this self-replication, our genes are subject to copying errors or "mutations", which are "ultimately the only possible source of genetic variation between individuals" (Penke et al. 2007, p. 552). Such copying errors consist of point mutations or single nucleotide polymorphism (SNPs, in which one of the four possible nucleotides in a base pair is substituted for another), copy-number variations (CNVs, duplications or deletions of base pair sequences), and rearrangements of larger chromosomal regions such as inversions or translocations. According to Penke et al. (2007), SNPs and duplication-type CNVs are likely the most common source of genetic variation among individuals because they can have phenotypic effects of any strength, including mild effects. Deletions, insertions, and larger rearrangements on the other hand are more likely to have very severe phenotypic effects. Although mutations can be phenotypically neutral or involve a functional improvement in the phenotype in relation to the environment, most tend to be harmful for an individual (Penke et al. 2007). Recently, genetic experiments revealed abundant CNVs in the human population, which may in part account for the existence of neuropsychiatric disorders such as autism and schizophrenia. Although the search for such susceptibility genes is complex, individuals with specific CNV syndromes may present themselves with known neuropsychiatric disorders. In addition to common genetic variants (e.g., the candidate gene *engrailed 2*) likely contributing to autism spectrum conditions, evidence accumulates that rare CNVs account for part of the genetic liability to autism spectrum conditions (Gupta and State 2007; Sebat *et al.* 2007; Christian *et al.* 2008; Cook and Scherer 2008; Glessner *et al.* 2009). Overall, recent data on autism spectrum conditions show that the disorder is the result of a mixture of inherited and new, or *de novo*, gene defects at many different loci, which vary among affected individuals. Moreover, evidence is accumulating that genes that influence specific processes (e.g., neuronal cell adhesion) and associated phenotypic traits are shared by different disorders: "the same genes may contribute to different disorders due to the overlap between these disorders' componential structures" (Grigorenko 2009, p. 127). For example, the same chromosomal regions (2q, 7q, and 13q) have been linked to both autism spectrum conditions and specific language impairment in different samples. Such findings are in line with a symptom-based rather than a holistic approach to psychiatric disorders. Indeed, high co-morbidity of and lack of clear separation between the current DSM-IV disorders is apparent (Regier *et al.* 2009). Rather than exemplifying discrete disease entities, mental disorders likely reflect a combination of symptoms or dysfunctional endophenotypes in varying degrees of severity (see also Chapter 7). Moreover, these specific symptoms may contribute to a variety of disorders.

9.3.3 **Mutational load model of neuropsychiatric disorders**

The polygenic nature and genetic heterogeneity of autism spectrum conditions is in line with the mutational load model of neuropsychiatric disorders that was recently put forward by Keller and Miller (2006a). Several different evolutionary explanations have been given for the variation in *common* mental disorders (at least, more common than would be expected from an evolutionary perspective), such as the hypothesis of ancestral neutrality or the theory of balancing selection (see Introduction). For example, frequency-dependent selection might explain the prevalence of pyschopathy (Mealey 1995). Keller and Miller's (2006a) mutational load model holds that the variation in common mental disorders is best explained by a polygenic mutation–selection balance. Their model aims to explain the evolutionary persistence of mental disorder susceptibility alleles. Mental disorder alleles are those regions of DNA (broadly defined to include both coding and noncoding regulatory regions) that differ between individuals and increase the risk of common mental disorders. The frequency of harmful single-gene or so-called Mendelian disorders (e.g., juvenile onset Parkinson's or achondroplastic dwarfism) can easily be explained by mutation–selection balance. More specifically, the frequency of these disorders can be explained by looking at the balance between genetic copying errors that turn normal alleles into harmful mutations, and selection that eliminates these mutations. Selection removes these mutations at a rate that is proportional to the fitness cost of the mutation. Because these mutations are so harmful to reproductive fitness, this typically results in a low equilibrium frequency of mutant alleles that have not yet been removed from the population by selection. Hence, the prevalence of these disorders is rare and does not pose an evolutionary paradox.

However, what about mental disorders such as schizophrenia, autism, and psychopathy? According to Keller and Miller (2006a), these are much more common, hundreds to thousands times more prevalent than would be expected from a single-gene mutation–selection model. Consequently, many researchers have rejected the mutation–selection model as a viable explanation of most mental disorders. However, Keller and Miller (2006a) argue against this conclusion by positing a multiple-gene or polygenic model instead. According to conservative estimates, the human brain carries an average of 500 mildly harmful mutations (Fay *et al.* 2001; Sunyaev *et al.* 2001). Because of variation in the number of mutations and the effect sizes of different mutations, the end result will be a continuous distribution across individuals with respect to most psychological dimensions. Individuals with a high mutation load will be at a higher risk of having mental disorders. The prevalence of common mental disorders therefore derives from their polygenic nature, or the fact that they are the result of many mutations with mild effects. Only those harmful mutations that have dramatic effects will be eliminated immediately. Mutations with mild effects on reproductive fitness will be removed more slowly and thus be more common. In line with this model, several studies have shown that children with autism carry a higher frequency of chromosomal abnormalities compared to normally developing individuals (Gupta and State 2007).

According to Keller and Miller:

> Everyone alive . . . has minor brain abnormalities that cause them to be a little bit mentally retarded, a little bit emotionally unstable, and a little bit schizophrenic.

> (Keller and Miller 2006a, p. 404)

Keller and Miller's model can in principle be extended to variation in personality traits in general, in which certain mental disorders are then seen as involving low-fitness extremes of particular personality traits or a combination of several low-fitness extremes reflecting different personality traits (see Buss 2006; Keller and Miller 2006b). The latter explanation appears well suited to integrate recent genetic and cognitive–behavioral findings in autism spectrum conditions. Penke *et al.* (2007) suggest that the severe nature of certain personality disorders might result from high mutational load, but may derive their specific characteristics from the combination of high mutational load with certain (extreme instances of) personality traits. Likewise, severe cases of autism spectrum conditions could be explained as involving high mutational load combined with extremely low empathy.

9.4 **Evolution of autistic traits: low empathy**

9.4.1 **Evolved gender differences in empathy: a differential selection pressure hypothesis**

Although Tooby and Cosmides (1990) reject heritable variation in personality traits as adaptive, they obviously do not claim that the complex psychological mechanisms that underlie personality traits are not adaptations. Indeed, human empathy is a psychological adaptation that is pivotal for, among others, the mother–child relationship, social bonding, and cooperation. Whether or not the normal variation in personality traits can be considered adaptive, both adaptive and nonadaptive views of normal variation in personality traits in the population as a whole allow for the possibility of evolved male/female differences in empathy. Male/female differences in empathy could be the result of differences in adaptive design because male/female social life differs in ways that potentially differentially affect the fitness payoffs of such a psychological mechanism. For example, Tooby and Cosmides (1990) hypothesize that a male versus female version of sexual jealousy likely evolved because of the different reproductive strategies of men versus women. Assuming that infidelity brings about different problems for men versus women (see Buss 1988), a qualitative difference, that is, a difference in kind and not just degree, in the psychological mechanism dealing with these problems might have evolved in men versus women. Similarly, specific social situations related to empathy are more apparent in male social settings versus female social settings and vice versa (e.g., warfare, the mother–infant bond, etc.), although empathy-related social settings might be more alike between men and women overall compared to sexual jealousy. Nevertheless, a difference in degree, and possibly even in kind, could be the result of evolutionary selection pressures.

Gender differences in empathy have been reported in numerous studies. These differences reflect averages and, obviously, individual men can have a more typically female cognitive style and individual women a more typically male cognitive style in terms of empathizing (Focquaert *et al.* 2007). As mentioned above, women show higher (self-reported) empathy compared to men (Baron-Cohen 2003; Rose and Rudolph 2006). Interestingly, a recent fMRI study found that men's empathic responses to the observation of painful stimulation are modulated by the perceived fairness of others (Singer *et al.* 2006).

Both men and women showed empathic responses towards fair opponents (fairness or unfairness was previously established in a sequential prisoner's dilemma game) in the anterior insula, fronto-insular cortex, and anterior cingulate cortex, although the latter was significant in women and borderline in men. There were no significant differences during painful stimulation of fair versus unfair players in empathy-related activity in pain-related brain areas in women. In contrast, men showed increased activity in the fronto-insular cortex for fair opponents, but not for unfair opponents. Moreover, men showed activation in reward-related areas of the brain when observing painful stimulation being given to unfair opponents. This study therefore suggests that it might be "easier" for men to decrease or "shut down" empathic responses towards unfair opponents. Moreover, a recent electrophysiological study (Han et al. 2008) measuring empathy-related event-related brain potentials (ERPs) found differences in both the early and late aspects of empathic brain processes between men and women. According to the authors, the early aspects reflect automatic processes underpinning emotional sharing (i.e., mirroring) and the late aspects reflect controlled processes underpinning the cognitive evaluation of others' pain. The early ERP effect was found to be comparable in men and women. Importantly, in women this ERP effect correlated with the subjectively perceived painfulness of the stimuli as well as one's own feeling of unpleasantness when observing the stimuli. The authors propose that the subjective feelings or conscious awareness of others' pain and one's own unpleasantness are more strongly determined by the early automatic component in women compared to men. This could imply a stronger linkage between a shared experiential state and the conscious awareness of other's feelings in women. The late ERP effect was found to be stronger in women compared to men. According to the authors, this might reflect more intense evaluation of the painful stimuli on behalf of the women. Overall, these findings suggest that empathizing in women is more strongly driven by mirroring processes (i.e., resulting from shared emotional states) compared to men, and while hypothetical, possibly implies that empathizing is a more cognitive driven process in men.

These gender differences in empathy might reflect evolved solutions to different social problem situations in men and women. Male social skills can be seen as adaptations to an evolutionary history of coalitional male–male competition and within coalitional dominance hierarchies, whereas female social skills most likely are adaptations to dyadic relationships focused on social and emotional support (e.g., the mother–infant bond). It has been shown that in many traditional societies men cooperate to form kin-based coalitions that compete with other male kin groups over ecological resources and reproductive control (Geary 1999; Geary et al. 2003). Strong empathic responses are potentially life-threatening when engaging in between-group competition (e.g., warfare), and might compromise one's dominance position and social status within the group (hence one's reproductive fitness). Moreover, the ability to easily "turn off" one's empathic reactions towards the "out-group" allows one to exclusively direct one's energy and resources to oneself and one's "in-group" (hence optimizing one's resources, as well as social status within the group). At the same time, this opens up the possibility of exploiting others to maximize one's own potential (in terms of both social status and resources). There is also genetic and anthropological evidence that the prototypical pattern during human evolution for females was to migrate and for males to stay in their birth group (e.g., Seielstad et al. 1998; Geary et al. 2003), although conflicting results have been found (see Handley and Perrin 2007). Consequently, it seems likely that the typical social setting during human evolution was different for males as opposed to females. According to Geary et al. (2003), male social strategies more likely reflect selection pressures acting on kin-based social relationships, whereas female social strategies more likely reflect selection pressures acting on dyadic nonkin social relationships in combination with mother–child interactions (Geary 1998). Relationships among kin are typically associated with high levels of cooperation and tolerance of nonreciprocal relationships, whereas relationships among nonkin tend to be less stable and more conflict-prone (Geary et al. 2003). The nature of male versus female evolutionary history suggest that selection pressures on empathy were different for males versus females, resulting at least in a difference in degree. Indeed, Nettle (2007) argues that sexual dimorphism in empathy suggests that, "though empathy is useful for both sexes, the fitness pay-offs for one sex are greater than the other" (p. 251). This might reflect the mother–infant bond that signals dependence and draws upon social support networks, or the fact that the consequences of social conflict might be more serious for females and especially their young, who may be endangered by aggression. Importantly, individuals with high empathy typically have harmonious interpersonal interactions and are able to avoid violence and interpersonal hostility. Such an ability is therefore extremely important with respect to the survival of one's offspring, although unconditional trust is, of course, almost never an adaptive strategy (Nettle 2006). Men have more to gain from personal status, even at the expense of social harmony (Nettle and Liddle 2008), hence "relaxing" selection pressures on empathy.

9.4.2 Evolutionary hypotheses about male/female social strategies

The hypothesis that high empathy is beneficial for females, but much less so for males during evolutionary history, and the sub-hypothesis that a low threshold to shut down empathy is beneficial for males, but much less so for females, is in line with recent evolutionary theories on male/female social strategies. We will briefly outline a number of theories on this topic, including (i) the tend-and-befriend hypothesis (Taylor *et al.* 2000), (ii) the reciprocity potential hypothesis (Vigil 2007), and (iii) the male-warrior hypothesis (Van Vugt *et al.* 2007).

(i) *The tend-and-befriend hypothesis.* Whereas "fight or flight" is considered to represent the prototypical response to stress in both men and women, Taylor *et al.* (2000) argue that behaviorally, women's responses to stress are marked by a pattern of "tend and befriend", a pattern that likely evolved to enhance reproductive success by affiliating with social groups, and especially other females, and by nurturing their own offspring. Tending is described as "quieting and caring for offspring and blending into the environment" (p. 412) and is well suited to avoid threats, whereas fight responses could potentially endanger oneself and one's offspring. Moreover, flight responses might be less effective when pregnant and caring for young offspring. Befriending is described as "the creation of networks of associations that provide resources and protection for the female and her offspring under conditions of stress" (p. 412). Taylor *et al.* (2000) discuss several empirical findings in animals and humans that support their theory, for example:

(a) Physical aggression is more prominent in men and largely confined to situations involving self-defence in women.

(b) As observed in animals, the effects of oxytocin and endogenous opioid peptides during stressful situations (e.g., threats) potentially mediate flight-response inhibition and might underlie "tending" mechanisms (e.g., maternal touching) in women. Oxytocin release during stressful situations enhances behavior (e.g., relaxation) that is antithetical to fight-or-flight responses. Research suggests that oxytocin's effects are more pronounced in female compared to male rats.

(c) Women are much more likely to seek affiliation with (same-sex) others under conditions of stress compared to men, and they will seek help and social support more often from women than from men, thus suggesting stronger befriending mechanisms in women. Tend-and-befriending mechanisms are undoubtedly linked to empathizing skills in humans. The stronger one's empathizing, the better one's tend-and-befriend mechanism will be. The need for tend-and-befriending possibly provided one of the adaptive problems leading to the evolution of higher empathizing in females compared to males.

(ii) *The reciprocity potential hypothesis.* Vigil (2007) investigated the concept of "reciprocity potential", which he claims is an important factor in reciprocal altruism and the ability to attract and maintain peer relationships. He argues that an individual's reciprocity potential can be displayed in two ways: (a) by displaying one's perceived capacities or (b) by displaying one's motivation to reciprocate. The first refers to indicators of health, financial potential, intelligence, and so forth. The second refers to an individual's perceived trustworthiness and is an indicator of the stability and security of a potential relationship. The best way to communicate one's motivation to reciprocate is through displays of empathy. Vigil (2007, 2008, 2010) provides preliminary evidence for the hypothesis that men will display more capacity cues, whereas women will display the investment component more strongly (i.e., one's motivation to reciprocate). Based upon an analysis of human evolutionary history as characterized by male-biased philopatry (i.e., remaining in or returning to one's birthplace) and kin-based male–male coalitional competition (e.g., Geary 1998; Geary *et al.* 2003), Vigil suggests that:

> exposure to and reliance upon more closely related kin may have relaxed the selective pressures for men to solicit and manipulate their social relationships through explicit displays of trustworthiness and hence the interpersonal investment component of reciprocity potential, in comparison to women.
>
> (Vigil 2007, p. 158)

whereas

> a disproportionate reliance upon more distantly related kin and non-kin among females may have selected for a greater propensity to advertise their trustworthiness and interchange more investment behaviors (e.g., vis-à-vis submissiveness displays, such as crying and expressed empathy).
>
> (Vigil 2007, p. 158–9)

Our hypothesis concerning evolved sex differences in empathy similarly claims that women during evolutionary history faced selection pressures that favor high empathy (or explicit displays of trustworthiness), whereas men did not. Moreover, men likely faced selection pressures that downplayed empathy.

(iii) *The male warrior hypothesis.* According to the male-warrior hypothesis (Van Vugt *et al.* 2007) men's social psychology and behavior is more intergroup oriented than women's. This is in line with a long evolutionary history of male–male coalitional competition and violent intergroup conflict, in which the benefits for men to engage in intergroup rivalry, in terms of mating opportunities and access to other valuable resources (e.g., food) sometimes outweighed the costs (e.g., serious injury, death). Research on violent intergroup competition in traditional societies (e.g., Yanomamö) and modern societies (e.g., street gangs) shows that men are much more likely to engage in such behavior compared to women (Van Vugt *et al.* 2007). High empathy vis-à-vis one's opponent or an inability to "shut down" empathizing temporarily is potentially life-threatening in case of violent intergroup conflict. Instead, low(er) empathy, combined with a low threshold to "shut down", is potentially "beneficial" during intergroup competition (in terms of fitness pay-offs).

9.4.3 Explaining autism: low empathy, high systemizing, and the mutational load model

Based on these evolutionary hypotheses and current experimental findings, high empathy, with a high threshold to "shut down", is more likely to have evolved during female evolutionary history, while low(er) empathy, with a low threshold to "shut down", is more likely to have evolved during male evolutionary history. On average, we should therefore find stronger mechanisms supporting empathy in women compared to men. Although further research needs to be done, current findings are in line with the possibility of stronger empathizing mechanisms in females. If this proves to be the case, it follows that low-fitness or maladaptive extremes involving extremely low empathy, as seen in high-functioning autism and Asperger syndrome, are more likely to occur in men. Moreover, possible detrimental effects of low(er) empathizing on males' social skills might be, at least to some extent, compensated by their systemizing skills, which are typically higher in men compared to women (Baron-Cohen 2003). Although contrary to Baron-Cohen's (2006) position (see above), such compensation appears to be the case for individuals with autism spectrum conditions, or at least for individuals with high-functioning autism or Asperger syndrome. The existence of compensatory mechanisms adds to the view that the latter cases of autism spectrum conditions might not be so devastating after all (Gernsbacher *et al.* 2006), although severe "autism is likely to be *evolutionary* harmful" (Keller and Miller 2006b, p. 433). Hence, the manifestation of maladaptive extremes of low empathy due to a combination of mutational load and differential selection pressures on the evolution of empathy in men versus women, possibly compensated for by normal or above average systemizing, might explain the prevalence of milder variants of autism, and the associated male/female ratio, whereas the prevalence of more severe cases of autism spectrum conditions is likely due to an overall higher mutational load additionally affecting cognitive traits. Skuse (2007) proposes that the clinical manifestations of autism spectrum conditions are influenced by an individual's general cognitive ability, in the sense that moderate-to-severe mental retardation "merely reveals autistic traits that were already present" (p. 393) and "symptomatic compensation occurs in many individuals of normal-range intelligence" (p. 392). Interestingly, Yeargin-Allsopp et al. (2003) found that as the rate of mental retardation increases in autism, the male/female ratio decreases, indicating that women, if diagnosed with autism spectrum conditions, are more likely to show severe forms of autism. High mutational load likely affects cognitive abilities (Penke et al. 2007) to the same extent in men and women and might explain why the male/female ratio drops in more severe cases of autism (i.e., because men and women are equally prone).

9.5 Conclusion

We set out to address the prevalence of autism spectrum conditions, and (extremely) low empathizing in particular, from an evolutionary psychiatric perspective. Autism spectrum conditions can be devastating to an individual's social life, which makes their prevalence puzzling from an evolutionary perspective. As mentioned in the introduction, heterogeneity is found for autism spectrum conditions at the behavioral, cognitive, neural, and genetic levels. A focus on specific traits or endophenotypes related to such conditions is therefore recommended. We focus on one aspect of autism spectrum conditions that is particularly important from an evolutionary perspective: empathy. Indeed, a lack of social "mirroring", reflecting (extremely) low empathizing skills, combined with normal or above average systemizing may lead to inappropriate and "artificial" social behavior. At least in individuals with high-functioning autism and Asperger syndrome, we argue that normal or above average systemizing skills may function as a compensatory strategy during social interaction. We suggest that evolved sex differences in empathy, namely low(er) empathy with a low threshold to "shut down" as an evolved strategy in men compared to high(er) empathy with a high threshold to "shut down" as an evolved strategy in women, provide part of the answer to the evolutionary puzzle surrounding autism spectrum conditions. Current findings on gender differences in empathy are in line with our hypothesis. A combination of mutational load effects and differential selection pressures on empathy in males versus females provides a possible explanation for the prevalence of autism spectrum conditions and the associated male/female ratio. In conclusion, we hypothesize that the autism continuum ranging from mild to severe psychopathology might reflect weaker versus stronger mutational load respectively, combined with maladaptive instances of low empathy (and compensatory strategies involving normal or above average systemizing in mild psychopathology).

Acknowledgments

We would like to thank the reviewers for useful comments to earlier drafts. Farah Focquaert is a postdoctoral research fellow of the Research Foundation–Flanders (FWO).

References

Adriaens, P.R. (2007) Evolutionary psychiatry and the schizophrenia paradox: a critique. *Biology and Philosophy*, 22, 513–28.

Baird, G., Simonoff, E., Pickles, A., Chandler, S., Loucas, T., Meldrum, D., and Charman, T. (2006) Prevalence of disorders of the autism spectrum in a population cohort of children in South Thames: the special needs and autism project (SNAP). *Lancet*, 368, 210–5.

Baron-Cohen, S. (2003) *The Essential Difference. The Truth About the Male and Female Brain*. Perseus, New York.

Baron-Cohen, S. (2006) The hyper-systemizing, assortative mating theory of autism. *Progress in Neuro-Psychopharmacology & Biological Psychiatry*, 30, 865–72.

Baron-Cohen, S. (2008) Theories of the autistic mind. *The Psychologist*, 21, 112–6.

Baron-Cohen, S., Knickmeyer, R. C., and Belmonte, M.K. (2005) Sex differences in the brain: implications for explaining autism. *Science*, 310, 819–23.

Bering, J.M. (2002) The existential theory of mind. *Review of General Psychology*, 6, 3–24.

Buss, D.M. (1988) The evolution of human intrasexual competition: Tactics of mate attraction. *Journal of Personality and Social Psychology*, 54, 616–28.

Buss, D.M. (2006) The evolutionary genetics of personality: Does mutation load signal relationship load?. *Behavioral and Brain Sciences*, 29, 441.

Carr, L., Iacoboni, M., Dubeau, M.C., Mazziotta, J.C., and Lenzi, G.L. (2003) Neural mechanisms of empathy in humans: A relay from neural systems for imitation to limbic areas. *Proceedings of the National Academy of Sciences*, 100, 5497–502.

Christian, S.L., Brune, C.W., Sudi, J., Kumar, R.A., Liu, S., Karamohamed, S., Badner, J.A., Matsui, S., Conroy, J., McQuaid, D., Gergel, J., Hatchwell, E., Gilliam, T.C., Gershon, E.S., Nowak, N.J., Dobyns, W.B., and Cook, E.H. (2008) Novel submicroscopic chromosomal abnormalities detected in autism spectrum disorder. *Biological Psychiatry*, 63, 1111-17.

Cook, E.H. Jr. and Scherer, S.W. (2008) Copy-number variations associated with neuropsychiatric conditions. *Nature*, 455, 919–23.

Dapretto, M., Davies, M.S, Pfeifer, J.H., Scott, A.A., Sigman, M., Bookheimer, S.Y., and Iacoboni, M. (2006) Understanding emotions in others: mirror neuron dysfunction in children with autism spectrum disorders. *Nature Neuroscience*, 9, 28–30.

Darwin, C. R. (1872) *The Expression of the Emotions in Man and Animals*. John Murray, London.

De Block, A. (2006) *Waanzin & Natuur. Darwin En de Psychiatrie*. Boom, Amsterdam.

Fay, J.C., Wyckoff, G.J., and Wu, C. (2001) Positive and negative selection on the human genome. *Genetics*, **158**, 1227–34.

Focquaert, F., Steven, M.S., Wolford, G.W., Colden, A., and Gazzaniga, M.S. (2007) Empathizing and systemizing cognitive traits in the sciences and humanities. *Personality and Individual Differences*, **43**, 619–25.

Focquaert, F., Braeckman, J., and Platek, S.M. (2008) An evolutionary cognitive neuroscience perspective on human self-awareness and theory of mind. *Philosophical Psychology*, **21**, 47–68.

Frith, U. and Happé, F. (1999) Theory of mind and self-consciousness: What is it like to be Autistic? *Mind & Language*, **14**, 1–22.

Geary, D.C. (1998) *Male, Female: The evolution of Human Sex Differences*. American Psychological Association, Washington.

Geary, D.C. (1999) Evolution and developmental sex differences. *Current Directions in Psychological Science*, **8**, 115–20.

Geary, D.C., Byrd-Craven, Hoard, M.K., Vigil, J., and Numtee, C. (2003) Evolution and development of boys' social behavior. *Developmental Review*, **23**, 444–70.

Gernsbacher, M.A., Dawson, M., and Mottron, L. (2006) Autism: Common, heritable, but not harmful. *Behavioral and Brain Sciences*, **29**, 413–14.

Glessner, J.T., Wang, K., Cai, G., Korvatska, O., Kim, C.E., Wood, S., Zhang, H.T., Estes, A., Brune, C.W., Bradfield, J.P., Imielinski, M., Frackelton, E.C., Reichert, J., Crawford, E.L., Munson, J., Sleiman, P.M.A., Chiavacci, R., Annaiah, K., Thomas, K., Hou, C.P., Glaberson, W., Flory, J., Otieno, F., Garris, M., Sooya, L., Klei, L., Piven, J., Meyer, K.J., Anagnostou, E., Sakurai, T., Game, R.M., Miller, J., Posey, D.J., Michaels, S., Kolevzon, A., Silverman, J.M., Bernier, R., Levy, S.E., Schultz, R.T., Dawson, G., Owley, T., McMahon, W.M., Wassink, T.H., Sweeney, J.A., Nurnberger, J.I., Coon, H., Sutcliffe, J.S., Minshew, N.J., Grant, S.F.A., Bucan, M., Cook, E.H., Buxbaum, J.D., Devlin, B., Schellenberg, G.D., and Hakonarson, H. (2009). Autism genome-wide copy number variation reveals ubiquitin and neuronal genes. *Nature*, **459**, 569–73.

Goldman, A.I. (2006) *Simulating Minds. The Philosophy, Psychology, and Neuroscience of Mindreading*. Oxford University Press, New York.

Grandin, T. (1995) *Thinking in Pictures and Other Reports from My Life with Autism*. Vintage Books, New York.

Grigorenko, E.L. (2009) At the height of fashion: What genetics can teach us about neurodevelopmental disabilities. *Current Opinion in Neurology*, **22**, 126–30.

Gupta, A.R. and State, M.W. (2007) Recent advances in the genetics of autism. *Biological Psychiatry*, **61**, 429–37.

Hadjikhani, N., Joseph, R.M., Snyder, J., and Tager-Flusberg, H. (2006) Anatomical differences in the mirror neuron system and social cognition network in autism. *Cerebral Cortex*, **16**, 1276–82.

Han, S., Fan, Y.m and Mao, L. (2008) Gender differences in empathy for pain: An electrophysiological investigation. *Brain Research*, **1196**, 85–93.

Handley, L.J.L. and Perrin, N. (2007) Advances in our understanding of mammalian sex-biased dispersal. *Molecular Ecology*, **16**, 1559–78.

Happé, F., Ronald, A., and Plomin, R. (2006) Time to give up on a single explanation for autism. *Nature Neuroscience*, **9**, 1218–20.

Hume, D. (1739–1740/2000) *A Treatise of Human Nature*. Oxford University Press, New York.

Keller, M.C. and Miller, G. (2006a) Resolving the paradox of common, harmful, heritable mental disorders: Which evolutionary genetic models work best? *Behavioral and Brain Sciences*, **29**, 385–452.

Keller, M.C. and Miller, G. (2006b) An evolutionary framework for mental disorders: Integrating adaptationist and evolutionary genetic models. *Behavioral and Brain Sciences*, **29**, 429–52.

Lahaie, A., Mottron, L., Arguin, A., Berthiaume, C., Jemel, B., and Saumier, D. (2006) Face perception in high-functioning autistic adults: Evidence for superior processing of face parts, not for a configural face-processing deficit. *Neuropsychology*, **20**, 30–41.

Lupski, J.R. (2008) Incriminating genomic evidence. *Nature*, **455**, 178–9.

Mandy, W.P.L. and Skuse, D.H. (2008) Research review: What is the association between the social-communicative element of autism and repetitive interests, behaviours and activities? *The Journal of Child Psychology and Psychiatry*, **49**, 795–808.

McIntosh, D.N., Reichmann-Decker, A., Winkielman, P., and Wilbarger, J.L. (2006) When the social mirror breaks: deficits in automatic, but not voluntary, mimicry of emotional facial expressions in autism. *Developmental Science*, **9**, 295–302.

Mealey, L. (1995) The sociobiology of sociopathy–An integrated evolutionary model. *Behavioral and Brain Sciences*, **18**, 523–41.

Minio-Paluello, I., Baron-Cohen, S., Avenanti, A., Walsh, V., and Aglioti, S.M. (2009) Absence of embodied empathy during pain observation in Asperger syndrome. *Biological Psychiatry*, **65**, 55–62.

Morrow, E.M., Yoo, S-Y., Flavell, S.W., Kim, T.K., Lin, Y.X., Hill, R.S., Mukaddes, N.M., Balkhy, S., Gascon, G., Hashmi, A., Al-Saad, S., Ware, J., Joseph, R.M., Greenblatt, R., Gleason, D., Ertelt, J.A., Apse, K.A., Bodell, A., Partlow, J.N., Barry, B., Yao, H., Markianos, K., Ferland, R.J., Greenberg, M.E., and Walsh, C.A. (2008) Identifying autism loci and genes by tracing recent shared ancestry. *Science*, **321**, 218–23.

Nettle, D. (2006) The evolution of personality variation in humans and other animals. *American Psychologist*, **61**, 622–31.

Nettle, D. (2007) Empathizing and systemizing: What are they, and what do they contribute to our understanding of psychological sex differences? *British Journal of Psychology*, **98**, 237–55.

Nettle, D. and Liddle, B. (2008) Agreeableness is related to social-cognitive, but not social-perceptual, theory of mind. *European Journal of Personality*, **22**, 323–35.

Penke, L., Denissen, J.J.A., and Miller, G.F. (2007) The evolutionary genetics of personality. *European Journal of Personality*, **21**, 549–87.

Regier, D.A., Narrow, W.E., Kuhl, E.A., and Kupfer, D.J. (2009) The conceptual development of DSM-V. *American Journal of Psychiatry*, **166**, 645–50.

Ronald, A., Happé, F., and Plomin, R. (2005) The genetic relationship between individual differences in social and non-social behaviours characteristic of autism. *Developmental Science*, **8**, 444–58.

Rose, A.J. and Rudolph, K.D. (2006) A review of sex differences in peer relationship processes: Potential trade-offs for the emotional and behavioral development of girls and boys. *Psychological Bulletin*, **132**, 98–131.

Rutherford, M.D. and McIntosh, D.N. (2007) Rules versus prototype matching: Strategies of perception of emotional facial expressions in the Autism spectrum. *Journal of Autism and Developmental Disorders*, **37**, 187–96.

Sebat, J., Lakshmi, B., Malhotra, D., Troge, J., Lese-Martin, C., Walsh, T., Yamrom, B., Yoon, S., Krasnitz, A., Kendall, J., Leotta, A., Pai, D., Zhang, R., Lee, Y.H., Hicks, J., Spence, S.J., Lee, A.T., Puura, K., Lehtimaki, T., Ledbetter, D., Gregersen, P.K., Bregman, J., Sutcliffe, J.S., Jobanputra, V., Chung, W., Warburton, D., King, M.C., Skuse, D., Geschwind, D.H., Gilliam, T.C., Ye, K., and Wigler, M. (2007) Strong association of de novo copy number mutations with autism. *Science*, **316**, 445–9.

Seielstad, M.T., Minch, E., and Cavalli-Sforza, L.L. (1998) Genetic evidence for a higher female migration rate in humans. *Nature Genetics*, **20**, 278–80.

Singer, T., Seymour, B., O'Doherty, J., Kaube, H., Dolan, R.J., and Frith, C.D. (2004) Empathy for pain involves the affective but not sensory components of pain. *Science*, **303**, 1157–62.

Singer, T., Seymour, B., O'Doherty, J., Stephan, K.E., Dolan, R.J., and Frith, C.D. (2006) Empathic neural responses are modulated by the perceived fairness of others. *Nature*, **439**, 466–9.

Skuse, D.H. (2007) Rethinking the nature of genetic vulnerability to autistic spectrum disorders. *TRENDS in Genetics*, **23**, 387–95.

Sunyaev, S., Ramensky, R., Koch, I., Lathe, W., Kondrashov, A.S., and Bork, P. (2001) Prediction of deleterious human alleles. *Human Molecular Genetics*, **10**, 591–7.

Taylor, S.E., Klein, L.C., Lewis, B.P., Gruenewald, T.L., Gurung, R.A.R., and Updegraff, J.A. (2000) Biobehavioral responses to stress in females: Tend-and-befriend, not fight-or-flight. *Psychological Review*, **107**, 411–29.

Tooby, J. and Cosmides, L. (1990) On the universality of human nature and the uniqueness of the individual: The role of genetics and adaptation. *Journal of Personality*, **58**, 17–67.

Van Vugt, M., De Cremer, D., and Janssen, D.P. (2007) Gender differences in cooperation and competition. The male-warrior hypothesis. *Psychological Science*, **18**, 19–23.

Vigil, J.M. (2007) Asymmetries in the friendship preferences and social styles of men and women. *Human Nature*, **18**, 143–61.

Vigil, J.M. (2008) Sex differences in affect behaviors, desired social responses, and accuracy at understanding social desires of other people. *Evolutionary Psychology*, **6**, 506–22.

Vigil, J.M., Geary, D.C., Granger, D.A., and Flinn, M.V. (2010) Sex differences in salivary cortisol, alpha-amylase, and psychological functioning following Hurricane Katrina. *Child Development*, **81**, 1228–40.

Wakefield, J.C. (2007) The concept of mental disorder: Diagnostic implications of the harmful dysfunction analysis. *World Psychiatry*, **6**, 149–56.

Wakefield, J.C. (2008) The perils of dimensionalization: Challenges in distinguishing negative traits from personality disorders. *Psychiatric Clinics of North America*, **31**, 379–93.

Yeargin-Allsopp, M., Rice, C., Karapurkar, T., Doemberg, N., Boyle, C., and Murphy, C. (2003) Prevalence of autism in a US metropolitan area. *Journal of the American Medical Association*, **289**, 49–55.

Chapter 10

The role of mood change in defining relationships: a tribute to Gregory Bateson (1904–1980)

John Price

In almost all group-living vertebrate species, relationships are asymmetrical in terms of power. The mechanism for creating and sometimes reversing asymmetry is ritual agonistic behaviour (threat and attack). In human beings the requisite asymmetry may also be produced by verbal means, as, too, may symmetry. Gregory Bateson included all these means of producing symmetry and asymmetry (words, threat, attack) in the term "defining the relationship", so that each asymmetrical (or complementary) relationship has a Definer and an Acceptor (who accepts the definition proposed by the Definer). In this chapter it is suggested that one evolutionary function of mood change is to facilitate the formation and reversal of complementarity, and another is to maintain complementarity once it has been established. Elevation of mood gives the Definer the courage, energy and forcefulness to impose a definition on a possibly reluctant Acceptor. Depression of mood enables an Acceptor to accept a definition which may deprive him of power and resources, and which in a normal mood state he would find unacceptable.

10.1 Introduction

The idea that mood changes relate to the gain or loss of territory or social rank has a history of at least 40 years (Price *et al.* 2007). Over the years, various formulations of this basic thesis have appeared, mostly written by clinicians who treat depressed patients every day (Price 1967, 1972, 1998, 2000, 2009; Gardner 1982; Price and Gardner 1995, 2009; Wilson and Price, 2006). In this chapter, I will review the many sources of inspiration for this hypothesis, and relate it to Gregory Bateson's work on communicating about the definition of human relationships.

The basic inspiration for the so-called "social competition hypothesis" of depression came from Darwin's theory of sexual selection (Price 1999). Darwin proposed that one sex selects members of the other sex for mating, and in so doing it rejects the rest. Even within each sex, there is selection and rejection. Darwin noted that animals, especially males, "drive away or kill their rivals" (Darwin 1871, p. 916) but he did not further pursue the fate of the unselected. The implication of this idea is that, in each generation since social life began, the population has been divided into those who have been selected, those who have not been selected, and also possibly those who have first been selected but then been de-selected (section 10.3).

A second source of inspiration came from comparative ethology (and, later on, behavioral ecology), which described the social structures that had evolved throughout the vertebrate sub-phylum to deal with the results of sexual selection. In group-living species, we were shown social hierarchies in which the selected occupied the senior positions while the unselected were pushed, often by means of fighting or agonistic behavior, into inferior ranks. For us, as psychiatrists, the marzipan on the cake was the fact that this fighting was largely ritualized, in that it took a symbolic form rather than lethal fighting. A corollary of the ritualization of fighting is that there must also be a ritualization of losing, and of the incapacity that accompanies losing in real fighting, such as being dead or seriously incapacitated. An animal that has been defeated has two main characteristics. First of all, it lies down on the ground. Second, it cannot get up. Both these qualities must be ritualized, but surprisingly the second quality was overlooked by the ethologists. They gave wonderful descriptions of the ritual submissive gestures that losers make to winners, but what about not being able to get up? It takes a psychiatric view to appreciate this ritual incapacity—an incapacity that we see in our depressed patients who are unable, for purely psychological or ritual reasons, to get up and carry on with their lives (section 10.4).

Further inspiration came from Paul MacLean's concept of the triune brain, providing the anatomical basis for the triune mind, or the old idea that the mind has three parts which operate relatively independently. We could see that fighting strategies could occur at all three levels, and that de-escalation at the higher level (in the form of voluntary surrender) could pre-empt or terminate de-escalation at one of the lower levels (in the form of depressed emotion or depressed mood). Moreover, MacLean's framework can account for both behavior based on intimidating the rival and also behavior designed to be attractive to the rival and to the social group as a whole (section 10.5).

Even though Bateson did not study animal hierarchies or depressed patients, his analysis of the nature of human relationships is of interest here for two reasons. First, he was one of the few people to study and describe relationships in terms of symmetry and asymmetry, and his definition of "complementary relationships", in terms of differential response to threat, seems to be an important clarification that has been neglected by social psychologists. Second, Bateson suggested that communication contains a "definitional" component in addition to the more obvious "informational" component, and this definitional component can be used to maintain asymmetry, or to create asymmetry, or even to reverse an asymmetrical relationship. In general, Bateson's ideas have been fully acknowledged in the family therapy literature, but as far as psychiatry is concerned he has been forgotten and his insights are unused. In part, this chapter is a personal tribute to Bateson, who has provided useful tools for my own thinking about power allocation in human beings and its relation to mood disorders (section 10.6).

To illustrate the arguments put forward in this chapter, I will first offer a story of a fairly typical depressive patient (section 10.2).

10.2 **The overthrown tyrant: a clinical case illustration**

A 55-year-old solicitor was referred to the psychiatric outpatient clinic after an overdose, at which time he gave a history of 3 months of major depression. He complained of sleeplessness, loss of interest in things, poor concentration and memory, poor appetite with the loss of half a stone in weight, tiredness, and suicidal thoughts. He had been off work for a month, and treated by his general practitioner with the antidepressant drug dothiepin. There was no previous history. He was a married man with two daughters; he had a good work record and was a moderate drinker. The depressed mood was associated with what the referring doctor called "obsessional thoughts" in which, when walking down the street, he felt irrational surges of anger against women who were pushing babies in prams. The anger was associated with images of assaulting them and injuring them. He was terrified that he was going to turn into a serial killer.

A diagnosis of major depressive illness was made. As he had not responded to dothiepin in 4 weeks, and was not getting any side effects, it was decided to double the dose. An arrangement was made to interview his wife, who confirmed the sketchy history given by the patient and added a rich background of family difficulties. She revealed that he had always been a tyrannical man, had dominated her, and had been severe with his two daughters. The younger daughter had a rebellious personality and there had been frequent rows between the father and this daughter. He had prevented the daughter taking a course of study, which she bitterly resented. After the daughters left home, relations with their father improved. The elder daughter married and had a miscarriage, and was told she was unlikely to have further children. The younger daughter married and had a son. This daughter would bring her child to visit the parents at weekends, and the father became devoted to his grandson. Gradually, however, the younger daughter started to take liberties with her father, make demands on him and in general to put him down. When he remonstrated with her and tried to resume his old bullying tactics, she stayed away for a few weekends. Eventually, she managed to induce a situation in which the visits of the grandchild were made conditional on her father's submissive behavior. This situation was tolerated for a while, but then she went beyond the bounds of what even the devoted grandfather was able to tolerate. On one occasion she said to her mother who was vacuuming, "Don't do that, Mum, let Dad do it, he's got nothing better to do." It was shortly after this episode that he became depressed.

This case represents a reversal of complementarity (an inversion of hierarchy) in that the father who had been dominant to his daughter now became subordinate to her. The daughter had found an effective weapon in her control of the grandchild, and the father had no defense against it. It was not a case of an elderly parent gratefully relinquishing the dominant role and leaving the child to take on the task of caring for the parent. The father suffered from what we once called an "involuntary subordinate strategy" (Gardner and Price 1999), and this was the outcome of a battle of wills; it was recognized by the medical profession as a major depressive illness.

The case also illustrates how easy it is for hostility to be transferred from one object to another. In this case the father had reason to be hostile to the daughter and her baby for frustrating his wishes, but he did not feel this hostility, instead he transferred it to mothers pushing their babies in the street. He could not express his hostility to his daughter because she was more powerful than he, and so he had what might be called free-floating hostility, which became attached to objects like his daughter but without her power. Transfer of hostility is, of course, common in the animal kingdom. When the alpha animal in a group threatens the beta animal, the latter does not respond with threat but rather with submission, and then in turn threatens the gamma animal in the group. The same can be seen in human military situations: the sergeant-major gives a "bollocking" to the sergeant, who bollocks the corporal, who bollocks the private (who may then take it out on the regimental cat). Whether the corporal feels consciously hostile to the sergeant or the private has not to my knowledge been studied.

Therapy in this case took the form of using the patient's depression (with associated readiness to take a back seat) to enable the mother to become more influential in the family. In joint interviews with the patient and his wife, she was encouraged to take a more assertive role, so that she was able to keep a reasonable peace between her husband and daughter, and when the latter tried to interfere she was able to say, "Please don't try to dictate who shall do what in my house!" and on the whole she did this so tactfully that the daughter was able to continue to bring the grandson for visits without attacking her father and so increasing his depression. In the end, father and daughter laughed about it together, saying they were both congenital tyrants, who needed to be kept in check by someone as amiable as the mother. The depression gradually resolved, but of course in the individual case it is not possible to say whether this was due to antidepressive medication, the passage of time, or the resolution of the family problem. In terms of social rank, the father started off dominant to his daughter, then became subordinate, and ended up equal; the mother started off subordinate to her daughter and ended up (benignly) dominant—which is a healthy form of family female hierarchy—and this reversal of rank did not induce depression in the daughter because it was managed very tactfully by the mother and was acceptable to the daughter.

Reversal of a dominance/subordinate relationship is not common, but it is probably easier to achieve than equality once any sort of asymmetry has already been established. I have seen several cases in marriage, when a previously subordinate wife gains confidence due to work experience or exercising authority over her children, and then becomes dominant to the husband. It is then the husband who gets depressed. Reversal also occurs when there is breakdown of dependent rank. I treated a woman who had been raised by her father to the number two position in the family hierarchy, over her mother and older sister. He did this to punish his wife for having the older sister while he was away at war. When the father died the mother and older sister took their revenge, the younger sister fell in rank and became depressed. I have also treated cases in which a grandparent raised a grandchild above the parents, with dire consequences when the powerful grandparent died.

10.3 **Darwin, Huxley, and sexual selection**

Darwin made it clear that natural selection is based on differential ability to deal with the physical environment, including predator and prey relations with other species, but at the same time he recognized that selection also occurs as a result of interactions with members of the same species. In *The Origin of Species* he wrote:

> This form of selection depends not on a struggle for existence in relation to other organic beings or the external conditions, but on the struggle between individuals of one sex, generally the males, for the possession of the other sex.

(Darwin 1859, p. 69)

In 1871 Darwin published *The Descent of Man and Selection in Relation to Sex*, which was devoted to a meticulous analysis of sexual selection. In this book he introduced the term for the first time, and he pointed out that sexual selection has two components. He wrote:

> Sexual selection depends on the success of certain individuals over others of the same sex, in relation to the propagation of the species (. . .). The sexual struggle is of two kinds; in the one it is between individuals of one sex, generally the male, in order to drive away or kill their rivals, the female remaining passive; whilst in the other, the struggle is likewise between the individuals of the same sex, in order to excite or charm those of the opposite sex, generally the females, which no longer remain passive but select more agreeable partners.
>
> (Darwin 1871, p. 916)

Darwin included both types of sexual selection under the same heading, but did not give them separate names. Julian Huxley (1938) introduced the term "intra-sexual selection" for the social process between members of the same sex, and he called mate choice "epigamic selection". Epigamic selection is a powerful amplifying device; if women would only mate with men who can sing in tune, the musical ability of the population would rapidly improve. Darwin concentrated on epigamic selection, rather than on intrasexual selection, and so have most of the biologists who have followed him. This, and the rather clumsy name, have probably shielded intrasexual selection (and the mainly nonlethal forms of social competition which subserve it) from the biological enquiry that it deserves. Huxley (1966) pays some attention to the fate of the unselected, pointing out that a significant proportion of adult birds fail to mate each year, and he wrote:

> [D]efeat in combat has far reaching general effects, birds though physically uninjured sometimes dying as a result, if not promptly removed from contact with other birds, and even when physically recovered losing the impulse to mate for the rest of the season. Conversely, successful threat-displays promote both general and sexual vigour.
>
> (Huxley 1966, p. 260)

As a result of Darwin's and Huxley's theorizing, we can speculate that in each generation of our ancestors, the population was composed of some who were selected, some who were not selected, and maybe some who were selected to start with and then were de-selected. This variation has not to my knowledge been used as the basis of personality study, but it would seem to account for at least some of the variation along the neuroticism/stability dimension of personality (Price 1969). One confusing thing about intrasexual selection is why anyone should choose to be unselected or de-selected—why hasn't the tendency to allow oneself to be "driven away" been bred out of the population? In the next section we will see that there are two answers to this, one from ethology and one from behavioral ecology.

10.4 Ritual agonistic behavior and ritual losing

One of the outstanding achievements of comparative ethology has been the demonstration of asymmetrical relationships between the members of almost all group-living vertebrates (Alcock 1989). Few species apart from man seem able to form a close, equal relationship with a member of the same sex. In the rest, relationships are defined as complementary by agonistic behavior. One animal threatens the other and if the other submits, the first animal has defined the relationship as one in which he is entitled to threaten the other, but the other is not entitled to threaten him. If the other one does not accept the definition, they adopt an escalating pattern of threat and fighting until one finally submits (or is dead) or one leaves the group (Huntingford and Turner 1987; Archer 1988).

In all species so far studied, this agonistic behavior has become ritualized so that outcomes are usually decided by threat rather than by fighting. The rituals adopted by different species vary, such as butting with the head, lashing with the tail, singing (in birds) and roaring (in stags), erection of gill pouches in fish, push-ups in lizards, but the overall framework is the same for all species. In most animal and human groups complementary relationships between those of different ages develop naturally during ontogeny, since the older members are larger and stronger than the young ones, and among a cohort of young ones there is often some fighting at adolescence, following which relationships tend to be stable. If two strange members of the same sex are put together, there is an agonistic encounter following which one becomes dominant to the other. Schjelderup-Ebbe showed this for domestic hens in 1935, when he described the confrontation between two strange hens. Three things could happen. Both hens could claim dominance, in which case they fought and the winner became dominant. Or one hen could claim dominance and the other not contest the issue, and automatically adopt a subordinate role. Or both could behave like subordinates, in which case one or the other would eventually realize that the dominant role was vacant and adopt it. Once formed, the asymmetry in the relationship was stable, and a reversal of asymmetry was associated with behavior disturbance (Schjelderup-Ebbe 1935; Price and Sloman 1987). The vast majority of animals develop relationships with strange conspecifics in the same way. A particularly clear account for hamadryas baboons is given by Kummer (1995).

Behavioral ecology is concerned with animal behavior from the point of view of its function, often applying mathematical models and using game theory (Krebs and Davies 1997). Unlike classical psychology, it sees behavioral variation in terms of alternative strategies, both life-long strategies, such as antisocial personality versus law-abiding personality (Troisi, 2005), and short-term strategies, such as the alternative fighting strategies of escalation and de-escalation (based on the primitive fight or flight response[1]). By utilizing game theory, behavioral ecology has explained the survival of apparently maladaptive strategies, such as allowing oneself to remain unselected or to be de-selected.

The capacity to "drive away or kill one's rivals" was not given a technical name until, in 1974, Geoffrey Parker introduced the term "resource-holding potential" (RHP). RHP is an intervening variable that is defined by its input and output. The input to RHP is whatever makes for success in fighting, such as size, strength, skill, and the availability of allies. The output from RHP is of two kinds: one relates to an immediate rival, with whom RHP is compared, giving a measure of relative RHP. If relative RHP is favorable, attack or other forms of escalation occur; if relative RHP is unfavorable, retreat or other forms of de-escalation occur. Undirected RHP is signaled by the general bearing of the individual, as in swaggering or furtive behavior. In humans, a fall in RHP is characteristic of depressive states, in which there is a general lowering of self-evaluation and pessimism about the likely outcome of any endeavor (Parker 1974).

The choice between escalation and de-escalation is also affected by the desirability of what is being fought about, technically known as the resource value (Krebs and Davies 1997). Obviously, the more valuable a thing is, the harder people are going to fight over it. Whereas RHP provides the "can", resource value provides the "will" to compete, and this may be over a particular issue or prize, or it may reflect general social ambition. A fall in resource value is characteristic of depression, in which nothing seems worthwhile and there is a generalized loss of energy.

The third variable important for the analysis of fighting behavior is ownership. Most animals win agonistic encounters on their own territory, and a hamadryas baboon, for example, will fight harder over a female belonging to his own harem. The sense of ownership is reduced in depression, as is the sense of entitlement. These three variables account for most of the symptomatology of depression, so that when we speak of depression, depressed mood, or clinical depression we refer to a state in which one or all of RHP, resource value, and sense of ownership are reduced.

It is useful to think of antagonistic encounters as occurring in two stages: a stage of assessment and a stage of engagement. The stage of assessment may

---

[1] Although fight and flight are well-known concepts in psychology, the wider categories of escalation (including not only fight but the active pursuit of goals) and de-escalation (including not only flight but submission and the relinquishing of goals) are not recognized. This is clear from the position of depression and anger in the classification of the emotions. In most systems anger and depression are classified together as "negative emotions" in contrast to the "positive emotions" of joy, exhilaration, and so on. But escalation/de-escalation theory recognizes anger as an escalatory emotion and depression as a de-escalatory emotion.

end with an amicable distribution of roles. One animal can see clearly that the other is bigger, stronger, and has more powerful allies, and so makes a signal of deference and/or submission. It is only if they are equally matched that a serious fight occurs, leading to the victory of one and the defeat of the other. So an animal can reach subordinate status either by backing off in the assessment stage or being defeated in the engagement phase. An animal that backs off in the assessment phase suffers no loss of RHP, so its relations with other animals are unlikely to be affected. However, if it is defeated in a ritual agonistic encounter, it loses RHP, so that it is no longer nearly equal to its former rival and its relation to other animals may be jeopardized.

10.5 A triune mind in a triune brain

The idea that the mind consists of two or more relatively independent parts has been around at least since the time of Plato. It has been most pithily expressed by Blaise Pascal in his well-known aphorism: "The heart has its reasons which reason knows nothing of." Ancient Eastern philosophers, whose ideas were largely promulgated in the West by Gurdjieff, used the metaphor of the cart, horse, and driver. The driver represented reason, or the rational mind, but he had only limited control over the horse, who represented the emotional mind (located in the heart), who in turn had limited control of the cart, representing the instinctive mind, located by some in the gut. Plato likened the three minds to different organs of state.

The work of the evolutionary neuroanatomist Paul MacLean has given support to the idea of the triune mind by his demonstration of a triune brain (MacLean 1990). Prior to MacLean, it was thought that over the course of evolution the brain had gradually grown in size, with the later additions on the whole controlling the earlier parts, largely by inhibition. MacLean pointed out that the forebrain had grown in three distinct stages, leaving three "central processing assemblies" which relatively independently respond to changes in the environment. First, the reptilian forebrain evolved from the fish and amphibian brains and concerned itself, as far as social relations went, with the courtship of the opposite sex and competition with the same sex by means of agonistic behavior. This brain is present in all reptiles, birds, and mammals, and in humans it occupies the basal ganglia or corpus striatum. Then, instead of a homogeneous accretion of additional brain volume, there developed a "paleomammalian brain", which dealt with mammalian social life, the family, the parent/offspring bond, play, and such social matters as were not part of reptilian life. This brain is situated in the limbic system. Not only did it deal with mammalian matters, but it also dealt with those problems that had been faced by reptiles and were also faced by mammals, such as the courtship of the opposite sex and competition with the same sex. In higher mammals there developed the neomammalian brain, which subserves what we recognize as rational thought and decision-making, and it brings these capacities to bear not only on modern problems such as technology and litigation, but also on the older problems that are addressed by the reptilian and paleomammalian brains, such as courtship and competition. This neomammalian brain is situated in the neocortex.

Thus we have three brains dealing with the same problems, and to some extent they co-operate, but also to some extent they act independently. They have different sources of information, they make different executive decisions, and they have different representations in awareness. This is quite a surprising situation, one that would not have been predicted, say, by an engineer accustomed to designing robots. The most surprising thing is that the rational brain, which appears to be the most sophisticated thinking machine ever to have evolved, has so little control over the two lower brains. One cannot will oneself to feel less depressed or less angry. The driver is not in control of the horse or the cart. It would have been easy for such control to have evolved, so the fact that it has not suggests that there is some advantage in having one or more relatively independent lower "central processing assemblies". In competitive relations with conspecifics, a decision frequently has to be made between escalation (fighting harder) and de-escalation (fleeing or submitting), and this decision appears to be made, relatively independently, by each of the three brains, sometimes sequentially, sometimes simultaneously. Possibly the rational brain, in order to maximize fighting ability, has delegated the contemplation of possible defeat to fail-safe mechanisms at the lower brain levels.

As noted earlier, decisions to escalate or de-escalate take place either simultaneously or consecutively at all three levels of the triune brain (see Table 10.1 for an overview of the options). At the rational, or neomammalian, level the decision is made consciously and voluntarily either to escalate by fighting harder or to back off. Escalation may take many forms, such as insulting or

Table 10.1 Escalating and de-escalating strategies at three brain levels: Agonistic competition

	Escalate		**De-escalate**
Rational, neomammalian level (isocortex)	Decide to fight on (stubbornness or courage)	or	Decide to back off (submission or escape)
Emotional, paleomammalian level (limbic system)	Feel assertive, angry, or hostile	or	Feel inferior (anxiety, depressed emotion)
Instinctive, reptilian level (basal ganglia)	Elevated mood	or	Depressed mood Anxious mood

attacking the opponent, obtaining a weapon, or recruiting allies. When de-escalating or backing off, the appeasement display may take the form of a graciously worded apology or a flowery speech of submission. At the emotional or limbic level, escalation takes the form of anger, indignation, and the exhilaration of combat, with its associated bodily changes. De-escalation at this level may recruit the dysphoric emotions of anxiety, depression, and the sense of being chastened. At the instinctive level, we hypothesize that escalation in the reptilian brain takes the form of elevated mood, giving the individual a prolonged increase in energy, optimism, and self-confidence. Since mood is pervasive and, from its origin in the reptilian brain, affects all the higher levels of the brain, in the human (and probably the chimpanzee) it will increase sociability with which to recruit allies. Conversely, de-escalation at the instinctive level takes the form of depressed mood and may include unfocused anxiety, fatigue, and a sense of physical disability. The appeasement display at this level communicates this impairment and disability to any rival or to society as a whole.

Methods of competition have become more complex over the course of evolution. Group living lengthened the duration of contests, so that even in apes a struggle for dominance may take several months to be resolved. In additon, instead of fleeing, as happens in territorial species, the loser could remain in the group with the winner of the contest, and this gave rise to appeasement or submissive behavior, which reflects the capacity to live in a subordinate social role. Anxiety and fear of the dominant individual, together with relatively low self-esteem and lowered mood, enabled the social hierarchy to maintain stability and prevent rebellion. At some stage in evolution, this stabilizing anxiety gave rise to a new way of relating to a higher-ranking individual: respect. The leaders of the group made themselves attractive to the group members instead of (or in addition to) intimidating them. Social rank was then determined by the choice of the group rather than by agonistic dyadic encounters. The new self-concept of social attention-holding power (Gilbert 1992) began to replace RHP, as group members evaluated themselves according to their power to attract interest and investment (such as votes or other forms of political support). Related to social attention-holding power is the concept of prestige, which is the extent to which the group is prepared to invest in the individual. Prestige competition was added to, but did not entirely replace, agonistic competition (Barkow 1991).

The capacity for escalation and de-escalation appears to have survived the switch to prestige competition, but takes different forms, at least at the upper two forebrain levels. At the highest level, pursuit of goals replaces the decision to attack, so that escalation consists of the adoption of new goals and de-escalation consist of giving up goals. The goals are usually ones that lead to prestige, if achieved. Also, on social occasions, escalation takes the form of self-assertion, such as standing up to speak and promoting one's own goals, whereas de-escalation takes the form of self-effacement and allowing other people's goals to take precedence in the group. At the emotional level, the escalation of prestige competition is less dramatic than the anger of agonistic competition; it takes the form of exhilaration, enthusiasm, and self-confidence. De-escalation reflects the fact that punishment comes from the group rather than from a dominant individual, so there is social anxiety, guilt, and shame. This is an appeasement display to the group, expressing contrition for breaking group rules or for failing to come up to group standards.

Finally, at the instinctive, reptilian level of the forebrain, little seems to have changed: elevation of mood represents escalation and depression of mood de-escalation. However, the information which leads to the activation of the strategy set is clearly different. Instead of measuring punishment received from the rival, the reptilian brain in some way monitors social standing in the group, and is sensitive to group approbation and disapprobation, to comparison of self with other group members, and with one's own aspirations, and to the knowledge of having failed the group in some way by not living up to its standards, or, having broken the group's rules, to the likelihood of being found out. Note that depressed and elevated mood are "all or nothing" things; whereas at the higher levels it is possible to escalate in some areas of life and de-escalate in others, in the reptilian brain the mood change is pervasive and affects all aspects of life, it is not situation dependent. This may reflect the pervasive change in the defeated reptile, who often loses his gaudy adult coloring and reverts to the dull brown or green of the adolescent coloration (Greenberg and Crews 1983).

The manifestation of escalation and de-escalation at the three brain levels are shown for agonistic competition in Table 10.1 and for prestige competition in Table 10.2.

Normally "ranking stress" or a "resource challenge" will activate only one or two of the three levels and then, if anger accompanies rational escalation, the individual is likely to win the conflict and the resource challenge is dealt with. Or, if chastened mood accompanies rational submission, the individual loses the conflict and becomes reconciled to the loss of whatever was at stake. However, two very human tendencies may lead to trouble. Our often implacable ambition and stubbornness may lead to prolonged escalation at the rational level in situations in which victory is extremely unlikely, and then the anticipation of losing may activate the reptilian level strategy set and select for de-escalation at that level. The resulting incapacitating depression makes winning even less likely, and a chronic situation results in which there is continued escalation at the rational level and continued de-escalation at the instinctive level. This is a common manifestation of depressed mood seen in

Table 10.2 Escalating and de-escalating strategies at three brain levels: prestige competition

	Escalate		De-escalate
Rational, neomammalian level (isocortex)	Adopt new goals, actively pursue existing goals, assert oneself	or	Give up goals, efface oneself
Emotional, paleomammalian level (limbic system)	Feel assertive, exhilarated, and enthusiastic	or	Feel inferior (shame/guilt/sense of failure, social anxiety)
Instinctive, reptilian level (basal ganglia)	Elevated mood	or	Depressed mood / Anxious mood

the clinic, as first pointed out by Edward Bibring (1953), who noted that his depressed patients were often clinging on to unrealizable goals.

The other human tendency is our desire to see fair play and our intolerance of injustice. This manifests at the emotional, limbic level, which seems finely tuned to evaluate the fairness of events and particularly of other people's actions. If we feel we have been treated unfairly we feel angry, and if this anger is ineffective in righting the situation, our reptilian strategy set may be activated and we have a mood change. If elevation of mood is selected, we may then have enough energy to right the wrong, but if depression is selected, the depressive incapacity then makes effective action even more impossible. Then, again, we get chronic reptilian de-escalation, which presents in the clinic as depressive illness.

10.6 Gregory Bateson: defining the relationship

In human beings the methods of analyzing relationships are more various, and whereas we have fighting and physical threats, like the raised fist or wagging finger, in most cases the method of ritual fighting is verbal, so that one could say that verbal fighting is the human species-specific form of ritual agonistic behavior. The type of verbal exchange is varied in both quality and quantity. There may be an exchange of insults, or shouting, or verbal abuse at the more primitive end of the scale. Then there may be reasoned argument to persuade the other person that they are at fault or in the wrong or otherwise in a one-down position. There may be direct assertions of control, such as "you are my daughter and therefore you do what I say". There may be more subtle attempts at gaining power, like being bossy and telling others what to do.

It may be difficult to tell where power lies in a relationship. Who decides, but who decides who decides? In my view, Bateson's concept of "defining the relationship" gives the best solution to the problem of identifying control in a relationship (see also Hinde 1987). In the 1950s Gregory Bateson, an English anthropologist, gathered together in Palo Alto a brilliant group of people who had skills in anthropology, psychiatry, family therapy, and communications engineering (Lipset 1980). Many ideas came from this group, such as the double-bind theory of schizophrenia and the concept of systemic versus linear thinking, and in fact they were the inspiration for the diaspora of family therapy around the world (Watzlawick *et al.* 1967; Bateson 1972; Watzlawick and Weakland 1977; Wilder-Mott and Weakland 1981). Out of this group grew the Mental Research Institute, which celebrated its 50th anniversary in 2009.

Focusing on human communication, Bateson had two important ideas. One is the idea that each communication between two people contains at least two components: one is an informational component and the other is a command or definitional component, which confirms or defines the nature of the relationship between the two communicators. The other is the idea that relationships can be either symmetrical or complementary along a number of dimensions, but particularly in terms of relative power. In a symmetrical relationship the expression of power is responded to by the expression of power (what Bateson called "more of the same"), whereas in a complementary relationship the expression of power is responded to by a reduction in the expression of power or by the expression of submission (what Bateson called "less of the same, or more of something different"). In a complementary relationship the definitional communications of one member are accepted by the other, resulting in an asymmetrical relationship in which one member is dominant or one-up, and the other subordinate or one-down (Sluzki and Beavin 1965). Since the basic difference is who defines and who accepts the definition provided by the other, I will call the dominant or one-up member the Definer and the subordinate or one-down member the Acceptor, although I do not think the Bateson group used the terms in this sense. The definition of a relationship contains influence from three sources: from each of the two members of the relationship and from outside the relationship; this situation may be illustrated by a series of Venn diagrams (see Appendix 10.1).

The daughter's statement in the case described at the beginning of the chapter, "Let father do the vacuuming" has an informational component that concerns the allocation of housework, but it also has two definitional components, one being that the daughter is entitled to allocate the task of vacuuming to her father, indicating that she is the Definer in relation to her father, and the other being that she is entitled to allocate housework in her mother's house, indicating that she is the Definer in relation to her mother. It is a very powerful assertion of power. The father must have been made angry by being told to vacuum by his daughter, but he could not afford to be angry because of the daughter's hold over him, and so he developed a de-escalating strategy at the reptilian level of the forebrain, which manifested itself as depressive illness. This enabled him to accept her definition of their relationship, or at least not to oppose it. He retired from the arena as a medical casualty. Hostility is often redirected to a safer target (usually down the hierarchy), and in this case the father redirected his anger to women pushing their small children along in the street.

In many such cases the Acceptor is not aware of being manipulated into the role of Acceptor. The fate of being maneuvered into the role of Acceptor may still be associated with depression, even if the statements used appear boosting rather than putting down, and even if the role as defined is not at all depressing. This is particularly the case when one person uses forms of speech that are associated with use by superior people to inferior people, especially when such usages are complimentary, such as bestowing praise in a situation in which an inferior would not normally praise a superior. In these cases the Definer appears well-disposed to the other, who does not consciously realize that he or she is being maneuvered into the role of Acceptor. However, at an unconscious level the effect is felt, and then the escalation/de-escalation strategy set is accessed, and the Acceptor finds him/herself becoming irrationally angry or inexplicably depressed. In general, depression occurs when submission is unacceptable or impossible (Dixon 1998; Gilbert 2001). Often submission is acceptable, or even welcomed, and then depression does not occur, as when the daughter in our case study accepted the dominance of her mother.

The offering of a unilateral definition of a relationship comes into the category of a catathetic signal, defined as a signal which lowers the RHP of the recipient unless returned in full measure (Price 1988). It is therefore like a blow or an insult, and is part of the repertory of ritual agonistic behavior. It is like a serve at tennis, which if returned leaves the two players equal, but if not returned leaves the receiver one-down.

Like a serve, the offer of a unilateral definition is not only a catathetic signal, but also a request for a reply, to enter into a negotiation (a rally), so that the outcome of the interaction is not something boring like an ace service, but a manifestation of repeated superior skill by the eventual winner. People do not like to win too easily, like the merchant who is disgusted if the buyer accepts the first price—he enjoys haggling. Kortmulder (1998) has pointed out that even fish have an appetite for a symmetrical encounter and may handicap themselves to get a more even "rally". It is more fun to beat someone who is near one's own level of skill than someone who cannot even return a serve. Of course, a negotiated definition can leave a couple with an equal relationship, which cannot happen with a tennis rally. In this sense, tennis is more similar to animal agonistic behavior than to human conflict.

It has been noted by ethologists that the general form and rules of ritual agonistic behavior are similar for all vertebrates, but that each species has a particular method of fighting. Offering definitions could well be the human species-specific form of agonistic behavior. It depends on language, which ties in with the fact that ritual agonistic weapons tend to become hypertrophied like the peacock's tail, and language is certainly hypertrophied in humans. Moreover, it does away with the problem which in humans, but not in animals, attends the use of "aggressive" acts such as hitting and insulting. This problem lies in the moral code that condemns fighting, and particularly a man hitting a woman. Therefore, if A attacks B and B does not retaliate, it could be that B is weaker than A, but it could also be that B has been trained not to settle differences by fighting, or, if A is a woman and B a man, B has been trained to believe that a man should not hit a woman. This moral training makes fighting a bad method of determining dominance in many situations, especially between the sexes. By not returning the blow for moral reasons, the courteous man loses RHP. This may be balanced by a gain in social attention-holding power as he contemplates his chivalrous behavior, but it seems likely that some damage is still done.

If offering a definition is a catathetic signal, like hitting or insulting, what are we to make of Bateson's suggestion that every communication contains a definitional (or command) component as well as an informational component? Can we deal with a situation in which every communication is like an insult or blow? One answer to this is given by Brown and Levinson (1987), who do indeed approach communication with the idea that every statement runs the risk of lowering the "face" of the recipient, and they demonstrate how this omniprevalent danger is counteracted in normal intercourse by forms of politeness and other subtle strategies.

Another answer lies in the fact of redundancy. Even if every statement defines the relationship, the vast majority of statements define the relationship in the way it has already been defined and agreed on by the two parties. In other words, the vast majority of definition statements are redundant and therefore do not come into Brown and Levinson's category of "face threatening acts" (FTAs). It is the unilateral definition statement that is an FTA (catathetic signal), in that it gives a definition that has not already been bilaterally agreed.

Unlike animals, a pair of humans has the choice of forming a symmetrical or a complementary relationship. This decision has to be made before they decide who is going to be one-up in the event of their forming a complementary relationship. Let us say that A and B have passed the assessment stage and have agreed that there is no disparity in RHP (social power) between them. Will they become friends on an equal basis or will they enter a trial of strength to compete for the one-up position? Let us make the assumption that friendship is based on mutual trust, and that the offer of friendship gives the other the option of abusing the trust and using the friendship to gain a one-up position. Then the two prospective friends are in a prisoner's dilemma situation (Pusey and Packer 1997). The possible outcomes, in order of payoff, are: (i) to be one-up by the abuse of trust, (ii) to be equal friends, (iii) to have to enter a fair fight for the one-up position, and (iv) to be one-down because of abuse of trust by the other.

Certain social arrangements help to maximize the chances of arriving at the second option (to be equal friends). The payoff from the first option (one-up by abuse of trust) could be reduced, either by lowering the advantage to be gained from the one-up position or by some form of social scrutiny so that reputation is damaged if the abuse of trust is made public (this is the situation which pertains in egalitarian hunter/gatherer societies). Alternatively, there is the possibility of playing a tit-for-tat strategy, so that incipient attempts to abuse trust can be detected and punished by the other before the one-up position is secured.

If the third option is chosen, the two faithless friends enter into a negotiation for the one-up position, which in animals takes the form of ritual agonistic behavior and in humans can take many different forms, including ritual agonistic behavior. The moves in this game can be described in terms of offered definitions of the relationship, and the one-up winner (the Definer) is the one whose definition is accepted by the one-down Acceptor. The various moves or bouts of this negotiation can be described by the dollar auction model (Editorial 1989). In a dollar auction, the winner of the auction gets the dollar, or whatever sum is being auctioned, less his bid, but the second-highest bidder also loses his most recent bid without getting any prize. Once involved in the exchange, each contestant gets into a situation of "too much invested to quit" and so cannot withdraw until he runs out of money (comes to the predetermined point at which he "gives up" in a war of attrition). This should be true if the costs of the engagement increase with each bid or with each bout of the ritual agonistic encounter. The selection of which runs out of money first can be described by the hawk/dove game (Maynard Smith 1974) in the form of the war of attrition, if we make hawks richer than doves. In the real-life negotiation, neither contestant knows how much money either he or the other has— they have to go on bidding until one runs out. The fact of being hawk or dove is a hidden component of RHP: the discovery of which contestant, if either, is a hawk, is what the engagement phase of the ritual agonistic encounter is all about.

According to this model, people entering an asymmetrical relationship may play three consecutive games: prisoner's dilemma, in which they both "defect," the dollar auction, in which they both have "too much invested to quit", and the hawk/dove game, which decides which of them will win the dollar and be one-up on the other. Perhaps it is not surprising that some people prefer to be hermits!

Finally, I think it is an open question whether a close, long-term human relationship can be both hedonic and symmetrical. Relationships can be so subtle that it may be difficult to decide just what is going on. Hinde (1987) realized this and pointed out that one not only has to know who makes the important decisions, but who decides who makes the important decisions, and who decides on who it is who decides, and so on in infinite regression. To observe who defines the relationship is probably the closest we can get, bearing in mind that it is possible for A to define B as the Definer.

10.7 Conclusion

In this chapter, I have touched on the complex issue of the balance of power in relationships, and have found the most satisfactory model in that provided by Gregory Bateson and his colleagues. In a regressive situation of "Who decides?" and "Who decides who decides?", and "Who decides who decides who decides?" the most economical formulation is to ask who defines the relationship and who accepts the definition offered by the other, giving each complementary relationship a Definer and an Acceptor. Definition can be by any means available, including fighting, verbal abuse, nonverbal signals, definitive statements, and the use of words or behavior that are generally known to be used only by high-ranking to low-ranking people, even when such asymmetrical statements are friendly and supportive. We are surprised that, in spite of the brilliance of the Bateson team and their interest in ethology, they did not extend their discussions of symmetrical and complementary relationships to animals. This may well have been due to the fact that Schjelderup-Ebbe's original description of the "pecking order" in English was published as a book chapter as late as 1935 and at the time they were working (in the 1950s) the subject of social hierarchy was virtually taboo. It was believed that there was no phylogenetic relation between human and animal hierarchies (Tedeschi and Lindskold 1976, p. 496), and it was widely thought that hierarchy in animals was an artifact of captivity (Rowell 1974). The concepts of symmetry and complementarity have been largely restricted to the marital and family therapy literature (e.g., Rogers-Millar and Millar 1979). Human communication research has moved away from Bateson's ideas (Littlejohn and Foss 2008), so that we have obtained no input from relational dialectics (Baxter 1988) or from the study of speech acts; the dimensions of illocutionary force do not include symmetry versus complementarity (Searle 1969; Doerge 2006). Studies of social dominance orientation have been concerned with group effects and personality rather than mood changes (Sidanius and Pratto 2001). Work on the authoritarian personality has been concerned with social hierarchy, but has been mainly about the redirection of hostility onto political and racial groups which are perceived as inferior (Stone *et al.* 1993).

We have been accused of emphasizing the importance of competition at the expense of sexuality and affiliation. Freud based his evolutionary speculations on the effect of the ice-age on sexuality (De Block 2005), and Bowlby was rightly concerned with the importance of attachment for good human development (Bowlby 1980), but if these theorists had had at their disposal the knowledge we now have of animal behavior and evolutionary theory, they might well have concerned themselves more with the problems of interpersonal competition, which has been important in evolution for hundreds of millions of years, antedating the evolution of the family and complex sexual relationships. Sexuality and affiliation are of vital importance to mankind, but we would suggest that they were grafted onto already existing mechanisms for managing social symmetry and asymmetry.

Summarizing the theme of this chapter, the adaptive function of the capacity for mood change has been the creation and maintenance of asymmetry in relationships. This can be conceptualized at a number of different levels. Social asymmetry is predicted by Darwin's theory of sexual selection: who is selected and who is not; who, having been selected, is then de-selected? Social asymmetry is observed at the structural level of social hierarchy and territory: who is up and who is down, who is going up the hierarchy and who is going down, who has gained a territory and who has lost one? Social asymmetry is created at the level of ritual agonistic behavior: who has won and who has been ritually defeated? Social asymmetry is maintained in the population because it is an evolutionarily stable strategy according to the game theory models of behavioral ecologists. No-one wants to be unselected or low-ranking or defeated; these undesirable social roles can be made acceptable at the highest brain level with humility or at the lowest brain level with depressed mood.

After over 40 years of pondering on these matters, I think the best and most comprehensive theory is Bateson's idea of defining the relationship: who defines, and who is manipulated into the position of accepting an unacceptable definition? This covers all methods of creating asymmetry and even deals with death, in that we define our loving relationships as permanent, and when death offers another definition we are forced to accept it whether we like it or not. To prescribe for population mental health, we should ask parents and educators to encourage in the young the development of equal, reciprocal relationships, and to avoid the insidious maxim of "He who is not one-up is one-down" (Potter 1947). If asymmetry is inevitable, it should be managed at the highest mental level, to ensure that it is based on respect rather than fear and depression. Therefore, although Bateson may not quite have hit the target with his double-bind theory of schizophrenia, he may well unwittingly have provided an epistemology for the better understanding of affective disorders.

Appendix 10.1

Venn diagrams are used in this instance to illustrate the origin of the definition of a relationship between two people. Some or all (or none) of the definition originates from one person, some or all (or none) from the other person; these are represented by the nonoverlapping parts of the circles. The overlap represents that part of the definition agreed on by both people. The dark parts of the circles signify definition, the light parts lack of definition. The surrounding area represents the influence of people outside the pair; when dark it indicates that the person on that side is identified as Definer. The two circles can represent independent people or those in a particular relationship, such as husband and wife, brother and sister, or master and servant. In this case the circle on the left represents the husband and the circle on the right the wife. One should note that, although informative about the origin of definition, Venn diagrams have nothing to say about other important aspects of relationships. For instance, they say nothing about closeness or intimacy, which are characteristics of relationships which themselves may be the subject of conflicted definitions. More importantly for our own purpose, the diagrams do not give information about the acceptability of definitions proposed by one member and accepted by the other. Some definitions proposed by one member may be so acceptable that they almost constitute shared definition; others may be entirely unacceptable, but have to be accepted because there is no alternative. This difference is clearly important for the mental health of the Acceptor. It could be indicated in the diagrams by some form of hatching, but for present purposes it seems best to keep the illustrations reasonably simple.

I am indebted for the use of Venn diagrams to describe relationships to Piero De Giacomo, who used them in a somewhat different way (to describe current relating rather than the origin of the definition of relationships) (De Giacomo 1993; L'Abate and De Giacomo 2003).

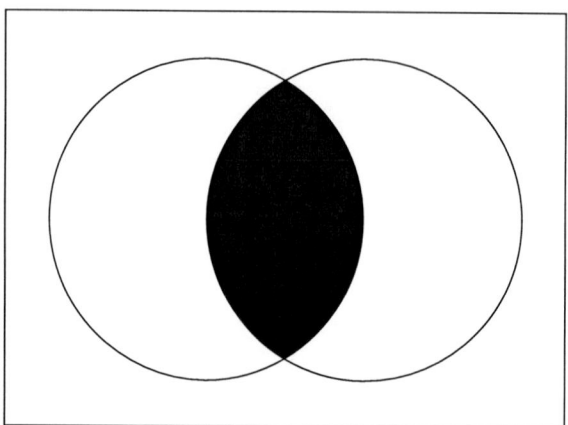

Fig. 10.1 A reciprocal, symmetrical marriage. The definition is mutually agreed between husband and wife.

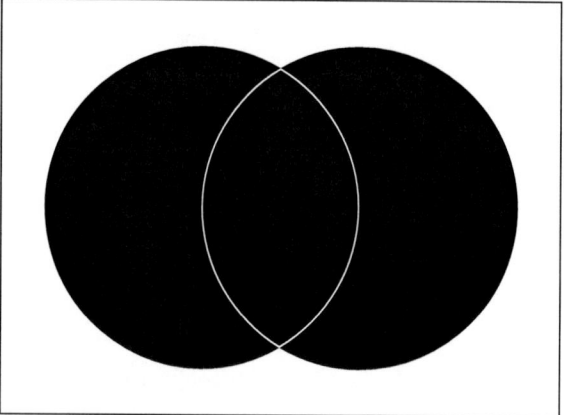

Fig. 10.2 A contested symmetrical marriage. Each is trying to define the relationship and there is only a certain amount of agreed definition.

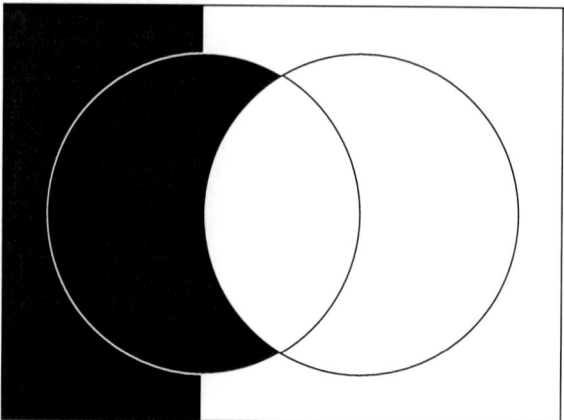

Fig. 10.3 A dominant husband in a male-dominated society. There is no shared definition and no contribution to the definition from the wife.

Fig. 10.4 A dominant wife in a male-dominated society.

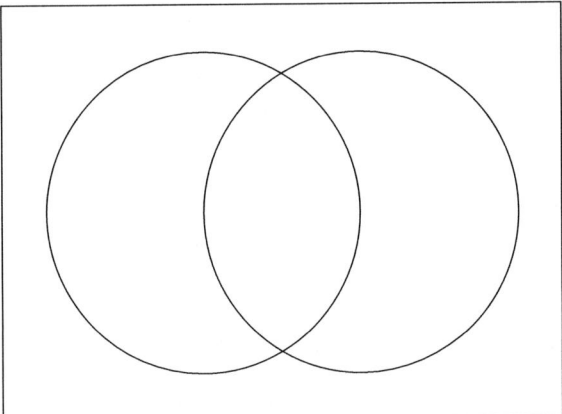

Fig. 10.5 Neither has attempted to define the relationship. This has been reported in the families of schizophrenic patients (Palazzoli *et al.* 1978).

References

Alcock, J. (1989) *Animal Behavior: An Evolutionary Approach* (4th edition). Sinauer Associates, Sunderland, MA.

Archer, J. (1988) *The Behavioural Biology of Aggression*. Cambridge University Press, Cambridge.

Barkow, J.H. (1991) Précis of *Darwin, Sex and Status: Biological Approaches to Mind and Culture*. *Behavioral and Brain Sciences*, 14, 295–334.

Bateson, G. (1972) *Steps to an Ecology of Mind*. Ballantine Books, New York.

Baxter, L.A. (1988) A dialectical perspective of communication strategies in relationship development. In S. Duck. (ed.), *Handbook of Personal Relationships*. Wiley, New York, pp. 257–73.

Bibring, E. (1953) The mechanisms of depression. In P. Greenacre (ed.), *Affective Disorders*; International Press, New York, pp. 309–16.

Bowlby, J. (1980) *Attachment and Loss*, volume 3. *Loss: Sadness and Depression*. Hogarth Press, London.

Brown, P. and Levinson, S.C. (1987) *Politeness: Some Universals in Language Usage* (2nd edition). Cambridge University Press, Cambridge.

Cronen, V.E., Johnson, K.M., and Lannamann, J.W. (1982) Paradoxes, double binds, and reflexive loops: an alternative theoretical perspective. *Family Process*, 21, 91–112.

Darwin, C. (1859) *The Origin of Species by Means of Natural Selection*. John Murray, London.

Darwin, C. (1871) *The Descent of Man and Selection in Relation to Sex*. John Murray, London.

De Block, A. (2005) Freud as an "evolutionary psychiatrist" and the foundations of a Freudian philosophy. *Philosophy, Psychiatry and Psychology*, 12, 315–24.

De Giacomo, P. (1993) *Finite Systems and Infinite Interactions: The Logic of Human Interaction and its Application to Psychotherapy*. Bramble Books, Norfolk, CT.

Dixon, A.K. (1998) Ethological strategies for defence in animals and humans: their role in some psychiatric disorders. *British Journal of Medical Psychology*, 71, 417–45.

Doerge, F.C. (2006) *Illocutionary Acts—Austin's Account and What Searle Made Out of It*. Tuebingen University Press, Tuebingen.

Editorial (1989) Of auctions, dilemmas and blood-letting: models of escalation behaviour. *Lancet*, 4, 1487–8.

Gardner, R. Jr. (1982) Mechanisms in major depressive disorder: an evolutionary model. *Archives of General Psychiatry*, 39, 1436–41.

Gardner, R. Jr. and Price, J.S. (1999) Sociophysiology and depression. In T. Joiner and J.C. Coyne (ed.), *The Interactional Nature of Depression: Advances in Interpersonal Approaches*. APA Books, Washington, DC, pp. 247–68.

Gilbert, P. (1992) *Depression: The Evolution of Powerlessness*. Lawrence Erlbaum Associates, London.

Gilbert, P. (2001) Evolution and social anxiety: the role of attraction, social competition and social hierarchies. *Psychiatric Clinics of North America*, 24, 723–51.

Greenberg, N. and Crews, D. (1983) Physiological ethology of aggression in amphibians and reptiles. In B.B. Svare (ed.), *Hormones and Aggressive Behavior*. Plenum Press, New York, pp. 469–506.

Hinde, R.A. (1987) *Individuals, Relationships and Culture: Links between Ethology and the Social Sciences*. Cambridge University Press, Cambridge.

Huntingford, F. and Turner, A. (1987) *Animal Conflict*. Chapman & Hall, London.

Huxley, J. (1938) The present standing of the theory of sexual selection. In G.R. de Beer (ed.), *Evolution: Essays on Aspects of Evolutionary Biology Presented to Professor E.S.Goodrich on his Seventieth Birthday*. Clarendon Press, Oxford, pp. 11–42.

Huxley, J. (1966) Introduction to "A discussion of ritualisation of behaviour in animals and man". *Philosophical Transactions of the Royal Society of London, Series B*, 251, 249–71.

Kortmulder, K. (1998) *Play and Evolution: Second Thoughts on the Behaviour of Animals*. International Books, Leiden.

Krebs, J.R. and Davies, N.B. (eds) (1997) *Behavioural Ecology: An Evolutionary Approach* (4th edition). Blackwell, Oxford.

Kummer, H. (1995) *In Quest of the Sacred Baboon*. Princeton University Press, Ewing, NJ.

Labette, L. and De Giacomo, P. (2003) *Intimate Relationships and How to Improve Them: Integrating Theoretical Models with Preventive and Psychotherapeutic Applications*. Praeger, Westport, CT.

Lipset, D. (1980) *Gregory Bateson: The Legacy of a Scientist*. Prentice-Hall, Englewood Cliffs, NJ. Reprinted in 1982 by Beacon of Boston and Fitzhenry and Whiteside of Toronto.

Littlejohn, S.W. and Foss, K.A. (2008) *Theories of Human Communication* (9th edition). Thomson Wadsworth, Belmont, CA.

MacLean, P.D. (1990) *The Triune Brain in Evolution*. Plenum Press, New York.

Maynard Smith, J. (1974) The theory of games and the evolution of animal conflicts. *Journal of Theoretical Biology*, 47, 209–21.

Palazzoli, S., Cochin, G., Prate, G., and Bascule, L. (1978) *Paradox and Counterparadox*. Aronson, New York.

Parker, G.A. (1974) Assessment strategy and the evolution of fighting behaviour. *Journal of Theoretical Biology*, 47, 223–43.

Potter, S. (1947/77) *The Theory and Practice of Gamesmanship: Or the Art of Winning Games Without Actually Cheating*. Penguin, London.

Price, J.S. (1967) Hypothesis: The dominance hierarchy and the evolution of mental illness. *Lancet*, ii, 243–6.

Price, J.S. (1969) The ritualisation of agonistic behaviour as a determinant of variation along the Neuroticism/Stability dimension of personality. *Proceedings of the Royal Society of Medicine*, 62, 1107–10.

Price, J.S. (1972) Genetic and phylogenetic aspects of mood variation. *International Journal of Mental Health*, 1, 124–44.

Price, J.S. (1988) Alternative channels for negotiating asymmetry in social relationships. In M.R.A. Chance (ed.), *Social Fabrics of the Mind*. Lawrence Erlbaum, Hove, pp. 157–95.

Price, J.S. (1998) The adaptive function of mood change. *British Journal of Medical Psychology*, 71, 465–77.

Price, J.S. (1999) Implications of sexual selection for variation in human personality and behavior. In J. van der Dennen and D. Smillie (eds), *The Darwinian Heritage and Sociobiology*. Greenwood Publishing Group, Westport, CT, pp. 295–308.

Price, J.S. (2000) Subordination, self-esteem and depression. In L. Sloman and P. Gilbert (eds), *Subordination and Defeat: An Evolutionary Approach to Mood Disorders and their Therapy*. Lawrence Erlbaum Associates, Mahwah, NJ, pp. 165–77.

Price, J.S. (2009) Darwinian dynamics of depression. *Australian and New Zealand Journal of Psychiatry*, **43**, 1–9.

Price, J.S. and Gardner, R. (1995) The paradoxical power of the depressed patient: a problem for the ranking theory of depression. *British Journal of Medical Psychology*, **68**, 193–206.

Price, J.S. and Sloman, L. (1987) Depression as yielding behavior: an animal model based on Schjelderup-Ebbe's pecking order. *Ethology and Sociobiology*, **8**, 85–98 (Supplement).

Price, J.S., Gardner, R., Wilson, D.R., Sloman, L., Rohde, P., and Erickson, M. (2007) Territory, rank and mental health: the history of an idea. *Evolutionary Psychology*, **5**, 531–54.

Pusey, A.E. and Packer, C. (1997) The ecology of relationships. In J.R. Krebs and N.B. Davies (eds), *Behavioural Ecology: An Evolutionary Approach* (4th edition). Blackwell, Oxford, pp. 254–83.

Rogers-Millar, L.E. and Millar, F.E. (1979) Domineeringness and dominance: a transactional view. *Human Communication Research*, **5**, 238–46.

Rowell, T.E. (1974) *Social Behaviour of Monkeys*. Penguin, Harmondsworth.

Schjelderup-Ebbe, T. (1935) Social behaviour of birds. In C. Murchison (ed), *Handbook of Social Psychology*. Clarke University Press, Worcester, MA, pp. 947–72.

Searle, J. (1969) *Speech Acts*. Cambridge University Press, Cambridge.

Sidanius, J. and Pratto, F. (2001) *Social Dominance: An Intergroup Theory of Social Hierarchy and Oppression*. Cambridge University Press, Cambridge.

Sluzki, C.E. and Beavin, J. (1965) Symmetry and complementarity: an operational definition and a typology of dyads. *Acta Psichiatrica y Psicologica de America Latina*, **11**, 321–30. Reprinted in P. Watzlawick and J.H. Weakland (eds) (1977) *The Interactional View*. W.W. Norton, New York, pp. 71–87.

Stone, W.F., Lederer, G., and Christie, R. (1993) Introduction: strength and weakness. In W.F. Stone, G. Lederer, and R. Christie (eds), *Strength and Weakness: the Authoritarian Personality Today*. Springer-Verlag, New York, pp. 3–21.

Tedeschi, J.T. and Lindskold, S. (1976) *Social Psychology: Interdependence, Interaction and Influence*. Wiley, New York.

Troisi, A. (2005) The concept of alternative strategies and its relevance to psychiatry and clinical psychology. *Neuroscience and Biobehavioural Reviews*, **29**, 159–68.

Watzlawick, P. and Weakland, J.H. (eds) (1977) *The Interactional View: Studies at the Mental Research Institute, Palo Alto, 1965–1974*. Norton, New York.

Watzlawick, P., Beavin, J., and Jackson, D.D. (1967) *Pragmatics of Human Communication: A Study of Interactional Patterns, Pathologies, and Paradoxes*. Norton, New York.

Wilder-Mott, C. and Weakland, J.H. (eds) (1981) *Rigor and Imagination: Essays from the Legacy of Gregory Bateson*. Praeger, New York.

Wilson, D.R. and Price, J.S. (2006) Evolutionary epidemiology of endophenotypes in the bipolar spectrum: evolved neuropsychological mechanisms of social rank. *Current Psychosis and Therapeutics Reports*, **4**, 176–80.

Chapter 11

From "evolved interpersonal relatedness" to "costly social alienation:" an evolutionary neurophilosophy of schizophrenia

Jonathan Burns

There is evidence that modern humans evolved a brain highly attuned and adapted to complex interpersonal relatedness. This 'social brain' is the substrate for an embodied understanding of 'mind' – a mind embedded in the physical matter of body, environment and social world. After Heidegger, Merleau-Ponty and Fromm, this philosophical stance better reflects the social origins of mental life than does the redundant dualism of Descartes. Schizophrenia is conceived as a disorder of social brain evolution in that it is characterised by what Eugene Bleuler termed an 'affective dementia.' Individuals with schizophrenia exhibit anatomical, functional and clinical evidence for social brain disorder. In this chapter, I describe this most human of maladies in terms of a 'phenomenology of social alienation' and, drawing on contemporary research data, make the case that schizophrenia represents a costly evolutionary trade-off in the emergence of embodied social consciousness.

11.1 Introduction

For more than a century, the psychopathological phenomenon termed "schizophrenia" has perplexed and frustrated clinicians and researchers alike. It has proved itself a vague and elusive concept, the subject of much controversy and little agreement. At every level of enquiry and at every historical stage, division and dissent have characterized schizophrenia discourse. Contemporary research in fields as diverse as genetics, brain imaging, neuropsychology, nosology, and epidemiology is characterized by multiple conflicting findings and increasingly regular contradictions of widely accepted "facts". For example the successful mapping of the human genome almost a decade ago has not led to the much-anticipated identification of the gene(s) responsible for schizophrenia. Well into the twenty-first century, schizophrenia remains an enigma to clinicians, researchers, and philosophers of mind.

But why is there so little agreement? Why is schizophrenia still such an enigma? Although there is no clear reason for this state of confusion, one key factor is the erroneous "philosophical paradigm" within which schizophrenia has been conceived. This paradigm can be roughly described as dualistic. I will argue that this critical misunderstanding lies at the root of our failure to make progress in research on the disorder and to establish a definitive model that has scientific validity and clinical utility. The dualistic paradigm is at least partially the product of several philosophical influences spanning nearly 2500 years. In his *Phaedo*, Plato argued that the human soul is immortal because it is obviously immaterial. Even the more "materialist" Aristotle claimed that the "intellect" must be immaterial for it to know anything. During the Middle Ages, philosophers constructed and defended many different and often quite sophisticated versions of Platonic and Aristotelian dualism. Rene Descartes (1596–1650), probably the most influential of the Enlightenment philosophers to consider the mind, argued for a rather radical dualism: substance dualism. He argued that there is only one idea that is always true: the idea that I am thinking, and therefore I am a thinking substance (a "res cogitans"). The Cartesian epithet "Cogito ergo sum" or "I think, therefore I am" became a cornerstone of the modern Western concept of mind. Descartes differentiated the body as a material entity, following the laws of physics, from the mind which is immaterial and devoid of physical qualities. This disembodiment of the mind was to have a powerful legacy, shaping humanity's concept of itself as a fundamentally split creature—a mortal body of base flesh, an oft unwilling host to an ethereal intangible entity called the mind/soul. There is still much debate in today's philosophy of mind whether or not a form of Cartesian dualism is a plausible philosophical position. However, the consensus view is that naïve interpretations of Descartes' dualism are untenable, even though such interpretations can still be found in religion. There, we witness the influence of Cartesian dualism. The supreme project of the Reformation was to disembody spiritual life by putting aside the aesthetic trappings of Catholicism and ridding inner spirituality of its messy earthly connections.

Some current philosophers of mind still defend an updated version of substance dualism (Smythies and Beloff 1989; Foster 1991), but sound as their arguments may be, they do not have many empirical implications. The questionable Cartesian views are the more naïve ones. Unfortunately, these

Cartesian views were the ones that had the most influence, both on folk psychology and on research in the cognitive and social sciences.[1] Likewise, the emergence of psychological theories of mind reflected the essential Cartesian divorce of body from mental and spiritual life. Freud, arguably the most influential scientist of the mind of the twentieth century, crafted a psychology of the individual. Unconscious drives, dynamics, and complexes are phenomena of the individual—alone in time and space—detached from the physical reality of the material and social environment.[2] The dualistic tradition influenced too the major phenomenologists upon whose work much of modern psychiatry was built, philosophers such as Edmund Husserl (1859–1938) and Karl Jaspers (1883–1969.) Even though Husserl rejects metaphysical (Cartesian) dualism and even though there are some paragraphs in Husserls *Ideas* on the reciprocal constitution of body and mind, he did divide the world of individual consciousness from the "world outside it" (Woodruff Smith 1995). Following the Husserlian lines, Jaspers provided a framework for conceptualizing psychopathology in terms of an internal individual psyche. Mood states, hallucinations, delusions, and thought disorders are descriptor terms relating to a single individual. "Mental states" are isolated from the world around—disembodied from phenomena outside individual consciousness.

The critical error of dualistic understandings of mind and its pathologies is this: individual experience is stripped of its interpersonal, social, and existential dimensions. Emotional and behavioral phenomena, both healthy and disordered, are detached from social and environmental context. Human beings are represented as isolated and solitary mental beings. Furthermore, "the mind" is somehow detached and independent of the physiological processes of brain, body, and world. It is an entity not subject to the laws of nature, to physics and biology. It is this Cartesian fallacy that has derailed our search for an understanding of schizophrenia that is consistent with the real world (Burns 2007.)

One might also argue that a naïve dualism is at least partly responsible for the virtual exclusion of evolutionary science from the biological revolution that characterized psychiatric research and discourse in the latter part of the twentieth century. While the Darwinian revolution has permeated and shaped almost every avenue of scientific research over the last 100 years, the study of "the mind" and its pathologies has not readily embraced evolutionary

---

[1] There is some empirical evidence that a Cartesian view is part of our innate folk psychology (see Bloom 2004; Bering 2006).
[2] Against this view one might concede that Freud's emphasis on Oedipal complexes and intrafamilial conflicts is evidence that he didn't discard the social environment as an important factor in pathogenesis. Furthermore, Sulloway (1992) argues that Freud was a materialist and someone who tried to give an evolutionary account of human motivation.

principles and concepts. If the mind is conceived as a disembodied entity, free of material laws, then it has nothing to do with the discoveries of Darwin and his successors. It may also be the case that the emerging biological paradigm in psychiatry was averse to a field of thought that had once flourished in its infamousy. In the decades following the publication of *The Origin of Species* (Darwin 1859), the principle of natural selection was applied to the human mind and intellect by Galton (1869) and others as a rationale for justifying social stratification and discrimination. This movement, known as "eugenics", inspired many racist and even genocidal political movements and policies such as *Aryan supremacy* and *apartheid*. In rejecting such morally reprehensible ideologies, scientists of the mind in more recent times have been careful to avoid the application of evolutionary principles to human psychology and behavior. Not all have, though. Edward O. Wilson (1975) published *Sociobiology* in 1975, arguing that evolutionary theory can illuminate the social behavior of humans, not just that of other creatures. He and his colleagues were immediately branded genetic determinists—the collective memory of the abuses of eugenics was perhaps still too fresh. In recent decades, however, a growing number of psychiatrists have embraced the evolutionary paradigm in their efforts to understand the nature of mental illness (see Stevens and Price 1996; McGuire and Troisi 1998; Brune *et al.* 2003; Burns 2007; Brüne 2009.)

11.2 **A philosophy of embodiment**

The adoption of an evolutionary perspective on the mind and its maladies necessitates an absolute rejection of a naïve Cartesian dualism. If our thinking is subject to the biological laws of evolutionary transformation, then it must exist as a phenomenon that is firmly embedded in the material substrate of brain, body, and world. For this reason, a philosophy of embodiment is key to any attempt to reinvigorate Darwinian thinking about the mind and psychopathology. Perhaps the father of such a philosophy was the German thinker Martin Heidegger (1889–1976.) As Bracken states so eloquently in his work *Trauma: Culture, Meaning and Philosophy*, "Heidegger's thought is a powerful antidote to the dominance of Cartesianism in the humanities and human sciences" (Bracken 2002, p. 9). Heidegger spoke of "being-in-the-world", meaning that we are not "in a world that is separate from ourselves . . . Rather, we allow a world to be by our very presence" (Bracken 2002, p. 88). Bracken explains that for Heidegger, the world exists *a priori*, that is, before, our human representation of it as thought. Bracken states: "Existence, in the sense of lived human existence, involved and embedded in the world, is the necessary precedent and the enabling condition of thought" (Bracken 2002, p. 91). What is in our minds is a construct derived from social and cultural information already present in the world around ourselves—our environment. Thus Heidegger divorces the Western concept of mind from its Cartesian origins and presents it as a manifestation of the dynamic interaction between individual and (socio-cultural) environment.

Equally important in overturning dualism and replacing it with a philosophy of embodiment was the French phenomenologist Maurice Merleau-Ponty (1908–1961). Macke (2007) writes of the significance of Merleau-Ponty's master work: "*The Phenomenology of Perception* succeeded in closing the book, so to speak, on the dialectic of body and mind that served as the fundamental, metaphysical puzzle of the second millennium" (Macke 2007, p. 403). Merleau-Ponty used the expression "body-subjects" to explain the bodily nature of consciousness. Our bodies, which perceive the world through sight and sound, through physical perception, give meaning to existence. It is through our bodies that we can act meaningfully on the world and the world can act meaningfully on us. Merleau-Ponty writes:

> The body is the vehicle of being in the world, and having a body is, for a living creature, to be interinvolved in a definite environment, to identify oneself with certain projects and be continually committed to them . . . Man taken as a concrete being is not a psyche joined to an organism, but the movement to and fro of existence which at one time allows itself to take corporeal form and at others moves towards personal acts. Psychological motives and bodily occasions may overlap because there is not a single impulse in a living body which is entirely fortuitous in relation to psychic intentions, not a single mental act which has not found at least its germ or its general outline in physiological tendencies. It is never a question of the incomprehensible meeting of two causalities, nor of a collision between the order of causes and that of ends. But by an imperceptible twist an organic process issues into human behaviour, an instinctive act changes direction and becomes a sentiment, or conversely a human act becomes torpid and is continued absent-mindedly in the form of a reflex. Between the psychic and the physiological there may take place exchanges which almost always stand in the way of defining a mental disturbance as psychic *or* somatic . . . Thus, to the question which we were asking, modern physiology gives a very clear reply: the psycho-physical event can no longer be conceived after the model of Cartesian physiology and as the juxtaposition of a process in itself and a *cogitation*. The union of soul and body is not an amalgamation between two mutually external terms, subject and object, brought about by arbitrary decree. It is enacted at every instant in the movement of existence.
>
> (Merleau-Ponty 1962, pp. 94–102)

Bracken and Thomas (2005) are clear that this world in which our consciousness is embodied is not just a physical material world of objects and places but is also a social world:

> Thus "embodiment" offers a valuable way of dealing with body mind dualism, but it also helps us to think our way around epistemological dualism (mind/society dualism.)

> Being a body-subject . . . also means that our being-in-the-world is just as much located in historic and cultural worlds as it is in the physical world.
>
> (Bracken and Thomas 2005, p. 127)

Individual consciousness therefore is not an entity independent of the social world. It is part of a greater "network" of interacting "minds" and bodies, invested with cultural and historical meaning—a "network" we call society or the social world. Thus, just as the "mind" is a social phenomenon, so the brain, which is the substrate for mental life, is a social organ embedded within a "network" of interacting brains. This brings us to the notion of a "social brain", a brain that is attuned to socially salient phenomena and is the product, both developmentally and in evolutionary terms, of forces emanating from the social world.

11.3 The evolution and development of the social brain

There is much evidence for the notion of a "social brain", emanating from fields as diverse as neuro-histology and neurophysiology to primatology and paleontology. This evidence covers the evolutionary history of the social brain, the developmental emergence of social behavior in individuals, and the growing field of social neuroscience. The social brain is a concept that has emerged in many literatures in a sort of "parallel" or "co-evolved" fashion. It is a concept that is the product of multiple and differing lines of evidence which, when gathered together, create a strong and robust "cable" of evidence. The philosopher of science, Wylie, has described a "cabling" methodology, whereby numerous strands of evidence are intertwined to construct a sound evidence base. Unlike arguments that form a logical "chain" of sequential links, the cabling method accommodates the inclusion of incomplete lines of evidence. The cognitive archaeologist Lewis-Williams argues that the cabling method is scientifically sound in that it is both *sustaining* (a parallel strand may compensate for a gap in another strand) and *constraining* in that it "restricts wild hypotheses that may take the researcher far from the [scientific] record" (Lewis-Williams 2002, pp. 102–3). With this methodology in mind, it is possible to consider the various strands of evidence for a social brain in modern *Homo sapiens*.

11.3.1 Evidence from primatology

Key questions in primate studies include: What differentiates primates from other mammals? What characteristic links primates together as a family? Perhaps most importantly, what behavioral characteristic of primates is responsible for driving the extraordinary evolution of the hominid brain?

In their 1953 paper "Social behaviour and primate evolution", Chance and Mead (1953, p. 417) were among the first to suggest that this characteristic might have been the need to master complex social dynamics. They wrote that "... the ascent of man has been due in part to a competition for social position. . ." Later, Humphrey argued that technology emerged as a consequence of and a solution to the problem of "time given up to unproductive social activity" (Humphrey 1976, p. 310). Social cohesion, he suggested, was essential for creating a context in which the learning of skills and knowledge critical for survival could occur. Forging social relationships is time-consuming and technological innovations (such as tool use) solved the problem of expediting survival-related tasks such as hunting and gathering. In his discussion of the emergence of social behavior, Byrne traces a gradual increase in the complexity of social functioning as one moves from phylogenetically more distant to nearer primate human relatives (Byrne 1999.) For example, in contrast to prosimians, the anthropoidea (monkeys and apes) use cooperation and alliances extensively, acquire dominance ranks, show long-lasting "friendships", devote substantial time to social grooming, engage in reconciliation behaviors, show knowledge of individual affiliations, and use techniques of social manipulation. It is in apes alone, however, that "theory of mind" (ToM) or representational intelligence is apparent.[3] Many authors (e.g. Premack and Woodruff 1978; Byrne 1999) have argued that nonhuman apes such as chimpanzees and orang-utans are capable of basic ToM skills, and date the origins of this social cognitive capacity to the period 16 to 5 million years ago.

11.3.2 Evidence from paleontology

Contrary to popular belief, enlargement in brain size alone was not solely responsible for the evolution of sophisticated modern human cognition. Regional enlargements of certain areas of the brain as well as the expansion of interconnected neural networks linking critical cortical and subcortical regions were almost certainly more important. Drawing on his analyses of hominid fossil endocasts and skulls, Holloway (1975) related the evolution of social behavior to brain reorganization (rather than brain enlargement). Interconnected brain regions recognized as key to social cognition and behavior were subject to greater-than-expected reorganization and expansion during

---

[3] It is important to note that some writers have critiqued the concept of ToM, arguing that it is based on a dualistic ontology. For example, Leudar and Costall (2004) maintain that the ToM framework adds little to the ideas of Chomsky and Grice, "inherits the traditional dualistic problem of other minds, tries to solve it, and ends up profoundly intellectualizing social interactions."

hominid evolution. For example, Holloway showed that endocasts from two *Australopithecine* fossils reveal an intermediate position of the lunate sulcus, between that of the human and that of the chimpanzee (Holloway 1983.) This position suggests, he argues, that the parietal association cortex (PAC) was significantly enlarged and reorganized as early as 3 million years ago. The right PAC has been shown to play an integral role in human social behavior and social cognition.

11.3.3 Evidence from comparative neuroscience

Modern neuroscience methods (e.g magnetic resonance imaging (MRI) and neuro-histology) have demonstrated that regions of the brain implicated in regulating social behavior have been subject to significant expansion and reorganization during the course of primate and hominid evolution. Comparative studies of extant primates have yielded important data in this regard. For example, Dunbar showed that neocortex size relative to whole brain size (the neocortex ratio) is correlated with social group size, the latter being a crude measure of social complexity (Dunbar 1992.) In hominoids, it is white matter (and specifically intra-hemispheric white matter) that increases relative to brain size, rather than grey matter (Hofman 1989; Rilling and Insel 1999.) This suggests that white matter connections linking the prefrontal and orbitofrontal cortices to posterior structures such as the superior temporal cortex and PAC have been subject to marked evolutionary change in the hominid line. This is borne out by other comparative research which shows greater-than-expected enlargement of temporal lobe white matter (Rilling and Seligman 2002), the corticobasolateral nucleus of the amygdala (Barton and Aggleton 2000), and the right intraparietal cortex (Gilissen 2001) as one moves from monkey to nonhuman ape to human. These interconnected structures underpin social cognition, ToM, and social behavior in humans.

11.3.4 Evidence from developmental neuroscience and psychology

Research into the development of social and language skills in infants reveals that "social consciousness" and interpersonal development is necessary before the emergence of "individual consciousness" and a sense of "self". George Herbert Mead (1863–1931), one of the founders of social psychology, argued that "meanings . . . arise in social interaction" and that "self-consciousness arises in the process of social experience" (Brothers 1997, p. 101). Brothers maintains that mothers teach "shared experience to their babies" through imitation games and that "the infant's basic faith in a shared world of subjectivity emerges from a matrix of physical interactions" (Brothers 1997, p. 77.)

Thus, individual consciousness is derived from collective meanings and, as Wittgenstein II argued, language is only meaningful in terms of its social context. Infants learn the meaning of language through their experience of the social world in which they are immersed. The developmental psychologist Vygotsky considered that the origins of our thoughts, beliefs, memories, and language lie in communal behavior and intersubjective discourse (Vygotsky 1978). This is close to Merleau-Ponty's description of the mind as an "embodied" phenomenon, constructed by and engaged in the physical world of the body and society (Merleau-Ponty 1962). Increasingly, infant research supports the notion that the development of an individual sense of self and identity depends on the primary development of a social sense of self and identity.

11.3.5 Evidence from neuro-histology

The identification of "mirror neurons" in the prefrontal cortex of both macaques and humans has helped to locate the neuronal basis of social cognition, mental state attribution, and possibly even empathy. Mirror neurons were first discovered in an area of the prefrontal cortex of macaque monkeys that is homologous to Broca's (speech) area in humans (Di Pellegrino *et al.* 1992). Mirror neurons discharge both when the individual performs a goal-directed action and when the individual observes another individual performing the same action. They therefore "mirror" an observed action within the motor cortex of the observer. In this way, they internally "represent" actions (Rizzolatti *et al.* 1996.) Moreover, it is the intention of the action that is represented rather than the action itself. Subsequent research demonstrated the existence of a mirror neuron system (MNS) within the human brain, comprising a cortical network including Broca's area, the prefrontal cortex, the superior temporal sulcus, and the posterior parietal cortex (Rizzolatti *et al.* 1996.) Gallese has argued that the MNS constitutes a neural basis for empathy and describes a shared intersubjective space across which the MNS maps a "multimodal representation of organism-organism relations" (Gallese 2003, p. 175.) This space is termed "the shared manifold of intersubjectivity." Thus one could argue that, at a cellular level, the MNS is evidence for embodied social behavior and social cognition (see also Chapter 9).

11.3.6 Evidence from social neuroscience

Neuropsychological and behavioral research (using various methods, such as functional imaging, electrophysiology, and animal studies) clarifies the structural and functional anatomy of the "social brain" as comprising the interconnected regions and circuits of the prefrontal, temporal, and limbic lobes and the parietal association cortex. Brothers describes the "social brain" as the higher cognitive and affective systems in the brain that underlie our ability to function as highly social animals and provide the substrate for social cognition, social behavior, and affective responsiveness (Brothers 1997.) Grady and Keightley (2002) consider social cognition as comprising face perception, emotional processing, ToM, self-reference, and working memory. In terms of face perception, neurons responding selectively to facial expression, eye gaze, and intended action have been identified in the fusiform gyrus, superior temporal sulcus, amygdala, and orbitofrontal cortex (Haxby *et al.* 2002.) The amygdala, with its important array of connections to both cortical and subcortical regions, functions as the brain's emotion regulation system, integrating emotional, motivational, and cognitive processes (Le Doux 1994.) Affiliative behavior, critical for the establishing and maintaining of social bonds, appears to be dependent on healthy functioning of the amygdala, orbitofrontal cortex, and temporal pole (Kling and Steklis 1976.) A host of evidence supports the central role of the prefrontal, orbitofrontal, and anterior cingulate cortices, superior temporal gyrus and parietal association cortex in ToM functioning (Abu-Akel 2003). In summary then, the "social brain" seems to consist of a network comprising the following structures and their interconnections: the prefrontal cortex, the orbitofrontal cortex, the anterior cingulate cortex, the superior temporal gyrus and cortex, the amygdala, and the parietal association cortex.

11.4 Schizophrenia and the evolutionary paradigm

In the introduction to this chapter, I discussed the apparent failure of a century of research efforts to identify the true nature of schizophrenia and to unravel its complex causation. This has much to do with the dominance of the Cartesian paradigm, but of equal importance is the lack of a unifying, biologically sound framework for conceptualizing the disorder. For reasons already discussed, the evolutionary paradigm was largely ignored in modern biomedicine's preoccupation with this enigmatic condition. Those who have ventured to understand schizophrenia in evolutionary terms have for the most part arrived at balanced polymorphism models, arguing that hidden advantages associated with the genotype serve to offset the evolutionary disadvantages associated with disease (e.g., Allen and Sarich 1988; Polimeni and Reiss 2003). Difficult to test empirically, these models have foundered on a basic principle of Darwinism, namely that a reproductive advantage has not been demonstrated in first-degree relatives (i.e., "carriers") of individuals with schizophrenia (see Burns 2006 and 2007 for a critique of balanced polymorphism models). A further problem with evolutionary analyses of this disorder is articulated clearly by Adriaens (2008, p. 1215) who argues that research evidence reveals that schizophrenia is in fact not a "natural kind", that is, it is not a "bounded and objectively real entity with discrete biological causes" and that therefore it "simply does not have an evolutionary history" (Adriaens 2008, p. 1215.) Is there then a way forward for understanding schizophrenia in evolutionary terms?

The latter and more damning critique of Adriaens is in fact the key to moving forward with this analysis. Schizophrenia as we know it in the current era is indeed a vague and indistinct phenomenon. As Adriaens (2008, p. 1215) states, it may indeed be no more than "a reified umbrella concept, constructed by psychiatry to cover a heterogenous group of disorders". Certainly the array of contradictory research findings emanating from genetics, brain imaging, neuropathology, and phenomenology suggests that we are dealing with a collection of similar but different conditions that have been grouped together by virtue of their sharing certain clinical symptoms. Importantly, these symptoms include hallucinations, delusions, and disorganized behavior and thought processes—the so-called "positive symptom" cluster. And it is here that the diagnostic and research problem lies. The century-old focus on these more flamboyant, dramatic, and disruptive manifestations has led clinicians and researchers alike into the frustrating cul-de-sac that is our current state of confusion about the exact nature of schizophrenia. These "symptoms" with which we diagnose schizophrenia are nonspecific in that they manifest in many other psychiatric disorders and they are inconsistent in that they are state, rather than trait, specific.

Is there then a "core" phenomenon that describes a 'natural kind'—a "bounded and objectively real entity"? Is there a "natural kind" that is phenotypically and genotypically substantiated by the evidence and is amenable to evolutionary analysis? Answering these questions requires us to step away from the diagnostic dogma[4] inherited from Kraepelin and Schneider and firmly adhered to by our profession for more than 100 years. Instead we need to look back to another patriarch of modern psychiatry, namely Eugene Bleuler.

11.5 "Interpersonal alienation" from the social world

Eugene Bleuler (1857–1939) was a Swiss psychiatrist with an unusual dedication to observing closely the symptoms of his patients in the Burghölzli Clinic

---

[4] Both Emil Kraepelin and Kurt Schneider emphasized what have now become known as the "positive" features of schizophrenia in terms of diagnosis. These symptoms include hallucinations, delusions, and disorganized thinking and behaviour—symptoms that are in fact nonspecific and commonly occur in psychiatric disorders other than schizophrenia.

near Zurich. He coined the term "schizophrenia" to describe what he considered a fundamental split or dissociation between inner thoughts and cognitions, and the emotional contact or engagement with the world. He described the core problem as an "affective dementia" (Bleuler 1923) in the sense that one experiences gross emotional and interpersonal detachment in an individual with schizophrenia (rather than irreversible degenerative disease.) He wrote that schizophrenia "is characterised by a specific kind of alteration of thinking and feeling, and with the relations with the outer world that occur nowhere else" (Bleuler 1923, p. 373). Beneath the more dramatic hallucinations and delusions, he argued, there exists a less obvious but core problem of "autistic" alienation from the social world. He described the core features of the condition in terms of "disturbances of association", a qualitative "affective disturbance", "ambivalence", and "autism".

Several successors to Bleuler have extended his focus on interpersonal alienation, describing hallucinations and delusions as a form of "exaggerated self-consciousness" or "hyperreflexivity" (Sass and Parnas 2003, p. 427) that is a secondary consequence of the primary problem, which is a loss of the "primal sense of vitality or vital connectedness with the world" (Bleuler 1923, p. 248.) Minkowski, a student of Bleuler's, viewed schizophrenia as a rupture between intellect and intuition, the former "associated with analysis and abstract reason" and the latter "based on . . . the vitality and temporal dynamism of experience as it is actually lived" (Sass 2001, p. 254). He defined schizophrenic autism as a loss of vital contact with reality. Blankenburg (1971, quoted in Sass 2001, p. 258) describes the core problem as a "loss of natural self-evidence" which, according to Sass (2001, p. 258), refers to "a loss of the usual common-sense orientation to reality . . . that normally enables a person to take for granted so many of the elements and dimensions of the social and practical world". These authors speak the same language, echoing Bleuler's original observation that the core problem of schizophrenia relates to the individual's sense of detachment and alienation from the "social self" and "social world".

If we return then to our earlier discussion of the work of Heidegger and Merleau-Ponty, we may observe the coincidence of these individuals' concept of a social self, originating from and embodied in a social world—"being-in-the-world"—and Bleuler's original identification of the core problem in schizophrenia being autistic alienation or disembodiment from "social self" and "social world". The construction therefore of a philosophy of mind that reflects the embodied nature of human consciousness and experience allows us to move away from the individualistic model of schizophrenia that has misguided our research agenda and towards a phenomenology of social alienation that is in keeping with growing evidence for social brain dysfunction in this disorder.

It also allows us to identify a core phenomenon or "natural kind" schizophrenia that is amenable to empirical investigation as well as to evolutionary analysis.

11.6 Schizophrenia as a social brain disorder

In recent decades there has accumulated a significant evidence base from a range of research methodologies supporting the notion of social brain dysfunction in schizophrenia. Neuropsychological, functional, and structural imaging and clinical approaches all confirm the centrality of social brain dysfunction in a syndrome that we might term "natural kind schizophrenia".[5] In other words, when social brain dysfunction is taken as the core diagnostic and pathological phenomenon, there exists a clinical entity that is empirically amenable to study as a "bounded and objectively real natural kind". The evidence for this claim is summarized below.

11.6.1 Evidence from ethological observation

Naturalistic observation of individuals with schizophrenia shows that specific problems with interpersonal nonverbal behavior are common to almost all cases. Brüne has reviewed this research in his book *The Social Brain: Evolution and Pathology* (Brüne *et al.* 2003.) He cites a number of common problems, including poor eye contact and reduced eyebrow raising, fewer upper face activities, fewer primary emotions, and more negative emotions, and lower scores on pro-social behavior, gesture, and displacement activities.

11.6.2 Neuropsychological evidence

Individuals with schizophrenia demonstrate multiple impairments of social cognition and theory of mind ability. These include eye gaze and facial affect recognition deficits, general emotion recognition deficits, impaired mentalization and "mindreading" ability, and social perception and attributional errors (Frith 1994.)

11.6.3 Evidence from functional imaging

Social cognition activation paradigms show functional abnormalities in a network comprising the dorsolateral-prefrontal cortex, orbitofrontal cortex, paracingulate and inferior parietal cortices, the superior temporal and lateral

---

[5] The clinical heterogeneity of the syndrome termed "schizophrenia" is such that many of the features of this phenomenon (e.g., hallucinations, delusions) transcend diagnostic boundaries and are apparent in a range of psychotic disorders. Arguably, therefore, there is a need to define a core psychopathological feature that is specific to schizophrenia, which might be termed the "natural kind schizophrenia".

fusiform gyri, and the amygdala (Frith *et al.* 1995; Fletcher *et al.* 1999.) This has been termed "functional dysconnectivity" in schizophrenia (Friston and Frith 1995). This research supports the argument that schizophrenia is a disorder of social brain functioning.

11.6.4 Evidence from structural imaging

MRI and diffusion tensor MRI show structural abnormalities of the same prefrontal, temporal, and parietal association areas as listed above (Lawrie and Abukmeil 1998; Sanfilipo *et al.* 2000) as well as disruption of white matter fasciculae connecting these regions (Burns *et al.* 2003). This has been termed "structural dysconnectivity" (Burns *et al.* 2003). Since these are the interconnected brain regions comprising the social brain, it is legitimate to locate the anatomical basis for schizophrenia in the social brain.

11.6.5 Clinical evidence

The negative symptoms of schizophrenia predict poor course and outcome in the disorder. In many ways one can argue that this cluster of symptoms is characteristic of core "natural kind schizophrenia". Negative symptoms incorporate many aspects of social cognition and behavior, such as affective responsiveness, motivation, and sociability. Both functional and structural imaging research shows that negative symptoms correlate with functional and structural deficits in prefrontal, parietal association, and limbic regions of the brain (Ross and Pearlson 1996; Sanfilipo *et al.* 2000). In other words, abnormal social cognition and social behavior in schizophrenia are associated with anatomical deficits in social brain structures.

11.7 Resolving the "schizophrenia problem" in evolutionary terms

The "schizophrenia problem" can be described in terms of the failure of a century's observations and research to converge on a "natural kind" that is empirically valid and meaningful as a biological phenomenon. I have argued that this frustrating lack of cohesion and agreement, and the resultant confusion and controversy aroused by the concept is in part a result of the dominance of Cartesian dualistic thinking about "the mind" and its pathologies. I have also argued that a general aversion to adopting an evolutionary perspective on schizophrenia (and other psychiatric disorders) has limited the search for gaining a fuller understanding of this very human phenomenon. Where an evolutionary approach has been adopted, a popular reliance on balanced polymorphism models has side-tracked efforts and rendered them open to serious criticism, especially from molecular and epidemiological vantage points. As Adriaens (2008) points out, the popular assumption of most evolutionary theorists—that there exists a paradox of constant prevalence in spite of obvious reproductive disadvantage—is in fact erroneous in the light of recent evidence suggesting marked epidemiological variation (McGrath 2006.)

Is it then possible to resolve the "schizophrenia problem" in evolutionary terms? I would argue that there are four key steps that together move us closer to understanding this elusive disorder as a "natural kind" with an evolutionary history as long as the very existence of our species:

1. Human psychological life, both in health and in sickness, needs to be understood in social terms. The redundant Cartesian framework needs to be replaced with a philosophy and phenomenology of embodiment that conceives of "the mind" as a manifestation of the dynamic interaction between individuals and the social world.

2. We need to move away from an individualistic model of schizophrenia, which has misguided our research agenda, and toward a phenomenology of social alienation that is in keeping with evidence for social brain dysfunction in this disorder. A return to Bleuler's original focus on "autistic alienation" from the "social self" and "social world" allows us to identify a "natural kind schizophrenia" that is amenable to both empirical and evolutionary analysis.

3. In adopting an evolutionary perspective on schizophrenia we need to move beyond a narrow Darwinian preoccupation with "natural selection" and its Malthusian message of "a struggle for existence" between individuals. The work of the Russian anarchist, Kropotkin, may provide a useful evolutionary vantage point for considering the origins of social behavior and schizophrenia. In his theory of *Mutual Aid* Kropotkin emphasized the evolutionary success of cooperative social behaviors in the struggle between organism and environment (Todes 1987; Gould 1997). Kropotkin wrote:

> If we . . . ask Nature: "who are the fittest: those who are continually at war with each other, or those who support one another?" we at once see that those animals which acquire habits of mutual aid are undoubtedly the fittest . . . Sociability is as much a law of nature as mutual struggle
>
> (Kropotkin 1902)

From this perspective, it is fair to state that the evolution of a highly sophisticated social brain was perhaps the defining event in the emergence of modern humans.

4. If "natural kind schizophrenia" is conceived in terms of a disorder of the evolved social brain in humans, it is legitimate to redefine it as a costly

by-product or consequence of social brain evolution (Burns 2006, 2007.) In other words, it exists and persists in our species because it is inextricably tied to the evolutionary process that gave rise to our social nature. Furthermore, if we assert that the multiple genes responsible for conferring vulnerability to schizophrenia are also responsible for the evolution and healthy functioning of the social brain, then we have no need to invoke problematic molecular models such as balanced polymorphism in resolving the schizophrenia paradox. Finally, the key to reconciling the "new epidemiology" of schizophrenia (variable prevalence and incidence) with an evolutionary genetic basis for the disorder is this: A spectrum of genetic vulnerability to schizophrenia exists in the population by virtue of the fact that these are "normal genes" responsible for neurodevelopment. Complex bidirectional gene–environment interactions operate throughout neurodevelopment to mediate expression of genetic vulnerability. Thus, a harmful social environment leads to increased expression of susceptibility genes, accounting for the epidemiological variability demonstrated in relation to factors such as urbanicity, migrant status, and socioeconomic disparity (Burns 2009).

In conclusion therefore, the development of a philosophy of mind that reflects the embodied social nature of human consciousness and experience allows us to move away from an individualistic model of schizophrenia and towards a phenomenology of social alienation that coincides with evidence for this being a social brain disorder. This helps us define a "natural kind schizophrenia" that is amenable to both empirical and evolutionary analysis. Evidence supporting the key place of sociality in the evolution of our species leads us to the realization that the existence of schizophrenia—a socially devastating clinical phenomenon—is inextricably linked to this social brain evolution. Importantly, the evolved genetic make-up that defines the unique social cognitive abilities of modern *Homo sapiens* also carries with it an inherent genetic vulnerability to harmful features of the social environment. Schizophrenia, therefore, is not just a costly by-product of social brain evolution in modern humans, but also a consequence of the unhealthy societies we create around us.

References

Abu-Akel A. (2003) A neurobiological mapping of theory of mind. *Brain Research and Brain Research Reviews*, **43** (1), 29–40.

Adriaens, P.R. (2008) Debunking evolutionary psychiatry's schizophrenia paradox. *Medical Hypotheses*, **70** (6), 1215–22.

Allen, J.S. and Sarich, V.M. (1988) Schizophrenia in an evolutionary perspective. *Perspectives in Biology and Medicine*, **32**, 132–53.

Barton, R. and Aggleton, J. (2000) Primate evolution and the amygdala. In: J. Aggleton (ed.), *The Amygdala: A Functional Analysis*. Oxford University Press, Oxford, pp. 480–508.

Bering, J.M. (2006) The folk psychology of souls. *Behavioral and Brain Sciences*, **29**, 453–98.

Blankenburg, W. (1971) *Der Verlust der Naturlichen Selbstverstandlichkeit: Ein Beitrag Zur Psychopathologie Symptomarmer Schizophrenien*. Ferdinand Enke Verlag, Stuttgart.

Bleuler, E. (1923) *Textbook of Psychiatry*, translated by A.A. Brill. George Allen and Unwin, London.

Bloom, P. (2004) *Descartes' Baby: How the Science of Child Development Explains What Makes Us Human*. Basic Books, New York.

Bracken, P. (2002) *Trauma: Culture, Meaning and Philosophy*. Whurr Publishers, London.

Bracken, P. and Thomas, P. (2005) *Postpsychiatry: Mental Health in a Postmodern World*. Oxford University Press, Oxford.

Brothers, L. (1997) *Friday's Footprint: How Society Shapes the Human Mind*. Oxford University Press, Oxford.

Brüne, M., Ribbert, H., and Schiefenhövel, H. (eds) (2003) *The Social Brian: Evolution and Pathology*. John Wiley and Sons, Chichester.

Burns, J.K. (2006) Psychosis: A costly by-product of social brain evolution in Homo sapiens. *Progress in Neuropsychopharmacology & Biological Psychiatry*, **30** (5), 797–814.

Burns, J.K. (2007) *The Descent of Madness: Evolutionary Origins of Psychosis and the Social Brain*. Routledge, Hove.

Burns, J.K. (2009) Reconciling "the new epidemiology" with an evolutionary genetic basis for schizophrenia. *Medical Hypotheses*, **72**, 353–8.

Burns, J., Job, D., Bastin, M.E., Whalley H., MacGillivray T., Johnstone E.C., and Lawrie S.M. (2003) Structural disconnectivity in schizophrenia: a diffusion tensor magnetic resonance imaging study. *British Journal of Psychiatry*, **182**: 439–43.

Byrne, R.W. (1999) Human cognitive evolution. In M.C. Corballis and S.E.G. Lea (eds), *The Descent of Mind: Psychological Perspectives on Hominid Evolution*. Oxford University Press, Oxford, pp. 147–59.

Chance, M.R.A. and Mead, A.P. (1953) Social behaviour and primate evolution. *Symposia of the Society of Experimental Biology*, **7**: 395–439.

Darwin, C. (1859) *The Origin of Species by Means of Natural Selection*. John Murray, London.

Di Pellegrino, G., Fadiga, L., Fogassi, L., Gallese, V., and Rizzolatti, G. (1992) Understanding motor events: a neurophysiological study. *Experimental Brain Research*, **91** (1): 176–80.

Dunbar, R.I.M. (1992) Neocortex size as a constraint on group size in primates. *Journal of Human Evolution*, **20**, 469–93.

Fletcher, P., McKenna, P.J., Friston, K.J., Frith, C.D., and Dolan, R.J. (1999) Abnormal cingulate modulation of fronto-temporal connectivity in schizophrenia. *Neuroimage*, **9**, 337–42.

Foster, J. (1991) *The Immaterial Self: A Defence of the Cartesian Dualist Conception of the Mind*, Routledge, London.

Friston, K.J. and Frith, C.D. (1995) Schizophrenia: a disconnection syndrome? *Clinical Neuroscience*, **3**, 89–97.

Frith, C. (1994) Theory of mind in schizophrenia. In A.S. David and J.C. Cutting (eds), *The Neuropsychology of Schizophrenia*, Lawrence Erlbaum Associates, Hove.

Frith, C.D., Friston, K.J., Herold, S., Silbersweig, D., Fletcher, P., Cahill, C., Dolan, R.J., Frackowiak, R.S., and Liddle, P.F. (1995) Regional brain activity in chronic schizophrenic patients during the performance of a verbal fluency task. *British Journal of Psychiatry*, **167**, 343–9.

Gallese, V. (2003) The roots of empathy: the shared manifold hypothesis and the neural basis of intersubjectivity. *Psychopathology*, **36**, 171–80.

Gilissen, E. (2001) Structural symmetries and asymmetries in human and chimpanzee brains. In: D. Falk and K.R. Gibson (eds), *Evolutionary Anatomy of the Primate Cerebral Cortex*. Cambridge University Press, Cambridge, pp. 187–215.

Gould, S.J. (1997) Kropotkin was no crackpot. *Natural History*, **106**, 12–21.

Grady, C.L. and Keightley, M.L. (2002) Studies of altered social cognition in neuropsychiatric disorders using functional neuroimaging. *Canadian Journal of Psychiatry*, **47**, 327–36.

Haxby, J.V., Hoffman, E.A., and Gobbini, M.I. (2002) Human neural systems for face recognition and social communication. *Biological Psychiatry*, **51**, 59–67.

Hofman, M.A. (1989) On the evolution and geometry of the brain in mammals. *Progress in Neurobiology*, **32**, 137–58.

Holloway, R.L. (1975) The role of human social behavior in the evolution of the brain. 43rd James Arthur Lecture on the Evolution of the Human Brain 1973. The American Museum of Natural History, New York.

Holloway, R.L. (1983) Cerebral brain endocast pattern of *Australopithecus afarensis* hominid. *Nature*, **303**, 420–2.

Humphrey, N.K. (1976) The social function of intellect. In P.P.G. Bateson and R.A. Hinde (eds), *Growing Points in Ethology*, Cambridge University Press, Cambridge, pp. 303–317.

Jaspers, K. (1962) *General Psychopathology*. Manchester University Press, Manchester.

Kling, A. and Steklis, H.D. (1976) A neural substrate for affiliative behaviour in nonhuman primates. *Brain and Behavioural Evolution*, **13**, 216–38.

Kropotkin, P. (1902) *Mutual Aid: A Factor in Evolution*. Heinemann, London.

Lawrie, S.M. and Abukmeil, S.S. (1998) Brain abnormality in schizophrenia. A systematic and quantitative review of volumetric magnetic resonance imaging studies. *British Journal of Psychiatry*, **172**, 110–20.

Le Doux, J.E. (1994) Emotion, memory and the brain. *Scientific American*, **270**, 50–7.

Leudar, I. and Costall, A. (2004) On the persistence of the "problem of other minds" in psychology: Chomsky, Grice and theory of mind. *Theory and Psychology*, **14** (5), 601–21.

Lewis-Williams, D. (2002) *The Mind in the Cave*. Thames & Hudson Ltd, London.

Macke, F.J. (2007) Body, liquidity, and flesh: Bachelard, Merleau-Ponty, and the elements of interpersonal communication. *Philosophy Today*, **51** (4), 401–15.

McGrath, J.J. (2006) Variations in the incidence of schizophrenia: data versus dogma. *Schizophrenia Bulletin*, **32**, 195–7.

McGuire, M. and Troisi, A. (1998) *Darwinian Psychiatry*. Oxford University Press, Oxford.

Merleau-Ponty, M. (1962) *Phenomenology of Perception*. Routledge & Kegan Paul, London.

Polimeni, J. and Reiss, J.P. (2003) Evolutionary perspectives on schizophrenia. *Canadian Journal of Psychiatry*, **48**, 34–9.

Premack, D. and Woodruff, G. (1978) Does the chimpanzee have a "theory of mind"? *Behavioural and Brain Sciences*, **4**, 515–26.

Rilling, J.K. and Insel, T.R. (1999) The primate neocortex in comparative perspective using magnetic resonance imaging. *Journal of Human Evolution*, 37, 191–223.

Rilling, J.K. and Seligman, R.A. (2002) A quantitative morphometric comparative analysis of the primate temporal lobe. *Journal of Human Evolution*, 42, 505–33.

Rizzolatti, G., Fadiga, L., Gallese, V., and Fogassi, L. (1996) Premotor cortex and the recognition of motor actions. *Brain Research and Cognitive Brain Research*, 3, 131–41.

Ross, C.A. and Pearlson, G.D. (1996) Schizophrenia, the heteromodal association neocortex and development: potential for a neurogenetic approach. *Trends in Neurosciences*, 19, 171–6.

Sanfilipo, M., Lafargue, T., Rusinek, H., Arena, L., Loneragan, C., Lautin, A., Feiner, D., Rotrosen, J., and Wolkin, A. (2000) Volumetric measure of the frontal and temporal lobe regions in schizophrenia: relationship to negative symptoms. *Archives of General Psychiatry*, 57, 471–80.

Sass, L.A. (2001) Self and world in schizophrenia: three classic approaches. *Philosophy, Psychiatry and Psychology*, 8 (4), 251–70.

Sass, L.A. and Parnas, J. (2003) Schizophrenia, consciousness and the self. *Schizophrenia Bulletin*, 29 (3), 427–44.

Smythies, J.R. and Beloff, J. (eds) (1989) *The Case for Dualism*. University of Virginia Press, Charlottesville, VA.

Stevens, A. and Price, J. (1996) *Evolutionary Psychiatry: A New Beginning*. Routledge, London.

Sulloway, F.J. (1992) *Freud, Biologist of the Mind: Beyond the Psychoanalytic Legend*. Harvard University Press, Cambridge, MA.

Todes, D.P. (1987) Darwin's Malthusian metaphor and Russian evolutionary thought, 1859–1917. *Isis*, 78, 537–51.

Vygotsky, L.S. (1978) *Mind in Society: The Development of Higher Psychological Processes*. Harvard University Press, Cambridge, MA.

Wilson, E.O. (1975) *Sociobiology: The New Synthesis*. The Belknap Press of Harvard University Press, Cambridge, MA.

Woodruff Smith, D. (1995) Mind and body. In B. Smith and D. Woodruff Smith (eds), *The Cambridge Companion to Husserl*. Cambridge University Press, New York, pp. 323–93.

Index

abnormal regulation mechanism 182
abnormality, functional 242
absence of reinforcement (extinction) 67
acalculia 169
Acceptor (subordinate/one-down member) 277–8, 280, 281, 283
acoustic cues 78
acquisition phase 74, 76–7
Active Listening 128
adaptation 2, 94, 97
  see also environment of evolutionary adaptedness; function, dysfunction and adaptation
adaptationist explanation/views of depression 119, 130
adaptationist framework 8, 11–13, 69, 71, 82–4, 135
adaptive functioning, level of 180
adaptive traits 7, 227
adaptiveness 147
adaptors, body-focused 128
addiction 176
adjustment disorders 207
Adolphs, R. 58
adornment, cultural differences in 70
Adriaens, P.R. 298–9, 303
affect(ion) 242
affective disorders 119
  see also depression
Affiliation 128
affiliative behavior 298
alcohol abuse 199
alienation, autistic 303
Allen, N.B. 204, 206
allergies 158
allies and status 190
Almazan, M. 97–8, 101
Alzheimer dementia 10
American Medico-Psychological Organization 177
amitryptyline 128
Amundson, R. 230
amusia 169
amygdala 43, 47, 56, 57–8, 298, 302
amygdalo-prefrontal network 59
ancestral neutrality hypothesis 250
Andreasen, N. 218
Andrews, P.W. 204, 206
anger 186, 193, 271n

anhedonia 204
animal hierarchies 281
anterior cingulate cortex 298
anti-Darwinism 143
antidepressant medication 212–13
antisocial personality disorder 133
anxiety 38, 59, 184, 186, 188, 193–4, 209, 275
  point 59
appeasement display 274, 275
Aristotle 144–7, 148, 151, 156, 290
arousal 77
artifacts 70, 80, 145
Asperger syndrome 242, 244, 246, 247, 257
assertion in deviant behavior 123
Assertive Behavior 128
assessment stage 271, 279
associative learning 67
asymmetry 266, 270
  see also relationships, asymmetrical
atavism 4
attachment:
  ambivalent 122
  anxious avoidant 121
  anxious resistant 121
  avoidant 122
  disorganized 121–2
  insecure 132–3
  insecure early 131
  relationships, early 121–2
  secure 121, 131, 132
  theory 120, 133
attention 57
attentional capture 50
attraction, cultural differences in 70
Auden, W.H. 198
autism 6, 15, 27, 242, 251
  high-functioning 242, 246–7, 257
  spectrum disorders 91, 243
  see also Asperger syndrome; empathy and autism
autistic alienation 303
automatic characteristics 56
automaticity 43, 45, 46

Badcock, P.B.T. 204, 206
balancing selection theory 250
Baron-Cohen, S. 92, 244, 245–6, 248, 257
basal ganglia 272, 273, 274, 275, 276, 278

Bateson, G. 85
  see also mood change, role of in defining relationships
Bateson, P. 73
beauty, standards of 70
Beck Depression Inventory 201
behavior:
  agonistic 272, 276
  deviant 122–4
  interpersonal disturbed 128–31
  interpersonal nonverbal 301
  nonverbal 129, 130, 134
  observable 76, 124–8
  obsessive 248
  repetitive 243, 248
  restricted 243
  social 127, 130, 295, 296
  systems 76
behavioral disorders 176
behavioral ecology 18, 270
behavioral elements and abbreviations, registered 126–7
behaviorists 178
Bengalese finches 73, 75, 76
Benton Facial Recognition Task 101
bereavement 175, 180, 202, 206
Bereczekei, T. 80
Bering, J.M. 247
Bernard, C. 227
Bibring, E. 276
Binet, A. 66, 67
bio-psycho-social model 180
biogenetic law 5
biological criterion 152
biological defects 161
biological determinism 19
biological dysfunction 217
biological function 144–7, 217
biologically designed-defenses 161–5
biology, philosophy 151–2
bipolar disorder 191, 194
bird species 78, 82
  Bengalese finches 73, 75, 76
  chickens 73, 78
  Darwin's finches 220
  domestic hens 270
  ducks 73
  goslings 71
  greylag geese 121
  jackdaws 71
  Japanese quail 73
  mallard ducks 75
  zebra finches 73, 75, 76, 77
birth interval effect 81
birth order effect 81
Bischof, H.-J. 74, 76, 77
Bishop, S.J. 59
black box essentialist account of function 150–1

Blankenburg, W. 300
Bleuler, E. 289, 299–300, 303
Blu, J.J. 119
Boorse, C. 19–21, 149, 171, 217–18, 227
borderline personality disorder 133
Bouhuys, A.L. 124, 128–9, 130
Bowlby, J. 16, 120, 132–3, 282
Bracken, P. 293–4
brain types (B, E and S) 245
break-down model 8
breakdown hypothesis 40, 41
breakdown/lower threshold of module 52
Brothers, L. 296–8
Brown, G.R. 82, 86
Brown, P. 279
Brulde, B. 154
Brüne, M. 301
Buller, D. 18
Bullough, V.L. 65
Burgoon, J.K. 122
Byrne, R.W. 295

cabling methodology 294
Cahn, D.D. 122
Cain, A.J. 220
Carnelley, K.B. 122
Cartesian dualism 292, 302
Cartesian paradigm 298
castration anxiety 67
categorical depression/normality distinction, inductive evidence for 200–3
categorical model 211
cats and dogs 78
causal factors see ethology
central mechanisms 76, 77
Chance, M.R.A. 295
chickens 73, 78
childhood experiences 81
childhood trauma 67
chimpanzees 78
Clahsen, H. 97–8, 101, 105
clinical evidence 302
clothing artifacts 70
Clouser, K.D. 153
cognitive architecture see developmental disorders and cognitive architecture
cognitive associative mechanism 55
cognitive problem-solving 204–5
collective meanings 297
commmunicative impairments 243
communication 266, 277, 279
comparative neuroscience 296
competition:
  agonistic 273, 275
  interpersonal 282
complementarity (inversion of hierarchy) 267
complementary relationships 266, 277, 279, 281
conditionoing paradigm 51

configural cues 102, 104, 110–11, 113
consciousness, individual 296–7
consequential evidence 123
consolidation phase 74, 77
contact and deviant behavior 123
context and DSM criteria 180, 189
contextual evidence and deviant behavior 123
conventionalist approach 199
Cook, M. 52
Cooper, R. 232
corpus striatum 272
cortical areas 56–7, 59, 273, 276, 296, 298, 301–2
Cosmides, L. 38–9, 209, 252
Crichton Browne, J. 3–4
criminology 5
Cro-Magnon 24
cross-fostering experiments 133
crouching postures 134
cues, holistic relation between 110–11
cultural differences in attraction 70
culture-boundedness 18
Culver, C.M. 144
  see also distinct sustaining causes: Culver and Gert
Cummins, R. 149, 219, 224–6
Curio, E. 135

Darwin, C. 1–2, 6–7, 135, 143–4, 147–8, 152, 242, 268–9
  Descent of Man and Selection in Relation to Sex 265, 268, 282
  and teleology 148–50
  The Origin of Species 268, 292
Darwinian approach 26, 36
Darwinian fitness 182
Darwinian picture of natural order 220
Darwinian psychiatry *see* evolutionary psychiatry
Darwinian revolution 60, 291
Darwinian thinking 292
Darwinism, psychiatric 3–6
Darwin's finches 220
das kranke Tier (ailing animal) 23–7
Davey, G.C.L. 54, 55
Davies, P.S. 225–6, 229–31
de Jong, P.J. 52–4
de-escalation (fleeing or submitting) 270, 271, 273–4, 275–6, 277–8
de-selected in intrasexual selection 269, 271
dedicated neural circuit 43
defects, biological 161
defences and diseases, distinction between 182, 218
defensive reactions 161–5
Definer (dominant/one-up member) 277–8, 280, 281, 283
definitional (command) component of communication 266, 277, 279

degeneration theory 4–5, 6
DeLoache, J.S. 50
depression 7, 184, 188–9, 193–4, 223, 271
  categorical 200–3
  grief exclusion for 175, 180, 202, 206
  morbid 223
  see also dysfunctions and depression; ethology; normality, disorder and evolved function in the case of depression; major depressive disorder
Deruelle, C. 102–3
Descartes, R. 290–1
descriptive generality requirement 228
determinism, biological 19
development and sexual imprinting in animals 76–8
developmental disorders and cognitive architecture 26, 91–113
  epistemology of developmental dissociations 105–9
  psychopathologies 93–8
    cognitive architecture 93–4
    dissociations, role of in decomposition of the mind 95–6
    face recognition in Williams syndrome 101–4, 110–12
    massive modularity hypothesis 94–5
    role of developmental psychopathologies 96–8
    why developmental psychopathologies provide no evidence for modularity 98–105
  see also Karmiloff-Smith
developmental neuroscience 296–7
deviant behavior 122–4
Diagnostic and Statistical Manual of Mental Disorders 66
DSM-I 177
DSM-II 177, 187
DSM-III 177, 180, 200
DSM-IV 178, 193, 194, 200, 201, 202, 206, 250
DSM-IV-TR 36–7
DSM-V 191, 192, 194
  see also evolutionary foundations for psychiatric diagnosis: DSM-V validity
differential selection pressure hypothesis 252–4, 258
Dillard, J.P. 131
disability, pathological 159–61
disease and illness, distinction between 218
disgust 55, 186
  -evoking status 53
disordered mood 203
displacement activities 134
displacement model 8, 10

dissociations 98, 107, 108, 112
  double 101
  impure 105, 107, 108, 110, 113
  impure double 96
  impure single 95
  pure 99, 105, 108
  pure double 95
  pure single 95
  role in decomposition of the mind 95–6
distinct sustaining causes: Culver and Gert 152–61
  basic idea and *prima facie* counterexamples 153–5
  failure of 'rational beliefs or desires' clause to eliminate *prima facie* counterexamples 155–6
  hybrid account of dysfunction as harm with no distinct external sustaining cause 153
  instability of distinct sustaining cause intuition 156–7
  normal inability versus pathological disability 159–61
  statistical criterion, retreat to 157–9
dollar auction model 280
domain-general systems 94
domain-specific systems 94
domestic hens 270
dorsolateral-prefrontal cortex 301
dot-probe paradigm 50
dothiepin 266
Down syndrome 91, 100
dual inheritance theory 18
ducks 73
Dunbar, R.I.M. 296
duplication-type copy-number variations 249
dyadic nonkin social relationships 254
dying species argument 167–9
dysconnectivity, functional 301–2
dysfunction 19–23, 144
  -based approaches 207
  biological 217
  as harm with no distinct external sustaining cause 153
  see also dysfunctions and depression; harmful dysfunction analysis; function, dysfunction and adaptation
dysfunctions and depression 203–7
  cognitive problem-solving 204–5
  learning 205
  motivational disengagement 204
  proportionality criterion 207
  risk regulation 204
  social signaling 205
dyslexia 104, 169, 170

early attachment and parental rearing styles 133
early experiences, adverse 131–2

eating disorders 176
Egger, H.L. 174
Eimer, M. 58
embodiment of others' pain 247
emotional contagion 242
emotional disorders 179, 181, 186–9
emotional level 274, 276
emotional mind 272
emotional paleomammalian level (limbic system) 272, 273, 276
emotional responses 175, 182
emotions 175
  abnormal 174
  aversive 7
  circumstances giving rise to 191
  and evolution 183–6
  normal 174
  positive 184, 187–8, 271n
  see also negative emotions; phobias and cognitive complexity of human emotions
empathy and autism 241–59
  autism spectrum condition 242–4
  empathizing-systemizing in autism spectrum conditions 244, 245–6
  empathy 241–2
  evolutionary hypotheses about male/female social strategies 254–7
  evolved gender differences in empathy: differential selection pressure hypothesis 252–4
  genetics of autism spectrum conditions 248–52
    mutational load model of neuropsychiatric disorders 250–2
    susceptibility genes 249–50
    triad of impairments 248–9
  low empathy, high systemizing and mutational load model 257–8
  'mirroring', lack of 246–8
empirical evidence and developmental disorders 109
encapsulation 43, 45, 51, 56, 93–4
Encouragement 128–9, 131
engagement stage 271
ennoblement through degeneration 24n
enthusiasm 275
environment of evolutionary adaptedness (EEA) 15–18, 40, 42, 45, 69, 71, 209
environmental circumstances 180
environmental features 156
environmental stimuli 41
environmentally caused reaction 154
epigamic selection 269
epistemological dualism 293
escalation (fighting harder) 270, 271, 273–4, 275–6, 278
essentialism 27

Estabrooks, G. 24
ethology: ontogenetic and causal factors and depression 6, 117–35, 301
  adverse early experiences 131–2
  attachment relationships, early and parental rearing styles 121–2
  deviant behavior 122–4
  disturbed interpersonal behavior 128–31
  ethograms 123, 124
  Ethological Coding System for Interviews 124–5, 128
  evolutionary explanations 132–3
  observable behavior and depression, association between 124–8
  why psychiatry needs ethology 133–5
etiological theory of biological functions 21
eugenics 5, 292
event-related potentials 50, 103, 253
evolution 19–27, 94
  of mental disorders 6–10
  see also environment of evolutionary adaptedness
evolutionarily recent stimuli 55
evolutionary approaches, critique of 165–71
evolutionary concept 220
evolutionary considerations 68
evolutionary explanations of depression 132–3
evolutionary fear module theory 51
evolutionary foundations for psychiatric diagnosis: DSM-V validity 173–94
  basic flaws 181–3
  diagnosis and its discontents 174–6
  emotional disorders 186–9
  emotions 183–6
  from clinical diagnosis to DSM 176–9
  motivational structure analysis 190–1
  progress, dissatisfaction with 179–81
  towards an evolutionary foundation for psychiatric nosology 191–4
evolutionary function 220
evolutionary hypothesis 54n, 254–7
evolutionary psychiatry 6, 10–18, 38–42, 60
evolutionary psychology 10–18, 69
evolutionary relevant fear 44–5, 47
evolutionary-irrelevant stimuli 48, 55
evolutionary-relevant stimuli 48, 55–6
evolved function *see* normality, disorder and evolved function in the case of depression
exaptation objection 169–71
excesses 7
exhilaration 271n, 275
Extraversion 130

face perception 298
face recognition in Williams syndrome 101–4, 106, 110–12
face threatening acts 279
face-inversion effect 102
facial features, individual 110
fairness 253
  of events 276
fatigue 204
fear 17
  conditioning paradigm 44
  irrelevant 44
  module 42–5, 55, 56
  module, specific 51
  neural circuitry 47, 51
  relevant 44–5, 47
  stimuli, novel 45
  stimuli, ontogenetic 45, 47
  stimuli, phylogenetic 45, 47, 52
featural cues 111
female hierarchy 268
fetishism *see* sexual imprinting and fetishism
fever of unknown origin 182
filial imprinting 71–2
final causes (Aristotle) 146
fish species 78, 278
fitness 2, 14
five-factor model of personality 244
flight and deviant behavior 123
flight or flight response 255, 271
Flykt, A. 51
Fox, N. 52
Frances, A.J.F. 174
free rider argument 166–7
Freud, S. 5–6, 24–7, 67, 282, 291
Frey, L.R. 122
friendship 186, 279, 280
function 20, 22
  natural 165–6, 167
  see also function, dysfunction and adaptation; functional explanation
function, dysfunction and adaptation 216–36
  dysfunction 228–31
  dysfunction and role of science 231–5
  theories of function 219–28
    selectionist view 217–18, 219–24, 229
    systemic capacity view 217, 219, 224–8, 229, 231–2, 234
  two-stage view 217–19
functional explanation and philosophy of psychiatry 143–71
  Aristotle and biological functions, mystery of 144–7
  black box essentialist account of function 150–1
  Darwin and teleology 148–50
  designed-defense objection 161–5
  harmful dysfunction analysis 152
  Lucretius on natural selection 147–8
  Nordenfelt's critique of evolutionary approaches 165–71

functional explanation and philosophy of
  psychiatry (*Cont.*)
  philosophy of biology to philosophy of
    medicine 151–2
  see also distinct sustaining causes
functional magnetic resonance imaging
  (fMRI) 246, 252, 301–2
Futuyma, D.J. 69

gambling tasks 204
game theory 271
Gardner, C.O. 202
gaze aversion 134
gaze duration 127
Geary, D.C. 254
Geerts, E. 129–30, 131
gene-culture co-evolutionary theory 16
general expectancy bias 54
General Health Questionnaire
  (GHQ-12) 200–2
genes 187
genetic differences 84, 188, 191
genetic disorder 95
genetic quality, signs of in a partner 69
genetic traits 59n
Gert, B. 144
  see also distinct sustaining causes:
    Culver and Gert
Gilbert, P. 120
goal pursuit 185
goals, new 275
Godfrey-Smith, P. 22, 229
goslings 71
Gould, S.J. 11–12
Grandin, T. 247
Grant, E.C. 123
Grant, P. and R. 220
Gray, A. 148
greylag geese 121
grief exclusion for depression 175, 180,
  202, 206
Griffiths, P.E. 227, 230–1
Gross, L. 222
grove snail (*Cepaea nemoralis*) 220
guilt 186, 275

Haeckel 5
Hale, W.W. 129
hamadryas baboons 270
Hamilton Rating Scale for Depression 123
hand movements, body-focused 128
Happé, F. 248
harm 217
harmful dysfunction analysis 20–2, 144, 152,
  155–6, 158, 160–2, 165–71, 176, 203
hawk/dove game 280
head aversions 128
Heidegger, M. 292–3, 300

heritability 208, 243
  see also genetic
hierarchy:
  animal 281
  female 268
  inversion 267
  social 265, 274
Hinde, R.A. 281
Hippocrates 150, 151
HM (amnesiac) 95
Hoffer, A. 6
Hogan, J.A. 76, 82, 86
Holloway, R.L. 295–6
Holmes, A. 58
homogamy 79
homosexuality 12, 65, 71
Horwitz, A. 22, 176, 223
hostility, free-floating 267
hostility, transfer of 267
human nature 23–7
Hume, D. 241
Humphrey, N.K. 295
Husserl, E. 291
Huxley, J. 6, 268–9
hybrid view of the concept of mental
  disorder 19, 20–1
hypothalamic-pituitary-adrenocortical
  axis 132

identity 296–7
illiteracy 170
illusory correlations 44–5
imaging studies 246, 252, 301–2
Immelmann, K. 73, 75
imprinting 121
  see also sexual imprinting
inability, normal 159–61
incest 71
individual variation, challenge of 208–10
inferior parietal cortex 301
innateness 85–6
instinctive reptilian level (basal ganglia) 272,
  273, 274, 275, 276, 278
intentionality 25
International Coding Diagnoses 178
interpersonal alienation from the social
  world 299–301
interpersonal behavior, disturbed 128–31
interpersonal development 296–7
interpersonal space 127
intra-sexual selection 269
inversion effect 103
involuntary subordinate strategy 267
isocortex 273, 276
Israeli kibbutzniks 79

jackdaws 71
Japanese quail 73

Jaspers, K. 291
jealousy 189, 193, 252
Johnson, G. 57
Jones, I.H. 127
joy 271n
just-so stories 11–14

Kagan, J. 52
Karmiloff-Smith, A. on developmental disorders and cognitive architecture 92–3, 94–6, 98–101, 105, 108, 109–10
   Improved Argument 93, 108, 112
   Original Argument 93, 98–101, 105, 108, 109–10
   Premise 1 99–100, 105, 106
   Premise 2 99, 101–2, 104, 105, 106–7, 108, 109–12
   'strong reading' 106–7, 109–12
   'weak reading' 106, 109
Keller, M.C. 14, 119, 244, 250–1
Kendler, K.S. 201
Kendrick, K.M. 76, 78
Kitcher, P. 12, 225
Kortmulder, K. 278
Kraepelin, E. 120, 181, 299
Krafft-Ebbing, R. von 66
Kropotkin, P. 303
Kruijt, J.P. 73

Lahaie, A. 248
Laland, K.N. 82, 86
language:
   acquisition 246
   impairment 97, 100, 104, 250
   meaning 297
   processing 102
Lauder, G.V. 230
learning 205
   associative 67
   effortful 247
Ledoux, J. 47, 56
Levinson, S.C. 279
Lewis-Williams, D. 294
Lewontin, R. 11–12
life events 179, 192, 206–7, 210
life problem 190
Lilienfeld, S. 162
limbic level/system 272, 273, 274, 276
Little, A.C. 80
Lloyd, E.A. 221
LoBue, V. 50
Lorenz, K. 71–2, 121
lovemaps 68
Lucretius, T.C. 147–8

McGuire, M.T. 41–2, 124
McIntosh, D.N. 247

Macke, F.J. 293
MacLean, P. 265, 272
McNeil, D.G. Jr 164
magnetic resonance imaging (MRI) 302
   functional 246, 252, 301–2
major depressive disorder (MDD) 20–2, 178, 180, 199–203, 211, 266–7
male warrior hypothesis 256–7
male/female social strategies, evolutionary hypotheses about 254–7
malingering 177
mallard ducks 75
mamadryas baboons 271
mammals 78, 82
manic-depressive illness 199
Marino, L. 162
Marks, I. 39–40, 42, 52–3, 55
marriage:
   contested symmetrical 284
   dominant husband in male-dominated society 284
   dominant wife male-dominated society 285
   minor marriages in Taiwan 79
   reciprocal, symmetrical 284
massive modularity hypothesis 15–16, 18, 91–2, 94–6, 98, 100, 104–5, 107, 112
masturbation 65
mate-quality hypotheses 83
material resources 190
mates and offspring 190
Maudsley, Dr. H. 4, 5
Mayr, E. 6
Mead, A.P. 295
Mead, G.H. 296
medical/breakdown model 9
medicine, philosophy 151–2
Mendelian disorders 250
Mendelian genetics 6
Mental Research Institute 277
Merckelbach, H. 52–4
Merleau-Ponty, M. 293, 297, 300
Miller, G. 119, 244, 250–1
Miller, R. 128
Millikan, R.G. 219, 228
Mills, D.L. 103
mind-blind 245
mindreading (face-based) 247
Mineka, S. 43, 52
Minio-Paluello, I. 247
mirror neurons 297
mirroring processes 242, 246–8
mismatch model 8, 17, 40
modularity 109
   see also massive modularity hypothesis
Money, J. 68

mood:
  disordered 203
  disorders 26
  inter-individual variation 202
  intra-individual variation 202
  low 14, 184, 189, 205, 207
  normal low 199, 203–4
  sub-clinical low 201
  see also mood change
mood change, role of in defining
    relationships: Gregory Bateson
    tribute 264–85
  Bateson: defining the relationship 276–81
  Darwin, Huxley and sexual selection 268–9
  overthrown tyrant: clinical case
      illustration 266–8
  ritual agonistic behavior and ritual
      losing 270–2
  triune mind in a triune brain 272–6
  Venn diagrams 283–5
moral code 279
Morris, D. 67–8, 72
Morrow, E.M. 243
motivational disengagement 204
motivational structure analysis 190–1
motor mechanisms 76, 77
motor phase 74
Murphy, D. 9, 40–1, 53, 232
mutational load model of neuropsychiatric
    disorders 250–2, 257–8

Nagel, T. 1
natural function argument 165–6, 167
natural selection 1, 82, 83, 132, 147–8,
    188, 292
naturalists/naturalist account 19–21, 23
nausea and vomiting 165
Neander, K. 219
negative emotions 7, 174, 182, 184, 187–8,
    194, 271n
  see also anger; depression
neocortex 273
neomammalian brain 273, 276
Nesse, R.M. 7, 14, 17, 35–6, 38, 39–40, 42,
    52–3, 55, 59, 206
Nettle, D. 120, 204, 254
neural circuitry, specific 56
neuro-histology 297
neurological conditions 176
neuropsychiatric disorders, mutational load
    model of 250–2
neuropsychological evidence 301
Neuroticism 130, 209, 210
neutral stimuli 47
Nietzsche, F. 23–4
non-evolutionary-relevant stimuli 56
non-threatening target 46
nonverbal behavior 129, 134

nonverbal convergence 131
nonverbal interactions 122
nonverbal interpersonal processes 133
nonverbal support-giving behavior 129, 130
nonverbal support-seeking behavior 129–30
Nordenfelt, L. 144
  critique of evolutionary approaches 165–71
normal and abnormal, distinction
    between 179
normal species function 20
normality, disorder and evolved function in
    the case of depression 198–213
  categorical depression/normality
    distinction, inductive evidence
    for 200–3
  disorder versus complaint as basis for
    identifying depression 210–13
  individual variation, challenge of 208–10
  see also dysfunctions and depression
normality, residual 98–9, 107
normality/disorder threshold 209
normativists/normative accounts 19–21

object-recognition mechanism 72
observable behavior and depression,
    association between 124–8
observable behavior of interviewer 128
observable sexual behavior 76
obsessive behavior 248
Öhman, A. 43, 45, 52, 53, 56
ontogenetic conditions 49
ontogenetic contingencies 51
ontogenetic factors *see* ethology
ontogenetic threats 49–50
operationalized diagnoses 178
optimal threshold 208
oral-genital sex 65
orbitofrontal cortex 298, 301
Osmond, H. 6
outcomes, negative 44, 54
oversensitivities 7
overthrown tyrant: clinical case
    illustration 266–8
ownership and ritual agonistic behavior 271
oxytocin 255

P1 51n
pain 174
painful stimulation observations 252
paleomammalian brain 272, 273, 276
paleontology 295–6
panic disorder 41, 180, 189, 191
Pansa, M. 127
paracingulate cortex 301
parallel searches 46
paranoid personality disorder 7
paraphilias 65–6, 68, 71
parent-offspring conflict theory 133

parental features and sexual or partner preferences 80
parental rearing styles 121–2, 131
parental smoking habits during childhood and sexual attraction to smoking 81
parietal association cortex 296, 298, 302
Parker, G. 271
partner preferences 80, 82
partner recognition 76
Pascal, B. 272
Penke, L. 249
perceptual mechanism 76–7
Perrett, D.I. 80
Perring, C. 162–3
personal resources 190
personality:
　disorders 7, 124, 133, 176, 194
　five-factor model 244
　traits 242, 251
pessimism 204
Pessoa, L. 57
phenotype matching 80
philosophical anthropology 25–6
philosophical criticism 10–18
philosophical paradigm 290
phobias 17, 158
　accident 52
　adaptive 42
　age at onset 38n
　blood-injection-injury 54
　height 41
　large predators 53
　maladaptive 42
　mushrooms 53
　small-animals 54
　spiders 53–4
　see also phobias and cognitive complexity of human emotions
phobias and cognitive complexity of human emotions 35–60
　definition of phobia 36–7
　emotion, alternative conception of 55–9
　evolutionary psychiatry and phobias 38–42
　evolutionist's explanation of phobias 52–5
　fear, module for 42–5
　snakes and spiders vs syringes and guns 46–51
phylogenetic conditions 49
phylogenetic origin 55
phylogenetic stimuli 54
phylogenetic threats 49–50
physical protective responses 182
Pinker, S. 91–2, 95, 101
Plato 290
Plenge, M. 79
Plessner, H. 25
Polsky, R.H. 124
polygenic mutation-selection balance 250

positive symptom cluster 299
posture shifts 128
power, expression of 277
prefrontal cortex 59, 298, 302
prestige 274
　competition 275–6
Price, J. 8–9
primates 78, 121, 271
primatology 294–5
prisoner's dilemma 186, 280
proportionality criterion 207
prosopagnosia (inability to recognize faces) 15
　see also Williams syndrome
proximate approach to causes of behavior 134, 135
proximate causes of behavior 9, 118
pseudo-interaction paradigm 129
psychiatric Darwinism 3–6
psychiatric diagnosis see evolutionary foundations for psychiatric diagnosis
psychoanalysts 178
psychopathological conditions 120
psychopathologies 23–7, 93–8
psychopathy 251
psychoses 176, 223
psychosocial stress, severity of 180
psychotropic drugs 177

Raines, Dr. G. 177
Ranelli, C.J. 128
rapport 129
'rational beliefs or desires' clause 155–6
rational brain 273
rational choice theories 83
rational level 275
rational mind 272
rational neomammalian level (isocortex) 273, 276
rational thought and decision-making 273
rationality 27
realist approach 199
reason 272
recapitulationism 5, 6
reciprocity potential hypothesis 256
redundancy in relationships 279
regression 4
regurgitation response 165
relationships:
　asymmetrical 270, 280, 282
　complementary 266, 277, 279, 281
　symmetrical 277, 279, 281–2
　see also marriage
relativity 228
relaxation and deviant behavior 123
reproductive effort 190
reproductive potential, traits associated with 69

reptilian brain 272, 273, 274, 275, 276, 278
Research Diagnostic Criteria group 178
residual normality 98–9, 107
resource value 271
resource-holding potential 271, 278, 279, 280
respect 274
Restlessness 128
reversal of dominance/subordinate
    relationship 268
rigid species essences 147
risk regulation 204
ritual agonistic behavior 270–2, 278, 280
ritual agonistic weapons 279
ritual fighting and losing 265, 270–2
ritual incapacity 265
Rivermead Face Memory Task 101
rodents 78
Ronald, A. 248
rule-based strategy 247
Rupert, R. 228
Rutherford, M.D. 247

sadness 184, 209
    normal 223
Saxton, T.K. 84
Schaffner, K.F. 222
schizophrenia 25, 199, 211, 243, 285
    dysfunction 20–1
    as evolutionary puzzle 6
    mutational load model 251
    psychodynamic model 10
    susceptibility genes 249
    trade-off model 9
    see also schizophrenia, evolutionary
        neurophilosophy of
schizophrenia, evolutionary neurophilosophy
    of 289–304
    embodiment, philosophy of 292–4
    evolutionary paradigm 298–9
    'interpersonal alienation' from the social
        world 299–301
    resolution of 'schizophrenia problem' in
        evolutionary terms 302–4
    see also social brain, evolution and
        development of
schizophreniform disorder 124
Schjelderup-Ebbe, T. 270
Schneider, K. 299
Schutz, F. 73
seasonal affective disorder 128
Segrin, C. 131
selection *see* natural selection; sexual selection
selectionist view of biological
    function 217–18, 219–24, 229
selective association 45
selective characteristics 56
selectivity 43, 45

self-assertion 275
self-confidence 275
self-effacement 275
self-inflicted injuries 199
Seligman, M.E.P. 44n
senescence model 8, 10
sense of self 296–7
sensitive period 72, 74, 78–9
serial searches 46
sex differences 38n, 75–6
    in autism spectrum disorders 243–4, 258
    and empathizing 244–5, 252–4, 257–8
    in fetishism 66, 68
    in imprinting 75
sexual imprinting and fetishism 65–86
    adaptationism and sexual
        imprinting 82–4
    fetishistic preferences for artifacts 79
    historical background of fetishism 65–71
    normal fetishists 66
    pathological fetishists 66
    sexual conditioning 67, 81
    sexual imprinting in animals 71–7
        development 76–8
        sensitive period 74
        sex differences 75–6
        stability of imprinting 74–5
        what is learned 73–4
    sexual imprinting in humans 78–82
    sexual preference hypotheses 80, 84
    sexual preferences associated with the
        body 70, 81
    sexual preferences, evolution of 68–71
sexual selection 268–9, 282
shame 275
shared manifold of intersubjectivity 297
sheep and goats 76, 78
Sheppard, P.M. 220
single nucleotide polymorphisms 249
skin conductance responses 44
Skuse, D.H. 258
Smith, E.A. 17
smoke detector principle 39, 187, 213
social attention-holding power 274, 279
social behavior 127, 130, 295, 296
social brain, evolution and
    development of 294–8, 301–2
    clinical evidence 302
    comparative neuroscience 296
    developmental neuroscience and
        psychology 296–7
    ethological observation 301
    functional imaging 301–2
    neuro-histology 297
    neuropsychology 301
    paleontology 295–6
    primatology 294–5

social neuroscience 297–8
structural imaging 302
social cognition 102, 296, 298
impairments 301
social competition hypothesis 265
social consciousness 296–7
social effort 190
social hierarchies 265, 274
social impairments 243, 248
social interactions 77, 122, 131
social life 252
social neuroscience 297–8
social preference 74
social problems 253
social relationships 186, 254
social roles 145
social self 300
social settings 252, 254
social signaling 205
social support, lack of 122
sociobiology 10–18
somatic effort 190
sparseness and cognitive architecture 93–4
spatial configuration 102
Speaking Effort 128, 129, 131
species-specific appearance of parents and siblings 73
speech pauses 128
stabilization phase 74
startle response 165
statistical abnormality 242
statistical criterion 159–60
Statistical Manual for the Use of Institutions for the Insane 177
Stearns, S.C. 206
Steele, V. 66
Sterelny, K. 18
Stevens, A. 8–9
Stoller, R. 67
strange situation experiment 121
stress 122, 179, 209
structural dysconnectivity 302
structural imaging evidence 302
subcortical areas 56
submission 278
Submissive Behavior 128, 274
substance dualism 290
subtractivity or residual normality assumption 99
superior temporal and lateral fusiform gyri 298, 301–2
surprise 186
susceptibility genes 249–50
suspicion 186
symmetrical encounter 278
symmetry 83, 266
sympathy 242

system-blind 245
systemic capacity view of biological function 217, 219, 224–8, 229, 231, 232, 234
systemizing 244, 245–6, 257–8

Tager-Flusberg, H. 111–12
target position 46
Taylor, S.E. 255
teleology/ teleological explanation 145–7, 148–50, 151
teleonomy 149
temperamental traits 59n
temporal association area 302
tend-and-befriend hypothesis 255
Tengland, P.-A. 162
theoretical considerations 109–10
theory of mind (ToM) 295, 296, 298, 301
Thomas, M. 92, 98, 100–1, 104
Thomas, P. 293–4
Thomson, J.A. 204, 206
threat detection 47
threat superiority effect 46, 48–9, 51
threatening target 46
Tinbergen, N. 6, 117–18, 134, 135
Tooby, J. 38–9, 209, 252
top-down projections 57
trade-off (balancing selection) model 8–9
transcranial magnetic stimulation 247
triune mind in a triune brain 265, 272–6
Troisi, A. 41–2, 124, 128
trust 186, 280
two-stage view of mental disorder 217, 226, 231, 233, 235

ultimate causes of behavior 118, 134
unfairness 253
universality of preferences 83, 86
unpredictability 6
unselected in intrasexual selection 269, 271
untoward situations 194

value criterion 152
Venn diagrams 283–5
verbal fighting 276
Vigil, J.M. 256
violence 199
visual search paradigm 45, 46, 48
voluntary mimicry 247
Vos, D.R. 75, 77
vulnerability to mental disorders 25
Vygotsky, L.S. 297

Wachbroit, R. 222
Wakefield, J. 2, 19–23, 175–6, 203, 217–18, 219, 223

weeds and mental disorder
    analogy 232
Westermarck effect 79
white matter fasciculae 296, 302
Williams, G. 35–6, 38, 42, 59
Williams syndrome 91, 97–8, 100,
    101–4, 106, 110–12
Wilson, E. 11–12, 118, 292
Wittgenstein, L. 1

World Health Organization: *Global Burden of Disease* report (1996) 199
Wright, L. 219
Wynne, C.D.L. 119

Yeargin-Allsopp, M. 258
youth, traits associated with 69

zebra finches 73, 75, 76, 77